Ecological Studies, Vol. 181

Analysis and Synthesis

Ecological Studies

Volumes published since 2000 are listed at the end of this book.

H. BassiriRad (Ed.)

Nutrient Acquisition by Plants

An Ecological Perspective

With 46 Figures and 17 Tables

Springer

Prof. Dr. Hormoz BassiriRad
Department of Biological Sciences
University of Illinois at Chicago
845 West Taylor Street
Chicago, IL 60607
USA

ISSN 0070-8356
ISBN 978-3-642-06337-4 e-ISBN 978-3-540-27675-3

Springer is a part of Springer Science+Business Media
springeronline.com

Printed in Germany

Editor: Dr. Dieter Czeschlik, Heidelberg, Germany
Desk editor: Dr. Andrea Schlitzberger, Heidelberg, Germany
Cover design: *design & production* GmbH, Heidelberg, Germany

31/3150 YK – 5 4 3 2 1 0 – Printed on acid free paper

Preface

The adaptation and evolution of terrestrial plants depend, to a large extent, on their ability to acquire nutrients. Therefore, factors governing plant nutrient acquisition are central to our understanding of plant ecology and evolution. The mechanisms that control plant nutrient uptake and how plants respond to changes in the environment have helped our understanding of the forces that govern today's plant communities, and are also likely to be indispensable in models designed to predict how plant communities will respond to projected changes in the global climate. A modern and integrated treatment of these mechanisms is the motivation for the production of this volume. Even though there are other excellent books that deal with plant nutrition (e.g., Barber 1995; Marschner 1995; Tinker and Nye 2000), a number of major differences make the content of this volume very timely.

First, this volume does not deal with the broader concept of plant nutrition. Rather, the emphasis here is on factors that control plant nutrient acquisition. Recent methodological advances in the imaging of root growth, architecture and turnover, as well as isotope techniques, have rapidly filled the information gap that once hindered progress in our understanding of the patterns and mechanisms of nutrient acquisition by native terrestrial plants. This volume reflects on many of those mechanisms and patterns. Second, and perhaps more important, other books in this area often discuss nutrient uptake by plants from an agronomic perspective. In contrast, the focus of this volume is largely ecological. Consequently, many of the chapters discuss root system characteristics that have evolved to cope with conditions that are only rarely experienced by crop plants, e.g., heterogeneity of soil nutrients in time and space.

Given the importance of the factors that control available nutrients at the root surface, we devoted two chapters to soil nutrient bioavailability. Chapter 1 deals with soil physiochemical factors and Chapter 2 with biogeochemical nutrient cycling controlling the release of nutrients from soil organic reservoirs. Chapter 3 highlights soil spatial and temporal heterogeneity of nutrient bioavailability. Nitrogen fixation and its contribution to soil N availability are covered in Chapter 4. Chapter 5 addresses the regulation of nutrient uptake by internal plant demand and use efficiency. The focus of the remaining chapters is on root characteristics, particularly those that are to date understudied, e.g., kinetics of uptake (Chap. 6), architecture (Chap. 7), life span (Chap. 8), and mycorrhizal symbiosis (Chap. 9). Two further chapters deal with nutrient uptake and how it responds to key components of global change. Chapter 10 deals with changes in root zone temperature, whereas Chapter 11 considers plant nutrient uptake responses to changes in UV-B, ozone, and atmospheric CO_2 concentration. Finally, Chapter 12 presents a synthesis of the book, emphasizing the need for an integrated view of plant nutrient uptake based in particular on molecular biology and biogeochemistry.

Most chapters have a nitrocentric/nitrophilic perspective, reflecting the fact that the literature is dominated by studies of plant nitrogen uptake. This focus is justified, given the N limitation of primary productivity in many native systems. There is growing recognition, however, that the balance between the availability of N and other nutrients, e.g., N to P ratios (Aerts and Chapin 2000), or the availability of micronutrients (Hungate et al. 2004) may be more critical components of plant nutrition than N dynamics alone.

It was noted above that an adequate treatment of topics such as soil physical and chemical factors, which have received only limited coverage here, would require its own volume. Several other areas fall outside the scope of this book, but are indispensable to the understanding of nutrient intake by plants. For example, this volume covers very little about recent methodological approaches in field nutrient uptake studies. Advances in stable isotope technology and affordability have made stable isotopes a popular approach in assessing nutrient uptake. Also, ion uptake into the root, and its eventual transport to the shoot, are influenced by root anatomical characteristics, but this field remains incredibly understudied. Readers interested in this area of research are referred to McCully (1995), Clarkson (1996), and Peterson et al. (1999) for issues and perspectives on anatomical and structural aspects of roots that control nutrient uptake. The readers are also encouraged to examine a recent volume of the Ecological Studies Series edited by de Kroon and Visser (2003), whose content is very complementary to this book.

Finally, compiling this volume represents the efforts of more than just one person. This work would not have been possible without the input from my co-authors who were extremely patient with me. Due to an inevitable lag associated with the production of the book, some of the very newest literature may not appear in the various chapters. However, every effort has been made to reduce this citation gap to a minimum. In particular, I thank Vince Gutschick and John Lussenhop who provided numerous feedback on many issues pertaining to nutrient uptake by plants. Last, but not least, I owe much to Jacqueline Charette-BassiriRad who devoted considerable time helping me with various editorial aspects.

Chicago, March 2005 *Hormoz BassiriRad*

References

Aerts R, Chapin FS III (2000) The mineral nutrition of wild plants revisited: a re-evaluation of processes and patterns. Adv Ecol Res 30:1–67

Barber SA (1995) Soil nutrient bioavailability: a mechanistic approach, 2nd edn. Wiley, New York

Clarkson DT (1996) Root structure and the site of ion uptake. In: Waisel Y, Eshel A, Kafkafi U (eds) Plant roots – the hidden half. Marcel Dekker, New York, pp 483–510

de Kroon H, Visser EJ (2003). Root ecology. Ecological Studies, vol 168. Springer, Berlin Heidelberg New York

Hungate BA, Stiling PD, Dijkstra P, Johnson DW, Ketterer ME, Hymus GJ, Hinkle CR, Drake BG (2004) CO_2 elicits long-term decline in nitrogen fixation. Nature 304:129–129

Marschner H (1995) Mineral nutrition of higher plants, 2nd edn. Academic Press, New York

McCully M (1995) How do real roots work? Some new views of root structure. Plant Physiol 109:1–6

Peterson CA, Enstone ED, Taylor JH (1999) Pine root structure and its potential significance for root function. Plant Soil 217:205–213

Tinker PB, Nye PH (2000) Solute movement in the rhizosphere. Oxford University Press, Oxford, England

Contents

Contributors

BASSIRIRAD, Hormoz

Department of Biological Sciences, University of Illinois at Chicago, 845 West Taylor Street, Chicago, Illinois 60607, USA, e-mail: hormoz@uic.edu

BIELENBERG, Douglas G.

Department of Horticulture, 175 Poole Agricultural Center, P.O. Box 340319, Clemson University, Clemson, South Carolina 29634-0319, USA, e-mail: dbielen@clemson.edu

COMERFORD, Nicholas

Department of Soil and Water Science, University of Florida, 2169 McCarty Hall, P.O. Box 110290, Gainsville, Florida 32611-0290, USA, e-mail: nbc@mail.ifas.ufl.edu

EISSENSTAT, David M.

Department of Horticulture, Penn State University, 218 Tyson Building, University Park, Pennsylvania 16802, USA, e-mail: dme9@psu.edu

FINLAY, Roger D.

Department of Forest Mycology & Pathology, SLU, P.O. Box 7026, 750 07 Uppsala, Sweden, e-mail: Roger.Finlay@mykopat.slu.se

GLASS, Anthony D.M.

Department of Botany, Room #3529, 6270 University Boulevard, Vancouver, BC V6T 1Z4, Canada, e-mail: aglass@interchange.ubc.ca

GUTSCHICK, Vincent

New Mexico State University, Department of Biology, MSC 3AF, Las Cruces, New Mexico 88003, USA, e-mail: vince@nmsu.edu

KING, John S.

School of Forest Resources and Environmental Science, Michigan Technological University, 1400 Townsend Drive, Houghton, Michigan 49931, USA, e-mail: jsking@mtu.edu, johnking@fs.fed.us

LYNCH, Jonathan P.

Department of Horticulture, Penn State University, University Park, Pennsylvania 16802, USA, e-mail: jpl4@psu.edu

PREGITZER, Kurt S.

School of Forest Resources and Environmental Science, Michigan Technological University, 1400 Townsend Drive, Houghton, Michigan 49931, USA, e-mail: kspregit@mtu.edu

PRESCOTT, Cindy

Faculty of Forestry, University of British Columbia, 2714-2424 Main Mall, Vancouver, BC V6T 1Z4, Canada, e-mail: cindy.prescott@ubc.ca

PUSHNIK, James C.

New Mexico State University, Department of Biology, MSC 3AF, Las Cruces, New Mexico 88003, USA, e-mail: jpushnik@csuchico.edu

ROBINSON, David

Plant and Soil Science, School of Biological Sciences, Cruickshank Building, University of Aberdeen, Aberdeen AB24 3UU, Scotland, UK, e-mail: david.robinson@abdn.ac.uk

SPRENT, Janet

32 Birkhill Ave., Wormit Fife DD6 8PW, Scotland, UK, e-mail: JISprent@aol.com

VOLDER, Astrid

Department of Horticulture, Penn State University, 218 Tyson Building, University Park, Pennsylvania 16802, USA, e-mail: a-volder@tamu.edu,

1 Soil Factors Affecting Nutrient Bioavailability

N.B. COMERFORD

1.1 Introduction

The soil supports plants, is the physical determinant of root growth and extent, and is the main reservoir for plant-available water and nutrients. Therefore, the soil controls the availability of most essential plant nutrients. It regulates availability by means of biophysiochemical processes, which are functions of soil and plant properties. This chapter introduces the concept of soil nutrient bioavailability, and the soil factors that regulate it. It summarizes the processes that control nutrient release and movement in the soil, and points out the importance of selected plant root characteristics in nutrient acquisition.

The soil–plant system's capacity to supply/absorb nutrients is termed *soil nutrient bioavailability*, and is the ability of the soil–plant system to supply essential plant nutrients to a target plant, or plant association, during a specific period of time as a result of the processes controlling (1) the release of nutrients from their solid phase in the soil to their solution phase; (2) the movement of nutrients through the soil solution to the plant root–mycorrhizae; and (3) the absorption of nutrients by the plant root–mycorrhizal system (Comerford 1998).

Release of nutrients from the solid phase to the soil solution is controlled by the physiochemical processes of desorption and dissolution. It is also a biochemical process by way of mineralization. In both cases, soil and plant root–mycorrhizal characteristics help regulate the expression and dominance of these processes (Table 1.1). The movement of nutrients through the soil solution to the root–mycorrhizal surface is determined by mass flow and diffusion, which in turn are controlled by the interaction of soil and plant characteristics (Table 1.1). Finally, nutrient uptake at the root surface is dependent on the amount of root–mycorrhizal surface and its uptake characteristics.

The following discussion presents the three groups of processes outlined above. It concentrates on delimiting those soil characteristics important to defining nutrient bioavailability. Plant root growth, root function and nutri-

Ecological Studies, Vol. 181
H. BassiriRad (Ed.) Nutrient Acquisition by Plants
An Ecological Perspective

Table 1.1. Soil processes and the soil characteristics defining each process, along with a qualitative evaluation of the plant's influence on the process

Soil process	Defining soil characteristic	Potential plant influence on process
Sorption	Temperature (T)	Low, shading
	K_d of adsorption, pH	High, root exudates
	Solution ionic strength (μ)	None to low
Desorption/ dissolution	T	Low, shading
	K_d of desorption, pH	High, root exudates
	μ	None to low
Mineralization/ immobilization	Volumetric water content (θ)	High, transpiration
	T	High, shading
	Organic matter quality	High
	Enzyme concentration	High, phosphatase exudation
Mass flow	Hydraulic conductivity, θ	High, transpiration effect on θ
	Bulk density (ρ)	Low to medium, root action
	Pore-size distribution	Low to medium, root action
	Solution concentration (C_l)	Low to medium, root exudates
Diffusion	Impedance	High, transpiration effect on θ, root action on soil structure
	K_d	Low to medium, root exudates
	θ	High, transpiration
	ρ, pore-size distribution	Low to medium, root action
	C_l	Low to medium, root exudates

ent mineralization are covered in other chapters, so their treatment in this chapter will be confined to their influence on plant nutrient bioavailability, and their interactions with other components of the soil bioavailability system.

1.2 Release of Nutrients from the Soil Solid Phase

The key solid phase properties are (1) the quantity of labile nutrients, and (2) the manner of nutrient release to the soil solution. The term labile, in this context, is defined as nutrients in a plant-accessible form during a given time period. Maintenance of the soil solution's nutrient concentration is a central concept of soil nutrient supply, and will be referred to repeatedly in this chapter. It is central because (1) the quantity of nutrients in solution is low relative to a plant's demand and, therefore, must be replenished continuously; (2) practically all nutrient uptake occurs from the soil solution; and (3) nutrients must be in solution to move effectively to the plant's root–mycorrhizal surface, thereby giving the plant a larger soil volume from which it can extract nutrients. The solid phase maintains the soil solution nutrient concentration. Release from the solid phase to the soil solution occurs biochemically (mineralization and immobilization) or physiochemically (adsorption and desorption, and precipitation and dissolution).

Mineralization and immobilization are transformations of nutrients between their organic to inorganic forms. These transformations are functions of the soil temperature regime, soil aeration regime, soil water regime, and the quality of the organic matter from which the nutrient is mineralizing – because these factors control the population and activity of soil organisms (see Chap. 2, this Vol.). However, a subject that is not often considered is the comparative role that mineralization plays in regulating the soil solution concentration of N, P, and S. In extensively managed soil systems, mineralizable nutrients can be the dominant component of the ecosystem's total labile pool, particularly the N pool. However, contribution to the labile pool via mineralization is not necessarily the same as its regulatory influence on the soil solution concentration, especially when considering P and S. This question is crucial for conceptual modeling of the nutrient's soil solution concentration over time.

Reaction of nitrogen with the mineral–organic surface complex of soil is chemically distinct from that of P and S. When nitrate is the primary N species produced by mineralization, mineralization controls the soil solution's nitrate concentration. Since there is little to no sorption of nitrate onto the solid phase, this is the only option. However, ammonium, phosphate and sulfate can sorb to the soil mineral surface subsequent to mineralization. The greater the sorptive capacity of the soil, the greater should be the potential for the mineralized ion to be sorbed onto the mineral surface. Therefore, the role of mineralization should be to (1) control the soil solution concentration directly when the nutrient does not sorb onto the mineral soil, or (2) contribute to the soil solution concentration and to the nutrient concentration on the soil's mineral surface when the soil's sorption capacity is greater than zero. In the latter case, nutrient distribution between the solution and the mineral surface should be related to nutrient affinity for the mineral surface. While it can be shown that mycorrhizal hyphae can intercept mineralized P before other microorganisms, or before sorption to some clays (Joner and Jakobsen 1994), there is little relevant literature directly targeting this topic, even though the interaction between mineralization and sorption is an integral element of modeling bioavailability.

Adsorption–precipitation and desorption–dissolution reactions regulate the removal of nutrients from, or release into, the soil solution. Adsorption–desorption reactions are classified as either outer sphere or inner sphere (Soil Science Society of America 1996), which for the most part differentiates between cations (outer sphere) and anions (inner sphere), notably phosphate and sulfate. Anions participate in outer sphere reactions when the soil has a positive charge. Except for specific circumstances, however, this does not add much to our understanding of their bioavailability.

An outer sphere reaction is an electronegative attraction of charged nutrients to the exchange complex of the soil surface. Cation exchange is the sorption process when cations are involved, and anion exchange is the process when anions are concerned. Outer sphere reactions are rapid, have comparatively weak bonding between the nutrient and the soil surface, and do not have kinetic limitations for equilibrium. An inner sphere reaction occurs when the solute becomes part of the soil mineral surface, such as during ligand exchange. In this case, the bond is covalent, can exhibit strong hysteresis between sorption and desorption, and is considered not to have kinetic limitations to equilibrium, unless the time frame is on the order of minutes to hours.

Both sorption (adsorption–precipitation) and desorption are key to nutrient cycling in soils. They help describe the fate of nutrients added by mineralization and fertilization, and their subsequent influence on the soil solution concentration. Numerous studies describe sorption, while fewer studies address desorption (Singh and Jones 1976; Peaslee and Philllips 1981; Colombo et al. 1994; McDowell and Condron 2001; Koopmans et al. 2004), and even fewer detail the effect that desorption has on the soil solution concentration (Brewster et al. 1975; Raven and Hosner 1994; Bhatti and Comerford 2002; Barros Filho et al. 2004). This is regrettable, given the importance of desorption in understanding bioavailability. The problem is further complicated by the consensus that sorption and desorption are hysteretic, particularly for inner sphere reactions. Clearly, better understandings of desorption and the hystereses between desorption and sorption are required (Barros Filho et al. 2004).

Recognizing that desorption is a key to bioavailability, how is desorption best described? Unlike kinetically dynamic mineralization, desorption is described as an equilibrium or quasi-equilibrium reaction through the use of desorption isotherms. Remembering that the soil solution concentration is a key component to defining a nutrient's bioavailability, the most useful soil property derived from isotherms is the partition coefficient (K_d):

$$K_d = \mathrm{d}C_S/\mathrm{d}C_l \qquad (1.1)$$

C_S is the concentration of the labile nutrient on the solid phase, and C_l is the nutrient concentration in the soil solution. K_d is the rate of change of the labile nutrient on the solid phase with respect to the soil solution concentration, and is the first derivative of the adsorption or desorption isotherm. It describes the distribution of the labile form of the nutrient between the solid and solution phases, relates to the soil's ability to buffer a nutrient's solution concentration, and is key to describing a nutrient's ability to move in the soil solution (see below).

K_d is a function of the quantity of nutrient on the sorption complex, the pool size of the sorption complex, and the affinity of the nutrient for the sorption complex. Barros Filho et al. (2004) showed that, for soils with similar clay mineralogy, K_d can be predicted from the amount of nutrient sorbed and the clay content. The sorption complex is the cation exchange complex for outer sphere reactions, and the anion sorption complex for inner sphere reactions. For outer sphere reactions, K_d depends on the makeup of the other exchangeable cations on that complex, while release from the inner sphere depends on having anions with a higher affinity for covalent bonding with the soil surface (Fox et al. 1990; Lan et al. 1995).

Determining the quantity of mineralizable organic N, P, and S is covered in Chapter 2 (this Vol.), but note that the estimate is dependent on the time frame chosen. The quantity of labile inorganic nutrients has historically not been measured directly, but indexed by standardized soil extractions using acidic or basic solutions. These often correlate well with response to fertilization, or growth over a continuum of nutrient deficiency to sufficiency, but do not measure the total pool of labile inorganic nutrients.

The difficulty in estimating the quantity of labile inorganic nutrients differs between outer and inner sphere ions. For outer sphere inorganic nutrients, the amount of exchangeable cations can be a good first-order estimate, particularly if (1) the time frame is not too long, or (2) clay mineralogy does not promote rapid movement of interlayer solutes to the exchange complex. For example, exchangeable potassium is a good estimate of total labile potassium, unless (1) the clay mineralogy of soil allows for the presence of significant non-exchangeable K (Guzel et al. 2001), or (2) root chemistry dissolves interlayer aluminum compounds, releasing potassium tied up or occluded with the interlayer aluminum (Comerford et al. 1990).

In contrast, labile inner sphere nutrients are difficult to measure, since their release can also be a function of root chemistry. One labile pool of inorganic P available to all plants is that removed by simple equilibrium exchange. All that is necessary to promote P desorption is its removal from solution by uptake or leaching. Unfortunately, this is not a large component of the bioavailable pool. Ligand exchange and congruent dissolution are two ways by which inorganic P is released to solution beyond simple equilibrium exchange, but these require the chemical action of plant roots (see below).

1.3 Nutrient Movement Through the Soil Solution

Nutrients move to roots through the solution phase. Movement occurs by mass flow or diffusion, or both. These processes are soil state-dependent (Table 1.1). Mass flow (Q_{MF}) is defined as the quantity of nutrient flowing to the root system with the transpiration stream, or:

$$Q_{MF} = Q_W \; C_l \tag{1.2}$$

Q_W is the water flux into the root system. Darcy's law states that the hydraulic conductivity of soil facilitates water flow, while the motive force for water flow is the water energy gradient between the root and the soil. As soil water content decreases, not only does the energy status of water change, but also hydraulic conductivity decreases logarithmically, restricting soil water flow to the root. The quantity of nutrient flowing with the water is the nutrient supplied by mass flow. Therefore, the most obvious controls of mass flow is the water regime of the soil, or how the soil water content and its energy status change over time, and the desorption–mineralization processes that determine the soil solution concentration. This is followed by the pore-size distribution of soil, which determines hydraulic conductivity changes with soil water content. The best conditions for mass flow dominance are when (1) the soil sorption capacity for a nutrient is in the diminishing returns phase, and (2) water content does not seriously limit water flow. Under these conditions, nutrient delivery by mass flow is maximized. Note that the plant plays a role in determining the importance of mass flow through its effect on the water potential gradient driving mass flow, and its effect on the soil solution concentration (see below).

When mass flow does not meet the plant's nutrient demand, i.e., the amount arriving at the root system is less than that required by the plant, the nutrient concentration in the rhizosphere is reduced relative to the soil solution outside the rhizosphere – a nutrient gradient develops. Fick's law states that diffusive flux is a function of this concentration gradient:

$$Q_D = \left[\left(D_l \theta f(\theta)\right)/b\right] \mathrm{d}C_l/\mathrm{d}x \tag{1.3}$$

Q_D is diffusive flux, D_l the diffusion coefficient of the nutrient in water, θ the volumetric water content, f the impedance factor of the soil, x the distance from the root, and b the soil buffer power. This law illustrates that the initiation of diffusion is due to mass flow not meeting the plant demand. In other words, nutrient deficiencies or the soil's inability to support luxury consumption promote the onset of diffusion. A plant experiencing a nutrient deficiency must have diffusive flow supplying a portion of that nutrient.

Soil water content and its change over time (soil water regime) are other critical components of bioavailability. Their role in mass flow and diffusion is clear from Eqs. (1.2) and (1.3). Soil water content regulates diffusion through (1) determining the cross-sectional area available for diffusion (Eq. 1.2), (2) determining the path length of diffusion by controlling diffusive impedance (f), and (3) contributing to the retardation of nutrient movement, as indicated by the soil buffer power (b). The impedance factor is the ratio of the length of the straight-line path of nutrient movement to the actual path. Since its real-life range can be from less than 0.1 to greater than 0.4, it has the potential to change the diffusive flux by a factor of more than 4. The impedance factor is a function of bulk density and water content (Bhadoria et al. 1991; Kirk et al. 2003), so it is manageable through irrigation, physical soil management, and the action of plant roots and soil biota, such as earthworms.

The rate of nutrient flux by diffusion is retarded by any sorptive interaction that the nutrient has with the soil. This interaction is contained in the buffer power term of diffusive flow where buffer power is (Van Rees et al. 1990a):

$$b = \theta + \rho K_d \tag{1.4}$$

where ρ is the bulk density. The greater the interaction the nutrient has with the soil surface, the greater will be the b and K_d, and the slower will be the nutrient delivery to the plant root system via diffusion. When the K_d is near 0, as with nitrate in most soils, the buffer power is equivalent to the soil water content.

One question often asked is which nutrient is supplied by mass flow, and which by diffusion? Answering this question from the perspective given above, it should be clear that any nutrient could be supplied by either mechanism, if the soil conditions are appropriate. Nitrate supply is commonly considered to be by mass flow. If deficiencies occur, however, as in many natural terrestrial ecosystems, nitrate should have a significant diffusive component. Conversely, phosphate should be supplied by diffusive flow because of the high sorptive capacity of most soils. However, if the soil's sorption capacity is surpassed, then phosphate concentrations are high and mass flow is projected. Highly weathered, sandy soils can have low to zero P sorption capacity (Harris et al. 1996), which has contributed to surface water pollution when there have been high P inputs from agricultural practices (Harris et al. 1994).

1.4 Nutrient Uptake by the Root System

Root characteristics that define nutrient bioavailability are (1) the dynamics of root/hyphal growth and development, (2) root/hyphal morphology, (3) inter-root competition, and (4) root/hyphal uptake kinetics. These topics are covered in various chapters in this volume, in particular Chapters 4, 6, 7, 8 and 9. However, two points need to be emphasized here. First, nutrient uptake is, with few exceptions, sensitive to the amount of root/mycorrhizal surface area, and less sensitive to its uptake characteristics (Smethurst and Comerford 1993a, b; Barber 1995; Williams and Yanai 1996). This is particularly true for plants experiencing a nutrient deficiency.

Second, an underrated asset of root systems is the chemistry they add to the soil environment, particularly the rhizosphere. This can influence bioavailability by changing the nutrient release pattern directly, or by changing the microflora of the rhizosphere, which in turn influences nutrient dynamics in that zone. The major chemical changes induced by root systems that influence nutrient bioavailability are (1) soil pH change in the rhizosphere (Hinsinger 2001), (2) increased levels of phosphatase for P mineralization (Grierson et al. 2004), and (3) exudation of organic acids, particularly low molecular weight acids (Grierson et al. 2004). Soil pH changes occur via root respiration and the production of bicarbonate ion, or the maintenance of ion balance during nutrient uptake (Hinsinger et al. 2003; Shen et al. 2004). Nitrate uptake results in proton uptake or hydroxyl exudation. Both result in a pH increase. Ammonium uptake is balanced by proton exudation and results in a pH decrease. Since pH can influence the equilibrium concentration of nutrients in the solid phase, the result can be an increase in bioavailability, particularly of P, Fe (Hinsinger et al. 2003), and Zn (Loosemore et al. 2004).

Phosphatase production is correlated with reduced P in the soil solution. Phosphatase concentrations can be more than five times higher in the rhizosphere than in bulk soil (Radersma and Grierson 2004). Phosphatase exudation response to low solution P varies among species. Kamh et al. (2002) equated the greater exudation of some species to their greater capacity to absorb P at low solution concentrations. However, increased phosphatase activity does not result in increased P bioavailability unless there is sufficient organic P in a mineralizable form present in the rhizosphere.

Organic acid exudation by plant roots is linked to Al toxicity tolerance in plants (Ryan et al. 2001; Zhang et al. 2003; Kochian et al. 2004), and to phosphorus (Hoffland 1992; Nakamura et al. 2000; Ryan et al. 2001; Veneklaas et al. 2003; Radersma and Grierson 2004; Shen et al. 2004; Grierson et al. 2004), manganese (Gherardi and Rengel 2004) and iron (McCluskey et al. 2004) bioavailability. By increasing the desorption of these ions from less soluble compounds, the soil K_d most likely changes along with the definable pool of labile nutrients. Soil solution concentration, mass flow, and the potential for diffusive flux can be increased. It appears to be an adaptive mechanism where low nutrients can affect the gene expression of carbon metabolism (Uhde-Stone et al. 2003), resulting in organic acid exudation. This is variable among species, and even among genotypes/cultivars (Ryan et al. 2001; Wouterlood et al. 2004; Liu et al. 2004). This variability represents an opportunity to genetically manage and deploy plants to take maximum advantage of the limited P bioavailability of P-deficient soils.

1.5 Modeling Nutrient Bioavailability

Table 1.1 indicates some of the soil variables that influence nutrient bioavailability. The most important tend to be root amount, mineralization regimes, soil water regimes, and soil partition coefficients–buffer powers. The complexity of nutrient bioavailability lends itself to computer modeling. The problem is multidimensional in both the number of variables and in space. Thus, typical mechanistic models have upward of ten input variables, and both soil depth and horizontal variability add to this complexity. Rengel (1993) and Tinker and Nye (2000) reviewed mechanistic uptake models. Tinker and Nye (2000) classified mechanistic bioavailability models into transient-state or steady-state models, which solved equations either analytically or numerically, and did or did not include inter-root competition. These models deal with single (Barber and Cushman 1981; Bar-tal et al. 1991; Smethurst and Comerford 1993a; Yanai 1994; De Willigen and van Noordwijk 1994a, b) or multiple nutrients (Bouldin 1989; Silberbush et al. 1993).

These models are all based on the mechanistic approach presented in this chapter. They consider nutrient release from the solid phase, movement to the root by mass flow–diffusion, and uptake at the root system surface. They are primarily soil-based models using the root system as a boundary condition, so they are within the soil "black box" often found in ecosystem models. Whether the models are analytically or numerically based, they all need impressive amounts of input data that describe the distribution/function of the root system, soil chemistry, and soil physics. They are best thought of in the context of submodels for plant growth that require inputs from carbon allocation and hydrology models.

A bioavailability evaluation is more complex than the simpler process used to evaluate soil fertility in the past, i.e., soil testing correlated with plant growth, nutrient uptake, or fertilizer response. The soil testing approach was key to the "green revolution". It continues to provide nutrient management recommendations, and has fueled agriculture for decades. Today, large-scale, multi-site field research is becoming excessively expensive, and nutrient management practices are requiring minimum nutrient inputs for maximum return on investment with minimum offsite effects. A more useful approach under these limiting circumstances is to use field testing to verify or debunk selected model predictions.

In more extensively managed ecosystems, understanding management inputs are not as crucial as understanding the bioavailability ramifications that accompany ecosystem changes, such as bioavailability changes related to forest harvesting, climate change, or acid deposition. Modeling plays a role under these conditions. Bioavailability models should allow researchers not only to estimate potential nutrient supplies under current conditions, but also to ask what will be the result of changing conditions. However, the use of models is limited by (1) the number of inputs, (2) the quality of some of the input data, and (3) their application to complex multi-species systems where roots of many species have different characteristics.

Sensitivity analyses reduce the number of inputs for which accurate data are required. In each specific species–soil scenario, a limited number of inputs control plant uptake (Smethurst and Comerford 1993b; Barber 1995; Williams and Yanai 1996). When the soil limits nutrient supply, root amount and growth rate, soil water content, soil diffusive properties (including K_d), and the solution concentration where influx equals efflux (C_{min}; see Chap. 6, this Vol.) tend to control plant uptake. When soil supply is high relative to demand, root uptake kinetics (except for C_{min}), root amount, and water influx to the root system are most important.

With regard to soil parameters, predicting desorption from soil characteristics (such as clay content) and its hysteresis with sorption need to be better understood (Barros Filho et al. 2004). The control of the soil solution concentration by mineralization and desorption needs to be clarified. Spatial variability of K_d and labile nutrient pools, along with methods to predict this variability, will improve model accuracy and precision. With regard to the plant itself, a better understanding of the root's Michaelis-Menten parameters, and of the dynamics of parameter change in the short and long term is necessary. For severe nutrient deficiencies, a much better understanding of C_{min} will be fruitful. More accurate measurement and prediction of root system dynamics are required.

Even with these limitations, the utility of nutrient bioavailability models has been shown (Nye and Tinker 1977; Van Rees et al. 1990b; Smethurst and Comerford 1993b; Barber 1995) under both greenhouse and field conditions, particularly when time frames are relatively short. Further development of mechanistic models based on bioavailability concepts is useful to improve nutrient management, gain a better understanding of nutrient cycling, and predict stress responses of extensively managed ecosystems. Their incorporation into plant growth models, such as TREGRO (Weinstein and Yanai 1994), or into P cycling components of models such as ANIMO, GLEAMS, and DAYCENT (Lewis and McGechan 2002) shines light into the "black box" that has often characterized soil processes.

1.6 Summary

Estimating the labile pool of nutrients ranges from being relatively easy to more difficult, and estimates depend on the time frame considered. Desorption and mineralization are the key processes that maintain a nutrient's solution concentration. In the case of desorption, the partition coefficient is the key descriptive parameter. Exchangeable cations are a reasonable estimate of the pool of nutrients controlled by outer sphere reactions, while the pool of inner sphere nutrients can depend on the action of roots.

Nutrient flux to the root system is by mass flow or diffusion, or both. The initiation of diffusion depends on mass flow not meeting the plant's demand, so it indicates nutrient deficiency, assuming no luxury consumption. Soil water content is crucial to defining both mass flow and diffusion, therefore, a soil with a high quantity of labile nutrient will have low bioavailability if the water regime is not conducive to delivering nutrients to the plant. The soil's diffusion coefficient, including the soil buffer power, is a key factor in the diffusion of nutrients to plant roots.

Root systems have the ability to increase the bioavailable pool of nutrients due to their influence on soil chemistry. This effect is principally in the root–mycorrhizal rhizosphere. It can have a profound effect on the mineralization of organic P, and the desorption/dissolution of inorganic P, Fe, Mn, and Zn from chemical species that would not normally be available. The capacity to increase nutrient bioavailability is variable among species, and even among genotypes/cultivars.

Modeling bioavailability is a logical approach to the complexity of the topic. Under most circumstances, a subset of input parameters control nutrient uptake and require accurate, quality input data. However, the subset may change with soil and species conditions. The quality of inputs needs to be improved, but in spite of these limitations, bioavailability models have been shown to be useful. Their incorporation into plant growth models has been successful, and has taken soil processes out of the black box in which the soil is so often confined.

References

Barber SA (1995) Soil nutrient bioavailability: a mechanistic approach, 2nd edn. Wiley, New York

Barber SA, Cushman JH (1981) Nutrient uptake model for agronomic crops. In: Iskander IK (ed) Modelling wastewater renovation land treatment. Wiley, New York, pp 382–409

Barros Filho NF, Comerford NB, Barros NF (2004) Phosphorus sorption, desorption and resorption in soils of the Brazilian Cerrado. Biomass Bioenergy (in press)

Bar-Tal A, Bar-Yosef B, Chen Y (1991) Validation of a model of the transport of zinc to an artificial root. J Soil Sci 42:399–411

Bhadoria PBS, Kaselowsky J, Claassen N, Jungk A (1991) Impedance factor for chloride diffusion in soil as affected by bulk-density and water-content. Z Pflanzenernähr Bodenkd 154:69–72

Bhatti JS, Comerford NB (2002) Measurement of phosphorus desorption from a spodic horizon using two different desorption methods and pH control. Comm Soil Sci Plant Anal 33:845–853

Bouldin DR (1989) A multiple ion uptake model. J Soil Sci 40:309–319

Brewster JL, Gancheva AN, Nye PH (1975) Determination of desorption isotherms for soil phosphate using low volumes of solution and an anion exchange resin. J Soil Sci 26:364–377

Buckley TN, Miller JM, Farquhar GD (2002) The mathematics of linked optimisation for water and nitrogen use in a canopy. Silva Fennica 36:639–669

Colombo C, Barron V, Torrent J (1994) Phosphate adsorption and desorption in relation to morphology and crystal properties of synthetic hematites. Geochim Cosmochim Acta 58:1261–1269

Comerford NB (1998) Soil P bioavailability. In: Lynch JP, Deikman J (eds) Phosphorus in plant biology: regulatory roles in molecular, cellular, organismic, and ecosystem processes. American Society of Plant Physiologists, Rockville, Maryland

Comerford NB, Harris WG, Lucas D (1990) Release of non-exchangeable potassium from a highly-weathered, forested quartzipsamment. Soil Sci Soc Am J 54:1421–1426

De Willigen P, van Noordwijk M (1994a) Diffusion and mass flow to a root with constant nutrient demand or behaving as a zero-sink. 1. Constant uptake. Soil Sci 157:162–170

De Willigen P, van Noordwijk M (1994b) Diffusion and mass flow to a root with constant nutrient demand or behaving as a zero sink. 2. Zero sink. Soil Sci 157:171–175

Fox TR, Comerford NB, McFee WW (1990) Phosphorus and aluminum release from spodic horizon mediated by organic acids. Soil Sci Soc Am J 54:1763–1767

Gherardi MJ, Rengel Z (2004) The effect of manganese supply on exudation of carboxylates by roots of lucerne (*Medicago sativa*). Plant Soil 260:271–282

Grierson PF, Smithson P, Nzigulheba G, Radersma S, Comerford NB (2004) Phosphorus dynamics and mobilization by plants. In: van Norodwijk M, Cadisch G, Ong CK (eds) Below-ground interactions in tropical agroecosystems: concepts and models with multiple plant components. World Agroforestry Centre (ICRAF), Nairobi, Kenya, pp 127–142

Guzel N, Buyuk G, Ibrikci K (2001) Non-exchangeable and exchangeable potassium status of soils in relation to clay mineralogy and other soil properties in Hilvan area of upper Mesopotamia in southeastern Anatolia. Commun Soil Sci Plant Anal 32:2887–2892

Harris WG, Wang HD, Reddy KR (1994) Dairy manure influence on soil and sediment composition – implications for phosphorus retention. J Environ Qual 23:1071–1081

Harris WG, Rhue RD, Kidder G, Brown RB, Littell R (1996) Phosphorus retention as related to morphology of sandy coastal plain soil materials. Soil Sci Soc Am J 60:1513–1521

Hinsinger P (2001) Bioavailability of soil inorganic P in the rhizosphere as affected by root-induced chemical changes: a review. Plant Soil 237:173–195

Hinsinger P, Plassard C, Tang CX, Jaillard B (2003) Origins of root-mediated pH changes in the rhizosphere and their responses to environmental constraints: a review. Plant Soil 248:43–59

Hoffland E (1992) Quantitative evaluation of the role of organic acid exudation in the mobilization of rock phosphate by rape. Plant Soil 140:279–289

Joner EJ, Jakobsen I (1994) Contribution by 2 arbuscular mycorrhizal fungi to P-uptake by cucumber (*Cucumis sativus* L.) from P^{32} labeled organic matter during mineralization in soil. Plant Soil 163:203–209

Kamh M, Abdou M, Chude V, Wiesler F, Horst WJ (2002) Mobilization of phosphorus contributes to positive rotational effects of leguminous cover crops on maize grown on soils from northern Nigeria. J Plant Nutr Soil Sci 165:566–572

Kirk GJD, Solivas JL, Alberto MC (2003) Effects of flooding and redox conditions on solute diffusion in soil. Eur J Soil Sci 54:617–624

Kochian LV, Hoekenga OA, Pineros MA (2004) How do crop plants tolerate acid soils? Mechanisms of aluminum tolerance and phosphorus efficiency. Annu Rev Plant Biol 55:459–493

Koopmans GF, Chardon WJ, de Willigen P, van Riemsdijk WH (2004) Phosphorus desorption dynamics in soil and the link to a dynamic concept of bioavailability. J Environ Qual 33:1393–1402

Lan M, Comerford NB, Fox TR (1995) Organic anions' effect on phosphorus release from spodic horizons. Soil Sci Soc Am J 59:1745–1749

Lewis DR, McGechan MB (2002) A review of field scale phosphorus dynamics models. Biosyst Eng 82:359–380

Liu Y, Mi GH, Chen FJ, Zhang JH, Zhang FS (2004) Rhizosphere effect and root growth of two maize (*Zea mays* L.) genotypes with contrasting P efficiency at low P availability. Plant Sci 167:217–223

Loosemore N, Straczek A, Hinsinger P, Jaillard B (2004) Zinc mobilization from a contaminated soil by three genotypes of tobacco as affected by soil and rhizosphere pH. Plant Soil 260:19–32

McCluskey J, Herdman L, Skene KR (2004) Iron deficiency induces changes in metabolism of citrate in lateral roots and cluster roots of *Lupinus albus*. Physiol Plant 121:586–594

McDowell R, Condron L (2001) Influence of soil constituents on soil phosphorus sorption and desorption. Comm Soil Sci Plant Anal 32:2531–2547

Nakamara Y, Nanzyo M, Yamasaki S (2000) Utilization of apatite in fresh volcanic ash by pigeonpea and chickpea. Soil Sci Plant Nutr 46:591–600

Nye PH, Tinker PB (1977) Solute movement in the soil-root system, Studies in ecology, vol 4. University of California Press, Berkeley

Peaslee EE, Phillips RE (1981) Phosphorus dissolution-desorption in relation to bioavailability and environmental pollution. In: Stelly M, Dowdy RH, Ryan JA, Volk VV, Baker DE (eds) Chemistry in the soil environment. Am Soc Agron Spec Publ 40:241–259

Radersma S, Griereson PF (2004) Phosphorus mobilization in agroforestry: organic anions, phosphatase activity and phosphorus fractions in the rhizosphere. Plant Soil 259:209–219

Raven KP, Hossner LR (1994) Sorption and desorption quantity-intensity parameters related to plant available soil phosphorus. Soil Sci Soc Am J 58:405–410

Rengel Z (1993) Mechanistic simulation models of nutrient uptake – a review. Plant Soil 152:161–173

Ryan PR, Delhaize E, Jones DL (2001) Function and mechanism of organic anion exclusion from plant roots. Annu Rev Plant Physiol Plant Mol Biol 52:527–560

Shen J, Tang C, Rengel Z, Zhang F (2004) Root-induced acidification and excess cation uptake by N_2-fixing *Lupinus albus* grown in phosphorus-deficient soil. Plant Soil 260:69–77

Silberbush M, Sorek S, Yakirevich A (1993) K^+ uptake by root systems grown in soils under salinity. I. A mathematical model. Transport Porous Media 11:101–116

Singh BB, Jones JP (1976) Phosphorus sorption and desorption characteristics of soil as affected by organic residues. Soil Sci Soc Am J 40:389–394

Smethurst PJ, Comerford NB (1993a) Simulating nutrient uptake by single or competing and contrasting root systems. Soil Sci Soc Am J 57:1361–1367

Smethurst PJ, Comerford NB (1993b) Potassium and phosphorus uptake by competing pine and grass: observations and model verification. Soil Sci Soc Am J 57:1602–1610

Soil Science Society of America (1996) Glossary of soil science terms. Soil Science Society of America, Madison, Wisconsin
Tinker PB, Nye PH (2000) Solute movement in the rhizosphere. Oxford University Press, Oxford
Uhde-Stone C, Gilbert G, Johnson JMF, Litjens R, Zinn KE, Temple SJ, Vance CP, Allan DL (2003) Acclimation of white lupin to phosphorus deficiency involves enhanced expression of genes related to organic acid metabolism. Plant Soil 248:99–116
Van Rees KCJ, Comerford NB, Rao PSC (1990a) Defining soil buffer power: implications for ion diffusion and nutrient uptake modeling. Soil Sci Soc Am J 54:1505–1507
Van Rees KCJ, Comerford NB, McFee WW (1990b) Modeling potassium uptake by slash pine seedlings from low potassium supplying soils of the southeastern coastal plain. Soil Sci Soc Am J 54:1413–1421
Veneklaas EJ, Stevens J, Cawthray GR, Turner S, Grigg AM, Lambers H (2003) Chickpea and white lupin rhizosphere carboxylates vary with soil properties and enhance phosphorus uptake. Plant Soil 248:187–197
Weinstein DA, Yanai RD (1994) Integrating the effects of simultaneous multiple stresses on plants using the simulation model TREGRO. J Environ Qual 23:418–428
Williams M, Yanai RD (1996) Multi-dimensional sensitivity analysis and ecological implications of a nutrient uptake model. Plant Soil 180:311–324
Wouterlood M, Cawthray GR, Turner S, Lamberes H, Veneklaas EJ (2004) Rhizosphere carboxylate concentrations of chickpea are affected by genotype and soil type. Plant Soil 261:1–10
Wright IJ, Reich PB, Westoby M (2003) Least-cost input mixtures of water and nitrogen for photosynthesis. Am Nat 161:98–111
Yanai RD (1994) A steady-state model of nutrient uptake accounting for newly grown roots. Soil Sci Soc Am J 58:1562–1571
Zhang XG, Jessop RS, Alter D (2003) Organic acid exudation associated with aluminium stress tolerance in triticale and wheat. Aust J Agric Res 54:979–985

2 Decomposition and Mineralization of Nutrients from Litter and Humus

C.E. Prescott

2.1 Introduction

In addition to soil physiochemical factors enumerated in Chapter 1 (this Vol.), the availability of nutrients to vegetation in natural ecosystems is also dependent on the nutrients being recycled from organic (largely vegetative) matter through the related processes of decomposition and nutrient mineralization. Decomposition refers to the physical breakdown of the material (usually measured as mass loss), while mineralization refers to the release of inorganic nutrients available for plant uptake. Decomposition is primarily a biological process resulting from enzymatic activities of soil microorganisms and influenced in a variety of ways by activities of the soil fauna (Visser 1985). This chapter highlights the factors that control rates of decomposition and nutrient mineralization. The focus is largely on forested ecosystems, and on nitrogen and phosphorus, because these elements are often in low availability relative to plant requirements in forest ecosystems.

2.2 Decomposition

Melillo et al. (1989) present a general model of the decomposition process from litter to humus in two phases. During the early stage, there is rapid loss of water-soluble components followed by rapid loss of cellulose from the litter. There is little loss, or possibly even a gain of insoluble decay products (collectively referred to as "lignin"). During the early phase, carbon is relatively available and nutrients are limiting, and there is immobilization of the limiting nutrient (usually N). Once the litter reaches the late stage of decomposition, it is considered to be humus and it is distinguished by a stabilized content and slow decay of all components. The late stage of decay is characterized by a net loss of lignin and net mineralization of N. Here, early- and late-stage

Ecological Studies, Vol. 181
H. BassiriRad (Ed.) Nutrient Acquisition by Plants
An Ecological Perspective

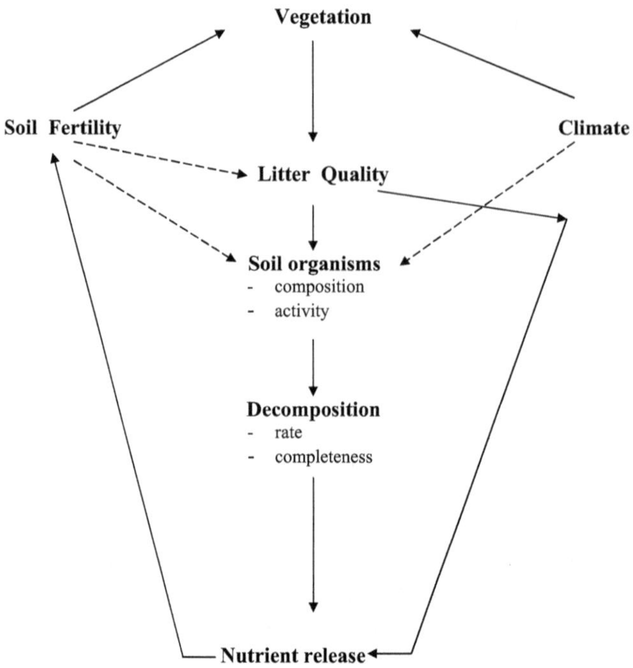

Fig. 2.1. A conceptual diagram depicting the factors that control rates of nutrient release from organic matter

decomposition will be discussed separately because the controlling factors and the methods for measuring rates differ.

An overview of the factors that control rates of nutrient release from organic matter is given in Fig. 2.1. According to this model, the species composition of vegetation is the primary determinant of the quality of litter produced in the ecosystem (i.e., its physical and chemical nature and decomposability). However, soil fertility and climate exert a tight control. The nature of the litter determines its palatability to soil organisms, and thus influences their composition and activity level. Soil biological activity is also influenced by soil fertility and climate (temperature and moisture). The activity of soil organisms determines the rate of decay and the completeness of decomposition. Finally, the rate of decay and concentrations of nutrients in the litter determine the rate of nutrient release, which creates a positive feedback to site fertility.

2.2.1 Early (Litter) Decomposition

Climate and litter quality are the main determinants of early rates of litter decomposition (Swift et al. 1979; Lavelle et al. 1993). Although the activity of soil organisms has been identified as a controlling factor (Lavelle et al. 1993; Couteaux et al. 1995), the rate of microbial activity should be considered as a mechanism by which the influences of climate and litter quality are realized (Fig. 2.1). Climate is the primary determinant of decay rate over broad geographical scales (Meentemeyer 1978; Berg et al. 1993; Aerts 1997). Decay will be uniformly slow in climates that are too cold, dry or wet to support a community of decomposers to operate at or near optimum (Couteaux et al. 1995). As a result, decay rates are generally high in the humid tropics and decline with increasing latitude or elevation, and are low in Mediterranean ecosystems, deserts and bogs (Swift et al. 1979). These differences are related to both climate and litter quality, as litter tends to be more recalcitrant in cold or dry environments (Aerts 1997). The relationship might best be described as climate determining the general range of decay rates, and litter quality determining the relative rates of litters within the range set by climate. For example, Berg et al. (1993) found that actual evapotranspiration (AET) best explained mass loss rates across a broad range of sites, but within a site most of the variation was explained by litter quality variables.

2.2.1.1 Moisture

In general, litter moisture contents in excess of 150% or below 30% (dry weight basis) tend to slow litter decomposition (Haynes 1986). Within this range, decomposition rates will increase with increasing moisture if temperature is adequate (Bunnell et al. 1977). Substantially greater rates of decay have been reported at moist sites compared to dry sites having similar elevation and temperature (Vitousek et al. 1994), and following irrigation of dry forests (Raison et al. 1990). Very high moisture will also severely constrain the activity of microbial communities (Bunnell et al. 1977; Couteaux et al. 1995). Schuur et al. (2001) reported a reduction in litter decomposition rates across a gradient of annual precipitation ranging from 2,200 to >5,000 mm, with a corresponding reduction in soil reduction–oxidation potentials. Improved aeration associated with the drainage of peat bogs has been shown to increase the rates of decomposition of litter (Lieffers 1988).

2.2.1.2 Temperature

It is very difficult to separate the effects of temperature on early decomposition from that of many other environmental factors, because soil temperature often co-varies with other factors that also affect decomposition. Moreover, decomposer organisms have a wide range of temperature optima (0–45 °C; Paul and Clark 1996), even though their activities often show a positive correlation with increased temperature (Swift et al. 1979). Vitousek et al. (1994) reported increased rates of decomposition along a decreasing elevation gradient in Mauna Loa, Hawaii, which was most strongly associated with temperature as opposed to precipitation or substrate quality. They found that a change of 10 °C in mean annual temperature was associated with a 4- to 11-fold increase in the decomposition of leaf litter of a dominant species at that site. Numerous manipulative experiments, including artificial warming of litter layers, have also shown that increased temperature results in higher rates of CO_2 evolution and mass loss (Q_{10}>2; PeterJohn et al. 1993; Hobbie 1996; Winkler et al. 1996).

Actual evapotranspiration, which incorporates both temperature and moisture, is usually the best predictor of decomposition rates across a range of climates. AET was correlated with mass loss of Scots pine needles along a climatic transect in Scandinavia (Meentemeyer and Berg 1986). AET was also correlated with first-year mass loss of a broad range of climatic conditions and leaf litters (Dyer et al. 1990), and with Scots pine needle litter along a transect from the sub-Arctic to the subtropics (Berg et al. 1993). In contrast, annual mean temperature best predicted first-year mass loss of root litter along a climatic transect from the Arctic circle to Germany (Berg et al. 1998). The lack of effect of moisture was attributed to these sites all being moist. Three-year mass loss across Canada was also best correlated with mean annual temperature (Moore et al. 1999). Inclusion of mean annual precipitation further improved the relationship, but AET was no better, indicating that summer moisture levels may not significantly constrain decomposition rates of litter in Canada.

2.2.1.3 Litter Quality

Decomposition rate is also determined by the chemical "quality" of the litter, which is largely a function of the relative proportions of major groups of C compounds in the litter. These groups, in order of decreasing decomposability, are sugars, cellulose, lignin, and phenols (Minderman 1968). Here, "lignin" is defined as the Klason lignin component, which is the acid-insoluble (AIS) residue that includes lignin, tannins, and cutin (Preston et al. 1997). The lignin content of the litter is often negatively correlated with the rate of decay (Meentemeyer 1978; Melillo et al. 1982; Tian et al. 1992; Van Vuuren et al. 1993). Initial concentrations of N or P in litter also often correlate with early decay rate (Witkamp 1966; Taylor et al. 1989b; Tian et al. 1992). Seneviratne (2000) suggested that N contents lower than 2 % limit the decomposition of tropical litters. Some of the apparent effect of N may actually be the result of low levels of polyphenols that usually accompany high concentrations of N in litter (Haynes 1986).

Relationships between decay rates and litter P concentrations have also been reported at sites where P availability is low due to either edaphic factors or N deposition (Aerts and de Caluwe 1989; Vitousek et al. 1994; Vesterdal 1999). Indices that incorporate both C chemistry and nutrient content, such as

C:N or lignin:N or C:P ratios, are often negatively correlated with early decay rates (Moore et al. 1999). In nutrient-poor or ericaceous systems, litter decay is most closely related to indices that also incorporate the concentrations of phenolic compounds, such as phenolics:N or P ratios (Aerts and de Caluwe 1989). Cutin may be more important in Mediterranean ecosystems where sclerophyllous tissues are more common. Thus, Gallardo and Merino (1993) found a significant negative correlation between cutin:N and decomposition in a dry, nutrient-poor Mediterranean ecosystem.

Other attributes of litter, such as toughness, also influence the rate of decay. For example, leaf toughness is a good predictor of the rate of litter mass loss in Mediterranean ecosystems (Gallardo and Merino 1993), and may be involved in the relationship between decay rates and AIS or cutin contents. Perez-Harguindeguy et al. (2000) found that leaf toughness or tensile strength was a good predictor of decomposition rate in a broad range of species. In the same study, the C:N ratio was also found to be a good predictor of decomposition rate, due largely to the fact that higher C:N values are often associated with compounds showing higher C enrichment, particularly lignin. The importance of toughness may also be implicit in the pattern of litter consumption by soil fauna, which is often delayed until litter becomes "softened" (Cortez 1998) or "conditioned" (Driebe and Whitham 2000) by microorganisms. Physical accessibility for microbes may also be an important determinant of decay rate, as suggested by the observed increase in microbial activity or decomposition once litter is ground (Fyles and McGill 1987).

There are three important caveats to keep in mind when assessing the relationship between litter "quality" and decomposition rate. First, the litter quality factor that best correlates with decomposition rates depends on the range of litter types examined. Across a broad range of substrates, C:N may be the best indicator of decay rates (Taylor et al. 1989a; Enriquez et al. 1993; Constantinides 1994; Seneviratne 2000); other factors assume greater importance within narrower ranges of substrates. Lignin:N appears to be a good predictor of decay rate in litters with initial lignin contents between 10 and 28 % (Taylor et al. 1989a, 1991). In litters with lignin contents <10 %, other factors limit decomposition; in litters with lignin contents >28 %, decay is limited by the high lignin content but a correlation may not be apparent, because decay rates will be uniformly low. Second, the factors that best correlate with decay rate depend on the length of the study. For example, Aerts and de Caluwe (1997) found that decay during the first 3 months was strongly related to P concentrations, whereas decay after 1 year was related more to concentrations of lignin and phenolics. Third, although the factor that is best correlated with decomposition rates can be used to predict decomposition rates, it does not necessarily elucidate the mechanism of action. Rather, it may simply be the factor that best integrates the array of physical and chemical characteristics of litter that determines decay rate. Given the high degree of autocorrelation among chemical characteristics of litter (Taylor et al. 1991; Parton et al. 1994), and their biological integration within tissues, the search for a single factor that controls decay rate may be a fruitless exercise.

Cornelissen et al. (2000) proposed that "functional attributes" of plants (i.e., "features that enhance the functioning of living plants and their leaves in their natural environment"; Cornelissen 1996), such as growth form, toughness and even autumn coloration, may be effective ways of predicting the relative rates of decay of diverse litter types. In a wide range of species and life-form groups, Cornelissen et al. (1999) found highly significant correlations between litter mass loss rates and leaf tensile strength. Similarly, Wardle et al. (1998) found that most of the variation in rates of leaf and root decomposition among 20 species of herbaceous plants could be predicted by plant traits such as vegetative growth characteristics, N concentrations, and palatability to an invertebrate herbivore.

2.2.1.4 Nitrogen Availability

The generally faster decay of N-rich litters suggests that litter decay rates would increase if their N content were increased through N fertilization or deposition, or would decrease if the N content declined as a result of elevated atmospheric CO_2 levels. In contrast, rates of decay have not been consistently altered by changes in the C:N ratio of litter resulting from N additions (Berg et al. 1987; Titus and Malcolm 1987; Hunt et al. 1988; Prescott et al. 1992; Cotrufo and Ineson 1995; Prescott 1995) or elevated CO_2 (Couteaux et al. 1991, 2000; Cotrufo et al. 1994; Kemp et al. 1994; Hirschel et al. 1997; Finzi et al. 2001). For example, mass loss rates of pine needle litter from fertilized plots were the same as those for unfertilized litter, and much slower than those for green needles, despite having N concentrations similar to that of green needles (Prescott 1995). Likewise, greater availability of N in the surrounding environment does not consistently increase the rate of litter decay (Fog 1988; Prescott 1995; Magill and Aber 1998). The responses to nutrient additions may depend on the C chemistry of the litters. Many of the reports of stimulated decay with N or P fertilization have been for litters with low lignin contents (<12 %; Cotrufo and Ineson 1995; Prescott 1995; Kwabiah et al. 1999), while reports of N fertilization suppressing decay have been for high-lignin litter (16–25 %; Hunt et al. 1988; Prescott 1995; Magill and Aber 1998). Greater stimulation of low-lignin litter compared to high-lignin litter by the same N addition rate has been reported by Hunt et al. (1988) and Hobbie (2000). Carreiro et al. (2000) found that chronic applications of aqueous NH_4NO_3 increased the decomposition of low-lignin (6 %) dogwood litter, but reduced the decay of high-lignin (26 %) oak litter. They also noted that cellulase activity increased in low-lignin litter, but the activity of lignin-degrading phenol oxidase declined substantially in the high-lignin litter in response to N addition. Thus, additional N may stimulate the decomposition of low-lignin litters, but may suppress the decomposition of high-lignin litters.

2.2.2 Completeness of Decomposition

The exponential decay model (Olson 1963) is often used to extrapolate rates throughout the entire process from early rates. However, there have been several reports of litter mass loss reaching an asymptote, rather than decaying exponentially, beginning with a laboratory study by Howard and Howard (1974). They found that mass loss from six leaf litters was better described by an asymptotic function than by an exponential function. Comparing the long-term decay of a variety of foliar and root litters, Melillo et al. (1989) and Aber et al. (1990) observed a two-phase decay sequence; during the first phase (up to about 80 % mass loss), fractional mass loss was relatively constant, whereas during the second phase mass loss was imperceptible. They concluded that use of the exponential decay model would greatly overestimate the rate of decay of phase-2 material (humus or soil organic matter). Berg (1991) suggested that there is a maximum decomposition limit, beyond which the rate of decay is immeasurably slow. This explains the deep accumulations of undecomposed litter in undisturbed, nutrient-poor pine forests on northern islands (Wardle et al. 1997), and the extreme age of soil organic matter (Berg et al. 1995; Paul and Clark 1996).

The completeness of decomposition is largely a function of the nature of the litter. Berg and Ekbohm (1993) found that the maximum decomposition limit of needle litter varied among species; lodgepole pine had a limit of nearly 100 %, while the limit value for Scots pine was about 84 % mass loss. Berg et al. (1996) found that limit values differed more among species than within a species, regardless of incubation site. This is consistent with other indications that the pattern of decay (i.e., the shape of the mass loss or N release curves) is determined by the species or type of litter, while other factors determine the rate of decay (Ladd 1983, cited by Anderson 1992; Prescott et al. 1993b; Hobbie 1996).

Contrary to expectations from the negative exponential decay model, there have been reports that high-quality litters have lower maximum decay limits and produce humus. This has been observed in N-fertilized versus unfertilized foliar litter (Berg and Ekbohm 1991; Prescott 1995; Cotrufo et al. 2001), green versus brown needles (Berg and Ekbohm 1991), broadleaf versus needle litter (Berg and Ekbohm 1991; Berg et al. 1996; Prescott et al. 2001), and high-quality (lignin:N=10) compared to low-quality (lignin:N=35) roots (Van Vuuren et al. 1993). Magill and Aber (1998) found that leaf litter in plots receiving chronic N additions entered the humus stage of very slow decay with 30–50 % of the original mass remaining, compared to 20 % typical of most litters. These observations suggest that higher-quality litter decays faster only until it reaches its maximum decay limit, after which it decomposes as slowly as low-quality litter. The tendency for high-quality litters to produce more humus may be related to their high initial N concentrations. A negative relationship between the N concentration in litter and its maximum decomposition limit has been demonstrated in several studies (Berg 1988; Berg et al. 1996). This may be the result of reactions of N compounds with aromatic substances in the soil, yielding recalcitrant humic compounds (Berg et al. 1996).

It should be noted that the reports of decomposition limits have thus far been from studies using litterbags or laboratory setups in which large soil fauna are excluded. Will (1967) reported an abrupt slowing of decay of pine needles in litterbags that was not apparent in the litter surrounding the litterbags, and considered this to be an artifact of litterbags. As suggested by Berg and Ekbohm (1991), the asymptotes may reflect the point at which agents other than microbial decomposers, such as soil animals, are needed for the process to continue. Soil macrofauna are known to have a large effect on the decomposition of litter once it has undergone initial microbial decay. Thus, the maximum decay limit may only apply to systems such as northern coniferous forests in which the soil fauna are not abundant, due to either poor environmental conditions or low-quality litter (Howard and Howard 1974; Berg 1991).

2.2.3 Late-stage Decomposition (Humus and Soil Organic Matter)

Relative to litter, humus is low in carbohydrates (cellulose, hemicellulose), high in large polyphenolic molecules, and has fairly stable N:P ratios of 15–17. Nitrogen concentrations are higher than in litter, but are highly variable among humus types. Late-stage decay is usually measured as rates of C mineralization (respiration) or N mineralization during incubation of soil or humus samples. The major factors controlling rates of humus and soil organic matter decay are moisture, temperature, organic matter quality and N availability, soil texture, and availability of exogenous labile C.

2.2.3.1 Moisture

Mineralization of soil organic matter generally increases with increasing moisture content up to field capacity, and then declines (Stanford 1974; Howard and Howard 1993; Paul and Clark 1996). In the organic horizons of a spruce-fir forest, Arnold et al. (1999) found negligible microbial biomass at low (20 %) moisture compared with higher moisture levels, regardless of temperature. Transplanting humus to a drier environment (400 mm less precipitation) slowed C mineralization, despite the mean air temperature being

0.8 °C warmer (Bottner et al. 2000). Irrigation of forests accelerated the decomposition of soil organic matter and the mineralization of C and N (Polglase et al. 1995), and stimulated N release from the forest floor (Kim and Burger 1997). The influence of moisture on the decomposition of soil organic matter appears to be greater at high temperatures (Douglas and Tedrow 1959; Zak et al. 1999). Drainage of boreal bogs has been shown to increase rates of respiration and decomposition of peat, at least at nutrient-rich sites (Martikainen et al. 1995; Laine et al. 1996). It is not clear if the greater respiration results from increased decomposition of the old peat, or from root-associated respiration or decomposition of new litter (Domisch et al. 2000). Hogg et al. (1992) found that C release from deep (>20 cm) peat remained low even when exposed to aerated conditions at 24 °C. They suggested that deep peat layers are resistant to decay, regardless of temperature and moisture conditions. They also found that peat samples that remained waterlogged did not increase respiration rates, regardless of temperature, demonstrating the overriding effect of excessive moisture on decomposition.

2.2.3.2 Temperature

When moisture conditions are adequate, rates of decomposition of humus and soil organic matter generally increase with increasing temperature (Paul and Clark 1996). Increasing the temperature in a boreal black spruce forest floor by 9 °C for three summers increased respiration and N mineralization rates, and decreased the mass on the forest floor, including the humus layer (Van Cleve et al. 1981). In a meta-analysis of 32 research sites in four biomes, Rustad et al. (2001) concluded that 2–9 years of experimental warming (0.3–6.0 °C) increased soil respiration rates by abut 20 %, and net N mineralization rates by about 46 %. Katterer et al. (1998) reviewed 25 studies and determined that within the range of 5–25 °C, a Q_{10} of 2 adequately described the relationship between temperature and CO_2 evolution from the soil. Soil and humus CO_2 fluxes from cold as opposed to warmer sites (<5 °C) are even more responsive to rising temperature, as they often exhibit Q_{10} values greater than 2 (Kirschbaum 1995; Niklinska et al. 1999; Bottner et al. 2000; Grogan et al. 2000).

In contrast to the many studies reporting a positive relationship between temperature and decomposition, Giardina and Ryan (2000) compared rates of soil C decomposition from 82 sites worldwide and found that rates were remarkably constant across a global mean annual temperature gradient. They point out that the increases in C mineralization often reported in laboratory studies tend to be short-lived, and so indicate either depletion of a small, active C pool, or a return to pre-disturbance (sampling) conditions. In contrast, long-term incubations and estimates of C turnover along natural temperature gradients show small or transient responses to increased temperature. They suggest that effects of warming on the release of C from soil is constrained by the low C quality of soil organic matter, along with binding to clay particles and insufficient moisture.

2.2.3.3 Humus Quality

Once litter has been humified, it is a much poorer substrate for decomposing organisms, so decomposition is slow regardless of litter type or location. Giardina et al. (2001) concluded that the C chemistry and turnover of soil organic matter was remarkably constant on a global scale, and suggested that the turnover of soil C was primarily constrained by the nature of the material itself. Nevertheless, differences in decay rates of humus or soil organic matter have been reported under the same environmental conditions, suggesting that there is sufficient variation in the chemical quality of these materials to affect their decay rates. Nadelhoffer et al. (1991) and Howard and Howard (1993) measured C mineralization from several soils under varying temperature and moisture regimes, and found that differences among soils were greater than their responses to temperature. Johansson et al. (1995) found that mass loss of highly decomposed pine needles was more closely related to their lignin content than to prevailing climatic conditions. The effect of lignin was strongest at sites with high AET (i.e., warm and wet sites). Berg and Staaf (1980) suggested that lignin concentration is the dominant factor controlling decomposition in the late stage, and others have found that rates of late-stage decay are inversely correlated to the initial concentration of lignin in the litter (McClaugherty et al. 1985; Constantinides 1994; Berg et al. 2000). Cutin, a component of leaf cuticles, may be a dominant component during late-stage decomposition (Gallardo and Merino 1993), as it is among the most recalcitrant of plant materials. High initial concentrations of N in litter may slow late-stage decay; Berg et al. (1987) found that N-fertilized pine needles decomposed faster initially but then slower compared with unfertilized needles. This could, however, also be interpreted as the fertilized needles reaching the humus stage sooner, such that their decay rate slowed sooner than that of the unfertilized needles. Finally, it has been suggested that the rate of late-stage decay may be inversely related to initial litter quality. In a long-term laboratory incubation study, Giardina et al. (2001) found that more C was released from soil organic matter under pine (which produced low-quality litter) than that under aspen (which produced high-quality litter).

2.2.3.4 Nitrogen Availability

There is some evidence that available N may actually reduce rates of humus decomposition. Reductions in microbial activity and greater forest floor accumulations have been reported following additions of N fertilizer to coniferous forest floors (Soderstrom et al. 1983; Nohrstedt et al. 1989). Magill and Aber (1998) found that chronic N additions slowed decomposition, and particularly "lignin" decay in foliar litter. Berg et al. (1987) noted that needle litter from N-fertilized trees decomposed more slowly during later stages than was the case for un-enriched litter. They attributed this to the inhibitory effect of high N levels on lignolytic enzyme production by white rot fungi, which are the only fungi that can degrade lignin completely. Pure culture studies with lignin-degrading fungi have demonstrated greater synthesis of lignolytic enzymes and accelerated lignin degradation at low inorganic N concentrations, and suppression of lignolytic enzymes at high N availability (Keyser et al. 1978; Reid 1983). In a field study with fresh leaf litter, Carreiro et al. (2000) also found reduced activity of lignin-degrading phenol oxidase after chronic applications of aqueous NH_4NO_3, but only in a high-lignin litter. They further suggest that N deposition or fertilization of forests with high-lignin litter will reduce the abundance or activity of white rot fungi, leading to slower late-stage decay and greater C storage. Coupled with the evidence for slower and less-complete decay presented above, it can be hypothesized that N deposition, or fertilization of ecosystems with high-lignin litter will lead to greater accumulations of humus or soil organic matter, and greater sequestration of C therein.

Availability of nutrients other than N may also influence rates of late-stage decay. Johansson (1994) found a positive correlation between Ca concentration (up to 0.5 %) and annual lignin decomposition, and suggested that Ca may regulate rates of late-stage decomposition (>65 % mass loss). Berg et al. (1996) found a positive relationship between the maximum decomposition limit and Mn^{2+} concentration in litter, which they attributed to the role of Mn^{2+} in the activity of lignin-degrading enzymes.

2.2.3.5 Soil Texture

In addition to the abiotic factors discussed above, soil texture also has a significant influence on rates of mineralization of soil organic matter (also discussed in Chap. 1, this Vol.). Fine-textured soils are usually associated with high contents of organic matter, and low rates of mineralization relative to coarse-textured soils (Burke 1989; Paul and Clark 1996; Koutika et al. 1999; Jobbagy and Jackson 2000). This has been attributed to the mineral particles physically protecting the organic matter from microbial decomposition. Mechanisms for this protection include adsorption of organic particles on clay minerals, entrapment in pores or aggregates inaccessible to microbes, and effects on microbial turnover processes (Oades 1988; Van Veen and Kuikman 1990). However, Giardina et al. (2001) noted that under similar moisture and temperature conditions in laboratory studies, there is often no clear effect of clay content on soil C mineralization rates, and the effect of clay content on N mineralization rates has also been inconsistent. As Van Veen and Kuikman (1990) and Giardina et al. (2001) point out, soil texture is related to many other characteristics of the soil and site, so the relationships between texture and soil organic matter may be related to factors other than physical protection in soil. Fine-textured soils tend to have greater water- and nutrient-holding capacities and support higher plant productivity, and hence greater inputs of organic matter to soil. Pastor et al. (1984) found a positive correlation between the silt and clay contents in subsoil and aboveground production along a soil fertility gradient in Wisconsin forests. Reich et al. (1997) found higher aboveground net primary production on finer-textured soils than on coarser-textured soils, regardless of forest type. In Douglas-fir forests, rates of aboveground litter input and clay content were both positively correlated with soil C and N (Prescott et al. 2000). Forests at sites with fine-textured soils may also support species that produce litter of relatively high quality (Pastor et al. 1984), which might be expected to lead to faster decay and less soil organic matter. However, as discussed above, there is some evidence that high-quality litter has a lower maximum decomposition limit, and produces soil organic matter that mineralizes more slowly. Therefore, the high-quality litter produced on fine-textured soils might actually lead to higher levels of soil organic matter, with slower rates of decay in the long term. Thus, while there may be a general relationship between soil clay and organic matter content, relationships between texture and rates of C and N mineralization, and the mechanisms behind the texture effect are less clear.

2.2.3.6 Labile Carbon

Availability of carbon substrates is thought to be more limiting than N during later stages of decay, when the readily metabolizable C in litter has been exhausted or transformed into recalcitrant humic materials. The possibility that humus decomposition (hence, nutrient availability) may be stimulated by available C was raised by Flanagan and Van Cleve (1983). Studies of lignin degradation also appear to support the hypothesis that humus decomposition is stimulated by available C. The white rot fungi that are most active in degrading lignin require other, more degradable C substrates as their primary energy source (Drew and Kadam 1979; Paul and Clark 1996). However, white rot fungi do not compete well with soil organisms, and may be restricted to woody debris (Paul and Clark 1996). Studies of fungal succession in decomposing litter have shown that generalist fungi such as *Penicillium*, *Trichoderma*, *Oidiodendron*, *Mortierella*, *Sporomiella*, and *Fusarium* are common in humus (Visser and Parkinson 1975; Soderstrom 1978; Paul and Clark 1996), whereas lignin decomposition specialists such as white rot fungi are rare in humus. Thus, it is not yet clear if the fungi that are common in humus are stimulated by degradable C (or inhibited by inorganic N).

Addition of fresh residues has been observed to stimulate decomposition and net release of N from indigenous soil organic matter in agricultural systems (Haynes 1986; Scott et al. 1996), and increases in tree growth have been reported following straw additions to N-poor forests (Prescott et al. 1995; Prescott and Zabek 1997). However, Haynes (1986) concluded that the "priming effect" tends to be short-lived and small relative to the amount of humus or soil organic matter present. Experimental additions of labile C to forest floors have produced inconsistent results (Bradley et al. 1997). Additions of readily available C in sugars and starch usually stimulate microbial activity and biomass production, but reduce N availability in forest floors because of greater immobilization in microbial biomass (Baath et al. 1978; Prescott and MacDonald 1994; Magill and Aber 2000a). However, long-term increases in N availability have been reported following additions of C to forests (Groffman 1999). Grazing by soil fauna appears to be critical in determining whether greater C inputs lead to greater immobilization or mineralization of N (Clarholm 1985; Ingham 1985; Couteaux et al. 1991).

The available carbon necessary for humus degradation could arise from dissolved organic carbon (DOC) in throughfall, litter leachate, or plant root exudates. Solutions passing through the organic layer are enriched in DOC, and leaf and root litter are important sources of easily metabolizable DOC for microbes in the forest floor and soil (Magill and Aber 2000b). Plant roots secrete a variety of carbon substrates that stimulate microbial activity in the rhizosphere (Helal and Sauerbeck 1984; Grayston et al. 1996), and several studies have suggested that relationships between root exudates, rhizosphere microflora and soil fauna have the potential to accelerate nutrient cycling (Clarholm 1985; Biondini 1988; Zhu and Ehrenfeld 1996; Priha et al. 1999). Hagvar and Edsberg (2000) observed faster and more complete decay of litter in close contact with fine roots during a 9-year study of unenclosed pine needle litter. However, Hendrickson and Robinson (1984) suggested that root exudates increase the activity of rhizosphere microorganisms but have little effect on humified material. Parmelee et al. (1993) and Bradley and Fyles (1995) found that roots stimulated populations of soil organisms and rates of N mineralization in mineral soil but not in humus. In many systems, ectomycorrhizal fungi will retain the C from plant roots, but might themselves serve as a source of C for other soil organisms (Rygiewicz and Andersen 1994; Grayston et al. 1996). However, Setala (2000) concluded that the role of ectomycorrhizal fungi in fueling the soil food web has been overestimated, whereas their role in decomposing organic matter has been underestimated.

2.3 Nutrient Mineralization

Patterns of nutrient release from decomposing litter do not always follow those for mass or C losses. While some nutrients (e.g., Mg^{2+}, Ca^{2+}, K^+, Na^+) are released as fast or faster than C, N and P are usually retained in the litter during the initial stages of decay (Swift et al. 1979). This net immobilization may lead to an increase in tissue nutrient content, indicative of net import of N or P into the litter (Staaf 1980a). Immobilization of N during early decay has often been reported, and results in a gradual increase in N concentrations, and hence a narrowing of C:N ratios during decomposition (Staaf 1980a; Vesterdal 1999). For example, the C:N ratio of red pine needle litter declined throughout a 77-month incubation and with depth in the profile, from 140 in fresh litter to 86 in humus to 17 in the mineral soil (Melillo et al. 1989). The pattern of P release is more variable, with reports of initial release (Lousier and Parkinson 1978; Peterson and Rolfe 1982; Rustad 1994; Monleon and Cromack 1996; Titus and Malcolm 1999) and initial immobilization (Gosz et al. 1973; Bartos and DeByle 1981; Staaf and Berg 1982; Berg and Laskowski 1997). The tendency for N or P to accumulate or be released varies with species and with site of incubation (MacLean and Wein 1978; Edmonds 1980; Bartos and DeByle 1981; Kelly and Beauchamp 1987; Tripathi and Singh 1992a), but appears to be closely related to the initial concentrations of these nutrients in the litter. In general, the initial rate of nutrient release is positively correlated with the initial concentrations of N or P (MacLean and Wein 1978; Aber and Melillo 1980; Edmonds 1980; Berg and Ekbohm 1983; Yavitt and Fahey 1986; Stohlgren 1988). In a review of tropical agricultural systems, Seneviratne (2000) found a positive linear relationship between N release and N concentrations in plant residues with <2 % N (C:N<27). Net immobilization dominated in plant residues with N concentrations up to 1.5 %, which is comparable to the values of 1.7–1.8 % reported by others (Aber and Melillo 1980; Constantinides 1994).

As a consequence of these differences in nutrient release patterns, the C:N and C:P ratios of litters tend to converge through the decay process. This pattern is illustrated in Vesterdal's (1999) comparison of N and P release from beech and spruce leaf litter from sites of differing fertility. Litter from the poorest site had the highest initial C:N ratio but immobilized N more rapidly than litter from the other sites, resulting in convergence of C:N ratios at 20–25 for birch leaves, and 28–35 for spruce needles after decomposing for 2.5 years. The pattern for P was even more striking, with C:P ratios of litter declining at the poor site as P was immobilized, and increasing at the rich sites as P was released, such that the C:P ratios converged at about 400 in birch leaves, and 500 in spruce needles after 2.5 years. Similar convergence of P contents has been reported in foliar litter of other species (Edmonds 1980; Berg et al. 1987; Rustad 1994; Aerts and de Caluwe 1997; Cotrufo et al. 1999), and for a wide range of litter types (Prescott et al. 1993b). The rapid initial losses of P probably reflect the tendency for extra P to be stored as inorganic P, which is rapidly leached from the litter early in the decay process. Laskowski et al. (1995) reported similar convergence of K^+ concentrations in a wide range of litter types as the material approached the humus stage (60 % mass loss). As Rustad (1994) points out, this convergence of nutrient contents is in keeping with the decay filter hypothesis (Melillo et al. 1989), which holds that as decomposition proceeds, litters of widely divergent initial chemistries are transformed into humus that has a relatively uniform chemistry.

The timing of nutrient release from agricultural residues can be largely predicted from the divergence of the residue's C:N or C:P ratios from the "critical ratio" for that element. The critical C:N is generally in the range of 25–30 (1.4–1.8 % N) for agricultural residues (Allison 1973; Haynes 1986). There is wider variation in critical C:N ratios of forest litter, with critical C:N generally increasing with increasing recalcitrance of litter. Critical C:N ratios in forest foliar litters have been reported as being 23 and 35 (Edmonds 1979, 1980), 29 (Lousier and Parkinson 1978), 34 and 31 (Gosz et al. 1973), 39 (Staaf and Berg 1977), 25 to 50 (McClaugherty et al. 1985), and between 25 and 59 in various tissues of bamboo (Tripathi and Singh 1992b). Higher critical C:N ratios have been reported in pine needles (37–80; Staaf and Berg 1977; Klemmedson 1985; Hart and Firestone 1992) and grass roots (50; Seastedt et al. 1992). Net N release was reported in tropical litters with C:N up to 109 (Thomas and Asakawa 1993) and 75 (Robbins et al. 1987). Critical C:N ratios are even higher in woody materials such as cones (167; Staaf and Berg 1977), twigs (>100) and branches (>300; Edmonds 1987). Hart (1999) measured net release of N from well-decayed wood with C:N ratios of 117; Prescott et al. (1993a) reported net N release from decayed wood with C:N ratios of 157, 107 and 70, but net immobilization in decayed wood with a C:N of 196. As Burger and Pritchett (1984) point out, the relationship between N concentration and net mineralization of organic N may only hold for the readily mineralizable part of the decaying material.

Critical C:P ratios for net P release from foliar litter has been reported as being 480 (Gosz et al. 1973), 350–450 (Rustad and Cronan 1988), 230 (Lousier and Parkinson 1978), 400 (Staaf 1980b), and >376 (Schlesinger and Hasey 1981). Edmonds (1980) found a large net release of P from Douglas-fir needle litter with a C:P ratio of 255, smaller net releases from alder (574) and Pacific silver fir (694), and net P immobilization in western hemlock (1,672).

Initial patterns of N and P accumulation and release may be best predicted by the N:P ratio of the litter. There is a general tendency for forest floor humus layers to have N:P ratios near 15 (Gosz et al. 1973; Lousier and Parkinson 1978; Vogt et al. 1986; Berg and Laskowski 1997). Thus, litters with initial N:P>15 are more likely to release N and retain P, while those with N:P<15 are likely to retain N and release P.

The exogenous availability of nutrients can influence patterns of uptake and release of N and P. Several studies have reported greater immobilization of N in litter placed on naturally N-rich or N-fertilized forest floors (Hirschfield et al. 1984; O'Connell 1986; Pastor et al. 1987; Berg and Tamm 1994; Chadwick et al. 1998) or mixed with N-rich litter (Staaf 1980b; Berg and Ekbohm 1983; Taylor et al. 1989b). Other investigators, however, have not found higher N immobilization at sites that are naturally N-rich (McClaugherty et al. 1985), N-fertilized (Prescott et al. 1992) or have N-fixers (White 1988). Vesterdal (1999) found greater immobilization of P in litter incubated at a site least enriched in P, but concluded that the incubation site did not influence N and P dynamics in litter as much as the initial N and P status of the litter.

Immobilization of N in litter is often explained as the consequence of the low concentrations of N relative to the requirements of decomposer organisms, hence as an indication of the extent to which they are limited by nitrogen. Berg and Soderstrom (1979) measured an increase in fungal biomass during decomposition of Scots pine needle litter that corresponded with the increase in absolute amount of N in the needles. Alternatively, increased uptake of N in fertilized or clear-cut forests may be related to higher availability of ammonium that can be chemically fixed (as ammonia) in litter and humus throughout the decay process, and has a particular affinity for fresh and low-N litters (Axelsson and Berg 1988). Accumulation of N also appears to be related to the tendency of N to become chemically immobilized during the humification process, particularly in high-lignin litters. The amount of N immobilized during decomposition is positively related to the initial lignin concentration of litter (Aber and Melillo 1982; Melillo et al. 1982; McClaugherty et al. 1985; Upadhyay and Singh 1985; White 1988). This increase has been attributed to the complexing of soluble polyphenols with protein (Berg and Theander 1984; Schlesinger 1985) as well as to the microbial synthesis of lignin-like materials (Paul and Clark 1996). Thus, increases in N content during decay could be related to either of these three processes, and so do not necessarily indicate the degree to which N "limits" decomposition.

The previous section has largely focused on gross mineralization, but in addressing nutrient availability to plants, immobilization may be extremely important. A substantial portion of the mineralized nutrients may be immobilized by microorganisms, and thus rates of net mineralization are often much lower than rates of gross mineralization. For example, in forest floor and surface soils of a boreal forest, Stottlemeyer and Toczydlowski (1999) determined that gross N mineralization rates were 23 times higher than net mineralization rates, and gross nitrification was 19 times higher than net nitrification rates. Competition between plant and microbes for mineralized nutrients is often intense (Kaye and Hart 1997). However, the nutrients immobilized in microbial biomass are re-mineralized when the cells lyse as a consequence of changing environmental conditions or in response to faunal grazing (Haynes 1986). Thus, on broader (annual) time scales, the rate of mineralization of nutrients in organic matter can be largely predicted from the initial characteristics of the litter, such as the %lignin, lignin:N or C:N ratios (Pastor et al. 1984; Wedin and Tilman 1990; Scott and Binkley 1997), or the nutrient content of annual litter fall (Reich et al. 1997; Prescott et al. 2000).

2.4 Summary

Many essential elements are made available to plants through the decomposition of organic matter and subsequent release in mineral form. The rate of litter decomposition is determined by temperature and moisture conditions, and the chemical and physical characteristics of the litter. Deviations from optimum in any one factor will slow decay, regardless of the levels of the other factors. Very low or very high moisture conditions will severely constrain decomposition. If moisture conditions are adequate, decomposition rates increase with increasing temperature; Q10 is usually near 2 but may be >2 at low temperatures.

The carbon chemistry and nutrient concentrations of the litter also influence the decay rate, and indices such as C:N or lignin:N ratios are good predictors of decay rate. The best predictor of decay rate varies according to the litter types examined, and the length of incubation. The C:N ratio is the best predictor across wide ranges of litters; lignin:N is a good predictor for litters with lignin concentrations between 10 and 28 %. Models that incorporate some measures of the physical nature and palatability of the litter are more robust in predicting decomposition rates of different litter types. Nitrogen availability also influences decay rate; N addition stimulates the decomposition of low-lignin litter (<12 %), but slows the decay of high-lignin (>18 %) litter. Alterations in C:N within a litter type do not affect decay rates as much as the same differences between litter types.

Contrary to the assumptions of the negative exponential decay model, mass loss often reaches an asymptote, with considerable mass remaining. This maximum decomposition limit is determined by the species or litter type, and appears to be lower in high-quality litter. Nitrogen-rich litter that decomposes fast initially may produce more humus that decays more slowly than does slowly decomposing, N-poor litter.

Decomposition of humus and soil organic matter is slow as a consequence of the low quality of the resource for decomposing organisms. Improvements in temperature and moisture conditions may increase decay rates to some extent, within the bounds set by substrate quality. Soil clay and organic matter contents are positively correlated, but the relationships between texture and rates of C and N mineralization are less clear. High concentrations of inorganic N may reduce rates of humus decomposition by inhibiting the activity of lignin-degrading enzymes. Labile C may stimulate these enzymes, and hence enhance humus decay, although the evidence is inconclusive.

Nutrients may be immobilized or released early in decomposition, depending on their initial concentration in the litter. Concentrations of N and P (hence, C:N and C:P ratios) tend to converge during decomposition, the nutrients being released from nutrient-rich litter and immobilized in nutrient-poor litter. The critical C:N and C:P ratios at which nutrient release begins vary among litter types, and generally increase with increasing recalcitrance of the litter. Exogenous N or P availability has a smaller influence on nutrient release than does their initial concentration in the litter.

References

Aber JD, Melillo JM (1980) Litter decomposition: measuring relative contributions of organic matter and nitrogen to forest soils. Can J Bot 58:416–421

Aber JD, Melillo JM (1982) Nitrogen immobilization in decaying hardwood leaf litter as a function of initial nitrogen and lignin content. Can J Bot 60:2263–2269

Aber JD, Melillo JM, McClaugherty CA (1990) Predicting long-term patterns of mass loss, nitrogen dynamics, and soil organic matter formation from initial fine litter chemistry in temperate forest ecosystems. Can J Bot 68:2201–2208

Aerts R (1997) Climate, leaf litter chemistry and leaf litter decomposition in terrestrial ecosystems: a triangular relationship. Oikos 79:439–449

Aerts R, de Caluwe H (1989) Aboveground productivity and nutrient turnover of *Molina caerulea* along an experimental gradient of nutrient availability. Oikos 54:320–325

Aerts R, de Caluwe H (1997) Nutritional and plant-mediated controls on leaf litter decomposition of *Carex* species. Ecology 78:244–260

Allison FE (1973) Soil organic matter and its role in crop production. Elsevier, Amsterdam

Anderson JM (1992) Responses of soils to climate change. Adv Ecol Res 22:163–210

Arnold SS, Fernandez I, Rustad LE, Zibilske LM (1999) Microbial response of an acid forest soil to experimental soil warming. Biol Fert Soils 30:239–244

Axelsson G, Berg B (1988) Fixation of ammonia (^{15}N) to *Pinus sylvestris* needle litter in different stages of decomposition. Scand J For Res 3:273–279

Baath E, Lohm U, Lundgren B, Rosswall T, Soderstrom B, Sohlenius B, Wiren A (1978) The effect of nitrogen and carbon supply on the development of soil organism populations and pine seedlings: a microcosm experiment. Oikos 31:153–163

Bartos DL, DeByle NV (1981) Quantity, decomposition, and nutrient dynamics of aspen litter fall in Utah. For Sci 27:381–390

Berg B (1988) Dynamics of nitrogen (^{15}N) in decomposing Scots pine (*Pinus sylvestris*) needle litter. Long-term decomposition in a Scots pine forest, VI. Can J Bot 66:1539–1546

Berg B (1991) Litter mass-loss rates and decomposition patterns in some needle and leaf litter types. Long-term decomposition in a Scots Pine forest, VII. Can J Bot 69:1449–1456

Berg B, Ekbohm G (1983) Nitrogen immobilization in decomposing needle litter at variable carbon:nitrogen ratios. Ecology 64:63–67

Berg B, Ekbohm G (1991) Litter mass-loss rates and decomposition patterns in some needle and leaf litter types. Long-term decomposition in a Scots pine forest, VII. Can J Bot 69:1449–1456

Berg B, Ekbohm G (1993) Decomposing needle litter in *Pinus contorta* (lodgepole pine) and *Pinus sylvestris* (Scots pine) monoculture systems – is there a maximum mass loss? Scand J For Res 8:457–465
Berg B, Laskowski R (1997) Changes in nutrient concentrations and nutrient release in decomposing needle litter in monocultural systems of *Pinus contorta* and *Pinus sylvestris* – a comparison and synthesis. Scand J For Res 12:113–121
Berg B, Soderstrom B (1979) Fungal biomass and nitrogen in decomposing Scots pine needle litter. Soil Biol Biochem 11:339–341
Berg B, Staaf H (1980) Decomposition rate and chemical changes in Scots pine needle litter. 2. Influence of chemical composition. Ecol Bull (Stockholm) 32:363–372
Berg B, Tamm CO (1994) Decomposition and nutrient dynamics of litter in long-term optimum nutrition experiments. ll. Nutrient concentrations in decomposing *Picea abies* needle litter. Scand J For Res 9:99–105
Berg B, Theander O (1984) Dynamics of some nitrogen fractions in decomposing Scots pine needle litter. Pedobiologia 27:261–267
Berg B, Staaf H, Wessen B (1987) Decomposition and nutrient release in needle litter from nitrogen-fertilized Scots pine (*Pinus sylvestris*) stands. Scand J For Res 2:399–415
Berg B, Berg MP, Bottner P, Box E, Breymeyer A, Calvan De Anta R, Couteaux MM, Esudero A, Gallardo A, Kratz W, Madeira M, Malkonen E, McClaugherty CA, Meentemeyer V, Munoz F, Piussi P, Remacle J, Virzo de Santo A (1993) Litter mass loss in pine forests of Europe and Eastern United States as compared to actual evapotranspiration on a European scale. Biogeochemistry 20:127–153
Berg B, McClaugherty CA, Virzo de Santo A, Johansson MB, Ekbohm G (1995) Decomposition of litter and soil organic matter – can we distinguish a mechanism for soil organic matter buildup? Scand J For Res 10:108–119
Berg B, Ekbohm G, Johansson M, McClaugherty C, Rutlgliano F, Virzo de Santo A (1996) Maximum decomposition limits of forest litter types: a synthesis. Can J Bot 74:659–672
Berg B, Johansson M, Meentemeyer V, Kratz W (1998) Decomposition of tree root litter in a climatic transect of coniferous forests in Northern Europe: a synthesis. Scand J For Res 13:402–412
Berg B, Johansson MB, Meentemeyer V (2000) Litter decomposition in a transect of Norway spruce forests: substrate quality and climate control. Can J For Res 30:1136–1147
Biondini M (1988) Carbon and nitrogen losses through root exudation by *Agropyron cristatum*, *A. Smithii* and *Bouteloua gracilis*. Soil Biol Biochem 20:477–482
Bottner P, Couteaux M-M, Anderson JM, Berg B, Billès G, Bolger T, Casabianca H, Romanyá J, Rovira P (2000) Decomposition of ^{13}C-labelled plant material in a European 65–40 latitudinal transect of coniferous forest soils: simulation of climate change by translocation of soils. Soil Biol Biochem 32:527–543
Bradley RL, Fyles JW (1995) Growth of paper birch (*Betula papyrifera*) seedlings increases soil available C and microbial acquisition of soil-nutrients. Soil Biol Biochem 27:1565–1571
Bradley RL, Fyles JW, Titus BD (1997) Labile C, humus form and N cycling in forests: concepts and methods. Recent Res Dev Soil Biol Biochem 1:63–76
Bunnell FL, Tait DEN, Flanagan PW, Van Cleve K (1977) Microbial respiration and substrate weight loss. I. A general model of the influences of abiotic variables. Soil Biol Biochem 9:33–40
Burger JA, Pritchett WL (1984) Effects of clearfelling and site preparation on nitrogen mineralization in a southern pine stand. Soil Sci Soc Am J 48:1432–1437
Burke IC (1989) Control of nitrogen mineralization in a sagebrush steppe landscape. Ecology 70:1115–1126

Carreiro MM, Sinsabaugh RL, Repert DA, Parkhurst DF (2000) Microbial enzyme shifts explain litter decay responses to simulated nitrogen deposition. Ecology 81:2359–2365

Chadwick DR, Ineson P, Woods C, Piearce TG (1998) Decomposition of *Pinus sylvestris* litter in litter bags: influence of underlying native litter layer. Soil Biol Biochem 30:47–55

Clarholm M (1985) Possible roles for roots, bacteria, protozoa and fungi in supplying nitrogen to plants. In: Fitter AH (ed) Ecological interactions in soil. Blackwell, Boston, pp 355–365

Constantinides M (1994) Nitrogen mineralization from leaves and litter of tropical plants: relationship to nitrogen, lignin and soluble polyphenol concentrations. Soil Biol Biochem 26:49–55

Cornelissen JHC (1996) An experimental comparison of leaf decomposition rates in a wide range of temperate plant species and types. J Ecol 84:573–582

Cornelissen JHC, Perez-Harguindeguy N, Diaz S, Grime JP, Marzano B, Cabido M, Vendramini F, Ceraolini B (1999) Leaf structure and defence control litter decomposition rate across species and life forms in regional floras on two continents. New Phytol 143:191–200

Cornelissen JHC, Perez-Harguindeguy N, Gwynn-Jones D, Diaz S, Callaghan TV, Aerts R (2000) Autumn leaf colours as indicators of decomposition rate in sycamore (*Acer pseudoplatanus* L.). Plant Soil 225:33–38

Cortez J (1998) Field decomposition of leaf litters: relationships between decomposition rates and soil moisture, soil temperature and earthworm activity. Soil Biol Biochem 30:783–793

Cotrufo MF, Ineson P (1995) Effects of enhanced atmospheric CO_2 and nutrient supply on the quality and subsequent decomposition of fine roots of *Betula pendula* Roth. and *Picea sitchensis* (Bong.) Carr. Plant Soil 170:267–277

Cotrufo MF, Ineson P, Rowland AP (1994) Decomposition of tree leaf litters grown under elevated CO_2: effect of litter quality. Plant Soil 163:121–130

Cotrufo MF, Raschi A, Lanini M, Ineson P (1999) Decomposition and nutrient dynamics of *Quercus pubescens* leaf litter in a naturally enriched CO_2 Mediterranean ecosystem. Funct Ecol 13:343–351

Cotrufo MF, Ineson P, Roberts D (2001) Decomposition of birch leaf litters with varying C-to-N ratios. Soil Biol Biochem 27:1219–1221

Couteaux M, Mousseau M, Célérier M, Bottner P (1991) Increased atmospheric CO_2 and litter quality: decomposition of sweet chestnut leaf litter with animal food webs of different complexities. Oikos 61:54–64

Couteaux MM, Bottner P, Berg B (1995) Litter decomposition climate and litter quality. Trends Ecol Evol 10:63–66

Couteaux M-M, Kurz C, Bottner P, Raschi A (2000) Influence of increased atmospheric CO_2 concentration on quality of plant material and litter decomposition. Tree Physiol 19:301–311

Domisch T, Finer L, Laiho R, Karsisto M, Laine J (2000) Decomposition of Scots pine litter and the fate of released carbon in pristine and drained pine mires. Soil Biol Biochem 32:1571–1580

Douglas LA, Tedrow JCF (1959) Organic matter decomposition rates in arctic soils. Soil Sci 88:305–312

Drew SW, Kadam KL (1979) Lignin metabolism by *Aspergillus fumigatus* and white-rot fungi. Dev Ind Microbiol 20:153–161

Driebe EM, Whitham TG (2000) Cottonwood hybridization affects tannin and nitrogen content of leaf litter and alters decomposition. Oecologia 123:99–107

Dyer ML, Meentemeyer V, Berg B (1990) Apparent controls of mass loss rate of leaf litter on a regional scale. Scand J For Res 5:311–323
Edmonds RL (1979) Decomposition and nutrient release in Douglas-fir needle litter in relation to stand development. Can J For Res 9:132–140
Edmonds RL (1980) Litter decomposition and nutrient release in Douglas-fir, red alder, western hemlock, and Pacific silver fir ecosystems in western Washington. Can J For Res 10:327–337
Edmonds RL (1987) Decomposition rates and nutrient dynamics in small diameter woody litter in four forest ecosystems in Washington, U.S.A. Can J For Res 17:499–509
Enriquez S, Duarte CM, Sand-Jensen K (1993) Patterns in decomposition rates among photosynthetic organisms: the importance of detritus C:N:P content. Oecologia 94:457–471
Finzi AC, Allen AS, DeLucia EH, Ellsworth DS, Schlesinger WH (2001) Forest litter production, chemistry, and decomposition following two years of free-air CO_2 enrichment. Ecology 82:470–484
Flanagan PW, Van Cleve K (1983) Nutrient cycling in relation to decomposition and organic matter quality in taiga ecosystems. Can J For Res 13:795–817
Fog K (1988) The effect of added nitrogen on the rate of decomposition of organic matter. Biol Rev 63:433–462
Fyles JW, McGill WB (1987) Decomposition of boreal forest litters from central Alberta under laboratory conditions. Can J For Res 17:109–114
Gallardo A, Merino J (1993) Leaf decomposition in two Mediterranean ecosystems of southwest Spain: influence of substrate quality. Ecology 74:152–161
Giardina CP, Ryan MG (2000) Evidence that decomposition rates of organic carbon in mineral soil do not vary with temperature. Nature 404:858–861
Giardina CP, Ryan MG, Hubbard RM, Binkley D (2001) Tree species and soil textural controls on carbon and nitrogen mineralization rates. Soil Sci Soc Am J 65:1272–1279
Gosz JR, Likens GE, Bormann FH (1973) Nutrient release from decomposing leaf and branch litter in the Hubbard Brook Forest, New Hampshire. Ecol Monogr 43:173–191
Gower ST, Son Y (1992) Differences in soil and leaf litterfall nitrogen dynamics for five forest plantations. Soil Sci Soc Am J 56:1959–1966
Grayston SJ, Vaughan D, Jones D (1996) Rhizosphere carbon flow in trees, in comparison with annual plants: the importance of root exudation and its impact on microbial activity and nutrient availability. Appl Soil Ecol 5:29–56
Groffman PM (1999) Carbon additions increase nitrogen availability in northern hardwood forest soils. Biol Fert Soils 29:430–433
Grogan P, Bruns TD, Chapin III FS (2000) Fire effects on ecosystem nitrogen cycling in a Californian bishop pine forest. Oecologia 122:537–544
Hagvar S, Edsberg E (2000) Vertical transport of decomposing spruce needles during nine years in a raw humus soil profile in southern Norway. Pedobiologia 44:119–131
Hart SC (1999) Nitrogen transformations in fallen tree boles and mineral soil of an old-growth forest. Ecology 80:1385–1394
Hart SC, Firestone MK (1992) Decomposition and nutrient dynamics of Ponderosa pine needles in a Mediterranean-type climate. Can J For Res 22:306–314
Haynes RJ (1986) Mineral nitrogen in the plant-soil system. Academic Press, Toronto
Helal HM, Sauerbeck DR (1984) Influence of plant roots on C and P metabolism in soil. Plant Soil 76:175–182
Hendrickson OQ, Robinson JB (1984) Effects of roots and litter on mineralization processes in forest soil. Plant Soil 80:391–405
Hirschel G, Korner C, Amone JA (1997) Will rising atmospheric CO_2 affect leaf litter quality and *in situ* decomposition rates in native plant communities? Oecologia 110:387–392

Hirschfield JR, Finn JT, Patterson WA (1984) Effects of *Robinia pseudoacacia* on leaf litter decomposition and nitrogen mineralization in a northern hardwood stand. Can J For Res 14:201–205

Hobbie SE (1996) Temperature and plant species control over litter decomposition in Alaskan tundra. Ecol Monogr 66:503–522

Hobbie SE (2000) Interactions between litter lignin and soil nitrogen availability during leaf litter decomposition in a Hawaiian montane forest. Ecosystems 3:484–494

Hogg EH, Lieffers VJ, Wein RW (1992) Potential carbon losses from peat profiles: effects of temperature, drought cycles, and fire. Ecol Appl 2:298–306

Howard PJA, Howard DM (1974) Microbial decomposition of tree and shrub leaf litter. 1. Weight loss and chemical composition of decomposing litter. Oikos 25:341–352

Howard DM, Howard PJA (1993) Relationships between CO_2 evolution, moisture content and temperature for a range of soil types. Soil Biol Biochem 25:1537–1546

Hunt HW, Ingham ER, Coleman DC, Elliott ET, Reid CPP (1988) Nitrogen limitation of production and decomposition in prairie, mountain meadow, and pine forest. Ecology 69:1009–1016

Ingham RE (1985) Interactions of bacteria, fungi, and their nematode grazers: effects on nutrient cycling and plant growth. Ecol Monogr 55:119–140

Jenny H, Gessel SP, Bingham FS (1949) Comparative studies of decomposition rates of organic matter in temperate and tropical regions. Soil Sci 68:419–432

Jobbagy EG, Jackson RB (2000) The vertical distribution of soil organic carbon and its relation to climate and vegetation. Ecol Appl 10:423–436

Johansson MB (1994) Decomposition rates of Scots pine needle litter related to site properties, litter quality, and climate. Can J For Res 24:1771–1781

Johansson MB, Berg B, Meentemeyer V (1995) Litter mass loss rates in late stages of decomposition in a climatic transect of pine forests. Long-term decomposition in a Scots pine forest 9. Can J Bot 73:1509–1521

Katterer T, Reichstein M, Andren O, Lomander A (1998) Temperature dependence of organic matter decomposition: a critical review using literature data analyzed with different models. Biol Fert Soils 27:258–262

Kaye JP, Hart SC (1997) Competition for nitrogen between plants and soil organisms. Trends Ecol Evol 12:139–143

Kelly JM, Beauchamp JJ (1987) Mass loss and nutrient changes in decomposing upland oak and mesic mixed-hardwood leaf litter. Soil Sci Soc Am J 51:1616–1625

Kemp PR, Waldecker D, Reynolds JF, Virginia RA, Owensby CE (1994) Effects of elevated CO_2 and nitrogen fertilization pretreatments on decomposition of tallgrass prairie leaf litter. Plant Soil 165:115–127

Keyser PT, Kirk TK, Zeikus JG (1978) Lignolytic enzyme system of *Phanaerochaete chrysosporium*: synthesized in the absence of lignin in response to nitrogen starvation. J Bacteriol 135:790–797

Kim DY, Burger JA (1997) Nitrogen transformations and soil processes in a wastewater-irrigated, mature Appalachian hardwood forest. For Ecol Manage 90:1–11

Kirschbaum MUF (1995) The temperature dependence of soil organic matter decomposition, and the effect of global warming on soil organic C storage. Soil Biol Biochem 27:753–760

Klemmedson JO (1985) Needle decomposition and nutrient release in ponderosa pine ecosystems. For Sci 31:647–660

Koutika L-S, Chone T, Andreux F, Burtin G, Cerri CC (1999) Factors influencing carbon decomposition of topsoils from the Brazilian Amazon Basin. Biol Fert Soils 28:436–438

Kwabiah AB, Voroney RP, Palm CA, Stoskopf NC (1999) Inorganic fertilizer enrichment of soil: effect on decomposition of plant litter under subhumid tropical conditions. Biol Fert Soils 30:224–231

Ladd JN (1983) Decomposition of plant material in Australian soils. I. The effect of quantity added on decomposition and on residual microbial biomass. Aust J Soil Res 21:563–570
Laine J, Silvola J, Tolonen K, Alm J, Nykanen H, Vasander H, Sallantaus T, Savolainen I, Sinisalo J, Martikainen PJ (1996) Effect of water level drawdown in northern peatlands on the global climatic warming. Ambio 25:179–184
Laskowski R, Berg B, Johansson MB, McClaugherty C (1995) Release pattern for potassium from decomposing forest needle and leaf litter. Long-term decomposition in a Scots pine forest 9. Can J Bot 73:2019–2027
Lavelle P, Blanchart E, Martin A, Martin S, Spain A, Toutan F, Barois I, Schaefer R (1993) A hierarchical model for decomposition in terrestrial ecosystems: application to soils of the humid tropics. Biotropica 25:130–150
Lieffers VJ (1988) Sphagnum and cellulose decomposition in drained and natural areas of an Alberta peatland. Can J Soil Sci 68:755–762
Lousier JD, Parkinson D (1978) Chemical element dynamics in decomposing leaf litter. Can J Bot 56:2795–2812
MacLean DA, Wein RW (1978) Weight loss and nutrient changes in decomposing litter and forest floor material in New Brunswick forest stands. Can J Bot 56:2730–2749
Magill AH, Aber JD (1998) Long-term effects of experimental nitrogen additions on foliar litter decay and humus formation in forest ecosystems. Plant Soil 203:301–311
Magill AH, Aber JD (2000a) Variation in soil net mineralization rates with dissolved organic carbon additions. Soil Biol Biochem 32:597–601
Magill AH, Aber JD (2000b) Dissolved organic carbon and nitrogen relationships in forest litter as affected by nitrogen deposition. Soil Biol Biochem 32:603–613
Martikainen P, Nykanen H, Almeida-Cortez JS, Silvola J (1995) Changes in fluxes of carbon dioxide, methane and nitrous oxide due to forest drainage of mire sites of different trophy. Plant Soil 168:571–577
McClaugherty CA, Pastor J, Aber JD (1985) Forest litter decomposition in relation to soil nitrogen dynamics and litter quality. Ecology 66:266–275
Meentemeyer V (1978) Macroclimate and lignin control of litter decomposition rates. Ecology 59:465–472
Meentemeyer V, Berg B (1986) Regional variation in mass-loss of *Pinus sylvestris* needle litter in Swedish pine forests as influenced by climate and litter quality. Scand J For Res 1:167–180
Melillo JM, Aber JD, Muratore JF (1982) Nitrogen and lignin control of hardwood leaf litter decomposition dynamics. Ecology 63:621–626
Melillo JM, Aber JD, Linkins AE, Ricca A, Fry B, Nadelhoffer KJ (1989) Carbon and nitrogen dynamics along the decay continuum: plant litter to organic matter. Plant Soil 115:189–198
Minderman G (1968) Addition, decomposition and accumulation of organic matter in forests. J Ecol 56:355–362
Monleon VJ, Cromack K (1996) Long-term effects of prescribed underburning on litter decomposition and nutrient release in ponderosa pine stands in central Oregon. For Ecol Manage 81:143–152
Moore TR, Trofymow JA, Taylor B, Prescott C, Camire C, Duschene L, Fyles J, Kozak L, Krannabetter M, Morrison I, Siltanen M, Smith S, Titus B, Visser S, Wein R, Zoltai S (1999) Litter decomposition rates in Canadian forests. Global Change Biol 5:75–82
Nadelhoffer KJ, Giblin AE, Shaver GR, Laudre JA (1991) Effects of temperature and substrate quality on element mineralization in six arctic soils. Ecology 72:242–253
Niklinska M, Maryanski M, Laskowski R (1999) Effect of temperature on humus respiration rate and nitrogen mineralization: implications for global climate change. Biogeochemistry 44:239–257

Nohrstedt H-O, Arnebrant K, Baath E (1989) Changes in carbon content, respiration, ATP content, and microbial biomass in nitrogen-fertilized pine forest soil in Sweden. Can J For Res 19:323–328

Oades JM (1988) The retention of organic matter in soils. Biogeochemistry 5:33–70

O'Connell AM (1986) Effect of legume understorey on decomposition and nutrient content of eucalypt forest litter. Plant Soil 92:235–248

Olson JS (1963) Energy storage and the balance of producers and decomposers in ecological systems. Ecology 44:322–331

Parmelee RW, Ehrenfeld JG, Tate RLI (1993) Effects of pine roots on microorganisms, fauna, and nitrogen availability in two soil horizons of a coniferous forest spodosol. Biol Fert Soils 15:113–119

Parton WJ, Schimel DS, Ojima DS, Cole CB (1994) A general model for soil organic matter dynamics. In: Bryant RB, Arnold RW (eds) Sensitivity to litter chemistry, texture and management. Soil Sci Soc Am Spec Publ 38:137–167

Pastor J, Aber JD, McClaugherty CA, Melillo JM (1984) Aboveground production and N and P cycling along a nitrogen mineralization gradient on Blackhawk Island, Wisconsin. Ecology 65:256–268

Pastor J, Stillwell MA, Tilman D (1987) Little bluestem litter dynamics in Minnesota old fields. Oecologia 72:327–330

Paul EA, Clark FE (1996) Soil microbiology and biochemistry. Academic Press, San Diego

Perez-Harguindeguy N, Diaz S, Cornelissen JHC, Vendramini F, Cabido M, Castellanos A (2000) Chemistry and toughness predict leaf litter decomposition rates over a wide spectrum of functional types and taxa in central Argentina. Plant Soil 218:30

PeterJohn WT, Melillo JM, Bowles FP, Steudler PA (1993) Soil warming and trace gas fluxes: experimental design and preliminary flux results. Oecologia 93:18–24

Peterson DL, Rolfe GL (1982) Nutrient dynamics and decomposition of litterfall in floodplain and upland forests of central Illinois. For Sci 28:667–681

Polglase PJ, Tompkins D, Stewart LG, Falkiner RA (1995) Mineralization and leaching of nitrogen in an effluent-irrigated pine plantation. J Environ Qual 24:911–920

Prescott CE (1995) Does nitrogen availability control rates of litter decomposition in forests? Plant Soil 168/169:83–88

Prescott CE, MacDonald MA (1994) Effects of carbon and lime additions on mineralization of C and N in humus from cutovers of western red cedar – western hemlock forests on northern Vancouver Island. Can J For Res 24:2432–2438

Prescott CE, Zabek LM (1997) Growth response and nutrient availability in western red cedar plantations following amendment with fish-wood compost and straw. Can J For Res 27:598–602

Prescott CE, Corbin JP, Parkinson D (1992) Immobilization and availability of N and P in the forest floors of fertilized Rocky Mountain coniferous forests. Plant Soil 143:1–10

Prescott CE, MacDonald MA, Weetman GF (1993a) Availability of N and P in the forest floors of adjacent stands of western red cedar-western hemlock and western hemlock-amabilis fir on northern Vancouver Island. Can J For Res 23:605–610

Prescott CE, Taylor BR, Parsons WFJ, Durall DM, Parkinson D (1993b) Nutrient release from decomposing litter in Rocky Mountain coniferous forests: influence of nutrient availability. Can J For Res 23:1576–1586

Prescott CE, Kishchuk BE, Weetman GF (1995) Long-term effects of repeated N fertilization and straw application in a jack pine forest. 3. Nitrogen availability in the forest floor. Can J For Res 25:1991–1996

Prescott CE, Chappell HN, Vesterdal L (2000) Nitrogen turnover in forest floors of coastal Douglas-fir at sites differing in soil nitrogen capital. Ecology 81:1878–1886

Prescott CE, Zabek LM, Staley CL, Kabzems RD (2001) Decomposition of broadleaf and needle litter in forests of British Columbia: influences of litter type, forest type and litter mixtures. Can J For Res 30:1742–1750
Preston CM, Trofymow JA, Sayer BG, Niu J (1997) ^{13}C nuclear magnetic resonance spectroscopy with cross-polarization and magic-angle spinning investigation of the proximate-analysis fractions used to assess litter quality in decomposition studies. Can J Bot 75:1601–1613
Priha O, Lehto T, Smolander A (1999) Mycorrhizas and C and N transformations in the rhizospheres of *Pinus sylvestris*, *Picea abies* and *Betula pendula* seedlings. Plant Soil 206:191–204
Raison RJ, Khanna PK, Connell MJ, Falkiner RA (1990) Effects of water availability and fertilization on nitrogen cycling in a stand of *Pinus radiata*. For Ecol Manage 30:31–43
Reich PB, Grigal DF, Aber JD, Gower ST (1997) Nitrogen mineralization and productivity in 50 harwood and conifer stands on diverse soils. Ecology 78:335–347
Reid CPP (1983) Nitrogen nutrition, photosynthesis and carbon allocation in ectomycorrhizal pine. Plant Soil 71:415–432
Robbins CT, Mole S, Hagerman AE, Hanley TA (1987) Role of tannins in defending plants against ruminants: reduction in dry matter digestion? Ecology 68:1606–1615
Rustad LE (1994) Element dynamics along a decay continuum in a red spruce ecosystem in Maine, USA. Ecology 75:867–879
Rustad LE, Cronan CS (1988) Element loss and retention during litter decay in a red spruce stand in Maine. Can J For Res 18:947–953
Rustad LE, Campbell JL, Marion GM, Norby RJ, Mitchell MJ, Hartley AE, Cornelissen JHC, Gurevitch J (2001) A meta-analysis of the response of soil respiration, net nitrogen mineralization, and aboveground plant growth to experimental ecosystem warming. Oecologia 126:543–562
Rygiewicz PT, Andersen CP (1994) Mycorrhiza alter quality and quantity of carbon allocated below ground. Nature 369:58–60
Schlesinger WH (1985) Decomposition of chaparral shrub foliage. Ecology 66:1353–1360
Schlesinger WH, Hasey MM (1981) Decomposition of chaparral shrub foliage: losses of organic and inorganic constituents from deciduous and evergreen leaves. Ecology 62:762–774
Schuur EAG, Chadwick OA, Matson PA (2001) Carbon cycling and soil carbon storage in mesic to wet Hawaiian montane forests. Ecology 82:3182–3196
Scott NA, Binkley D (1997) Foliage litter quality and annual net N mineralization: comparison across North American forest sites. Oecologia 111:151–159
Scott NA, Cole CV, Elliot ET, Huffman SA (1996) Soil textural control on decomposition and soil organic matter dynamics. Soil Sci Soc Am J 60:1102–1109
Seastedt TR, Parton WJ, Ojima DS (1992) Mass loss and nitrogen dynamics of decaying litter of grasslands: the apparent low nitrogen immobilization potential of root detritus. Can J Bot 70:384–391
Seneviratne G (2000) Litter quality and nitrogen release in tropical agriculture: a synthesis. Biol Fert Soils 31:60–64
Setala H (2000) Reciprocal interactions between Scots pine and soil food web structure in the presence and absence of ectomycorrhiza. Oecologia 125:109–118
Soderstrom B (1978) Soil microfungi in three Swedish coniferous forests. Holarct Ecol 1:61–72
Soderstrom B, Baath E, Lundgren B (1983) Decrease in soil microbial activity and biomasses owing to nitrogen amendments. Can J Microbiol 29:1500–1506

Staaf H (1980a) Release of plant nutrients from decomposing leaf litter in a South Swedish beech forest. Holarct Ecol 3:129–136

Staaf H (1980b) Influence of chemical composition, addition of raspberry leaves, and nitrogen supply on decomposition rate and dynamics of nitrogen and phosphorus in beech leaf litter. Oikos 35:55–62

Staaf H, Berg B (1977) Mobilization of plant nutrients in a Scots pine forest moor in central Sweden. Silva Fenn 11:210–216

Staaf H, Berg B (1982) Accumulation and release of plant nutrients in decomposing Scots pine needle litter. Long-term decomposition in a Scots pine forest, II. Can J Bot 60:1561–1568

Stanford G (1974) Nitrogen mineralization-water relations in soils. Soil Sci Soc Am Proc 38:103–106

Stohlgren TJ (1988) Litter dynamics in two Sierran mixed conifer forests. 2. Nutrient release in decomposing leaf litter. Can J For Res 18:1136–1144

Stottlemyer R, Toczydlowski D (1999) Nitrogen mineralization in a mature boreal forest, Isle Royal, Michigan. J Environ Qual 28:709–720

Swift MJ, Heal OW, Anderson JM (1979) Decomposition in terrestrial ecosystems. University of California Press, Los Angeles

Taylor BR, Parkinson D, Parsons WFJ (1989a) Nitrogen and lignin content as predictors of litter decay rates: a microcosm test. Ecology 70:97–104

Taylor BR, Parsons WFJ, Parkinson D (1989b) Decomposition of *Populus tremuloides* leaf litter accelerated by addition of *Alnus crispa* litter. Can J For Res 19:674–679

Taylor BR, Prescott CE, Parsons WFJ, Parkinson D (1991) Substrate control of litter decomposition in four Rocky Mountain coniferous forests. Can J Bot 69:2242–2250

Thomas RJ, Asakawa NM (1993) Decomposition of leaf litter from tropical forage grasses and legumes. Soil Biol Biochem 23:1351–1361

Tian G, Kang BT, Brussaard L (1992) Biological effects of plant residues with contrasting chemical compositions under humid tropical conditions – decomposition and nutrient release. Soil Biol Biochem 24:1051–1060

Titus BD, Malcolm DC (1987) The effect of fertilization on litter decomposition in clear-felled spruce stands. Plant Soil 100:297–322

Titus BD, Malcolm DC (1999) The long-term decomposition of Sitka spruce needles in brash. Forestry 72:207–221

Tripathi SK, Singh KP (1992a) Abiotic and litter quality control during the decomposition of different plant parts in dry tropical bamboo savanna in India. Pedobiologia 36:241–256

Tripathi SK, Singh KP (1992b) Nutrient immobilization and release patterns during plant decomposition in a dry tropical bamboo savanna, India. Biol Fert Soils 14:191–199

Upadhyay VP, Singh JS (1985) Nitrogen dynamics of decomposing hardwood leaf litter in a central Himalayan forest. Soil Biol Biochem 17:827–830

Van Cleve K, Barney R, Schlentner R (1981) Evidence of temperature control of production and nutrient cycling in two interior Alaska black spruce ecosystems. Can J For Res 11:258–273

Van Veen JA, Kuikman PJ (1990) Soil structural aspects of decomposition of organic matter by micro-organisms. Biogeochemistry 13:213–233

Van Vuuren MMI, Berendse F, de Visser W (1993) Species and site differences in the decomposition of litters and roots from wet heathlands. Can J Bot 71:167–173

Vesterdal L (1999) Influence of soil type on mass loss and nutrient release from decomposing foliage litter of beech and Norway spruce. Can J For Res 29:95–105

Visser S (1985) Role of the soil invertebrates in determining the composition of soil microbial communities. In: Fitter AH (ed) Ecological interactions in soil. Blackwell, Boston, pp 297–317
Visser S, Parkinson D (1975) Fungal succession on aspen poplar leaf litter. Can J Bot 53:1640–1651
Vitousek PM, Turner DR, Parton WJ, Sanford RL (1994) Litter decomposition on the Mauna Loa environmental matrix, Hawai'i: patterns, mechanisms, and models. Ecology 75:418–429
Vogt KA, Grier CC, Vogt DJ (1986) Production, turnover, and nutrient dynamics of above- and belowground detritus of world forests. Adv Ecol Res 15:303–377
Wardle DA, Zackrisson O, Hornberg G, Gallet C (1997) The influence of island area on ecosystem properties. Science 277:1296–1299
Wardle DA, Barker GM, Bonner KI, Nicholson KS (1998) Can comparative approaches based on plant ecophysiological traits predict the nature of biotic interactions and individual plant species effects in ecosystems? J Ecol 86:405–420
Wedin DA, Tilman DT (1990) Species effects on nitrogen cycling: a test with perennial grasses. Oecologia 84:433–441
White DL (1988) Litter decomposition in southern Appalachian black locust and pine-hardwood stands: litter quality and nitrogen dynamics. Can J For Res 18:54–63
Will GM (1967) Decomposition of *Pinus radiata* litter on the forest floor. 1. Changes in dry matter and nutrient content. N Z J For Sci 10:1030–1044
Winkler JP, Cherry RS, Schelsinger WH (1996) The Q10 relationship of microbial respiration in a temperate forest soil. Soil Biol Biochem 28:1067–1072
Witkamp M (1966) Decomposition of leaf litter in relation to environment, microflora, and microbial respiration. Ecology 47:194–201
Yavitt JB, Fahey TJ (1986) Litter decay and leaching from the forest floor in *Pinus contorta* (lodgepole pine) ecosystems. J Ecol 74:525–545
Zak DR, Holmes WE, MacDonald NW, Pregitzer KS (1999) Soil temperature, matric potential, and the kinetics of microbial respiration and nitrogen mineralization. Soil Sci Soc Am 63:575–584
Zhu W, Ehrenfeld JG (1996) The effects of mycorrhizal roots on litter decomposition, soil biota, and nutrients in a spodosolic soil. Plant Soil 179:109–118

3 Integrated Root Responses to Variations in Nutrient Supply

D. Robinson

3.1 Introduction

Plants require 15 essential elements – C, H, O, N, P, K, Ca, Mg, S, Fe, Cl, B, Mn, Mo and Cu – plus several others such as Si, Co and Na (Marschner 1986, p. 5). In land plants, all of these elements except C and O are obtained largely from soil via roots and (usually) associated mycorrhizas (see Chap. 9, this Vol.). The elemental compositions of plants can vary within restricted limits. This variation has environmental and genetic components.

The first component reflects the nutrient-supplying characteristics of a particular habitat, the second, the evolutionary history of a species. Two examples illustrate this. First, plants growing on saline soils usually accumulate more Na in their tissues than they would when growing under non-saline conditions, but members of the Chenopodiaceae such as *Beta* accumulate more Na than do cereals and legumes under the same conditions (Marschner 1986, p. 536). Second, plants growing on calcium-rich soils tend to contain more Ca than those from acidic soils, but dicots contain more Ca per unit dry weight than do monocots when on the same soil (Rorison and Robinson 1984). Such variation in elemental composition may have functional significance. Na accumulation in leaves reduces the internal water potential, thereby partly compensating for the salt-induced increase in external water potential. By contrast, Ca^{2+} accumulation probably has little direct physiological significance because Ca^{2+} is required physiologically only in meagre amounts, and much of the Ca^{2+} in plant tissue is adsorbed onto the negatively charged surfaces of cell walls external to plasma membranes.

Ultimately, however, variations in plant elemental compositions are limited by the metabolic systems which have evolved, and by the range of ions and molecules which these systems require as substrates and cofactors in biochemical processes (MacDonald and Davies 1996; Elser et al. 2000; Magnani and Grace 2000). This requirement constitutes a plant's *demand* (see also

Ecological Studies, Vol. 181
H. BassiriRad (Ed.) Nutrient Acquisition by Plants
An Ecological Perspective

Chap. 4, this Vol., for a definition of plant nutrient demand) for an element, i.e. the amount of an element needed over a certain growth interval to maintain the plant's stoichiometry within its genetically determined, metabolically feasible limits. Should the supply of an element be insufficient to maintain a plant's demand, various morphological and physiological responses can be expressed which can compensate for the shortfall in availability. These responses range from alterations in the expression of genes coding for or regulating ion transporters in root plasma membranes to adjustments in the rate of whole-plant growth and allocation among different organs. Although physiological and developmental responses are usually studied separately, their functional significance can be understood properly only if they are considered as part of an integrated response system which operates on many organisational levels, from the molecular to the organismal (BassiriRad et al. 2001). This chapter discusses how such responses to variations in nutrient supply could be integrated within plants such that nutrient demands can be met even if external supplies are variable. It highlights many important gaps in knowledge that remain to be filled.

3.2 Spatial and Temporal Variations in Soil Nutrient Supplies

For plants growing in soil, variability in nutrient supply is normal. Soil is inherently heterogeneous at all scales of measurement. This heterogeneity is caused by interactions of climate, parent rock, topography, vegetation, biological activity, and the time over which these interactions occur during soil formation (Rowell 1994, p. 1; Schlesinger 1997). Because these processes and their interactions do not occur uniformly, their product – soil – is similarly non-uniform. This non-uniformity is exacerbated in agricultural soils by the placement of fertiliser pellets or organic residues within the rooting zone of crops. On uncultivated land, vegetation has a major influence on local nutrient availability by the production of litter and its subsequent decomposition by soil microbes. Animals also have major influences on soil nutrient heterogeneity by providing point inputs as excreta and corpses for microbial decomposition and mineralisation.

The result is a mosaic of soil in which nutrient availability can vary from point to point. A good example of this is the two-dimensional map produced by Jackson and Caldwell (1993). They showed that extractable phosphate (P_i) varied significantly across a 0.5 × 0.5 m plot in the Utah desert, a scale small enough to influence the P supply to different parts of the root systems of an *Artemisia tridentata* shrub at the centre of the plot. At a larger spatial scale (20 × 20 m), Gallardo et al. (2000) found that soil NO_3^- concentrations varied from <9 to 15 $\mu g\ g^{-1}$ and NH_4^+ concentrations from <25 to 95 $\mu g\ g^{-1}$ in a Dehesa savannah in Andalucia, Spain. Both ions showed distinct patchiness in their

Fig. 3.1. Contour plots of extractable NH_4^+ and NO_3^- measured in soil around an holm oak (*Quercus* sp.) tree in a Dehesa ecosystem in Spain (Gallardo et al. 2000). Figure used with permission of the author and Kluwer Academic Publishers, Dordrecht, The Netherlands

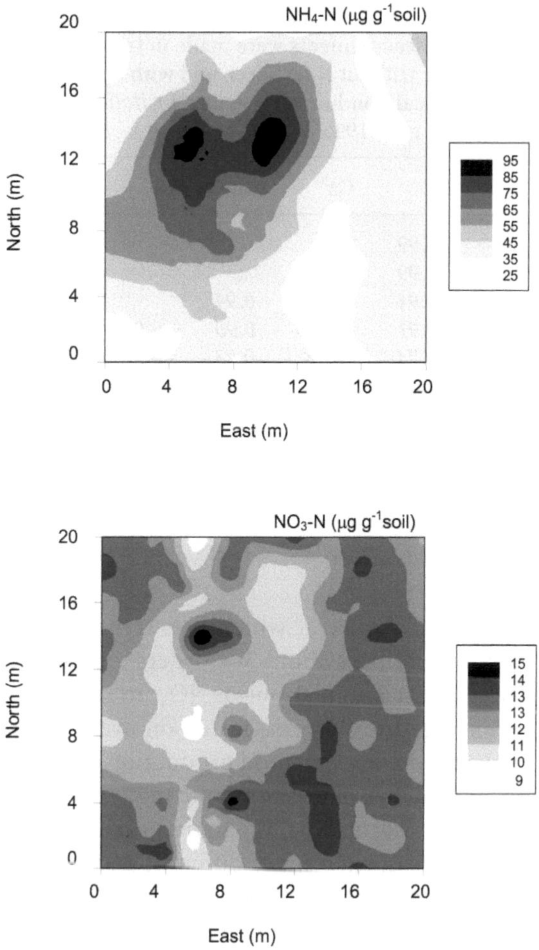

distributions around a single oak tree (*Quercus* sp.) within that area (Fig. 3.1), although it is notable that NO_3^-- and NH_4^+-rich patches did not always coincide. Solute concentrations in soil do not vary independently of each other because electrochemical reactions couple the concentrations of one ion to those of others which are present. Yanai et al. (1996) found strong positive correlations between the concentrations of major cations (Mg^{2+}, Ca^{2+}) and those of NO_3^- in a Scottish agricultural soil (Table 3.1). P_i concentrations were negatively correlated with those of cations. Concentrations of NH_4^+ and SO_4^{2-} were statistically independent of those of any other ions. This supports the impression given by Fig. 3.1 that a NO_3^- 'hotspot' in soil does not necessarily indicate one for NH_4^+.

Soil nutrient heterogeneity also has an important temporal dimension. The durability of a nutrient 'hotspot' in soil depends on the mobility of the

Table 3.1. Correlation matrix of major ions in solution extracted from a Scottish agricultural soil. Measurements were made in triplicate seven times over a 50-day period from soil with or without N fertiliser, and with or without growing *Zea mays* L. (maize) plants (*EC* electrical conductivity, *values in italics* correlations significant at $p<0.01$). (Data from Yanai et al. 1996)

	EC	Ca^{2+}	Mg^{2+}	Na^+	K^+	P_i	SO_4^{2-}	NO_3^-	NH_4^+
Ca^{2+}	*0.99*								
Mg^{2+}	*0.99*	*1.00*							
Na^+	*0.94*	*0.93*	*0.94*						
K^+	*0.91*	*0.90*	*0.90*	*0.96*					
P_i	*−0.74*	*−0.73*	*−0.73*	*−0.62*	*−0.54*				
SO_4^{2-}	0.04	0.08	0.06	0.01	−0.06	−0.08			
NO_3^-	*0.95*	*0.94*	*0.94*	*0.90*	*0.89*	*−0.65*	−0.17		
NH_4^+	0.01	0.02	0.03	0.04	0.02	0.01	−0.07	0.02	
H^+	*0.89*	*0.88*	*0.89*	*0.89*	*0.85*	*−0.66*	0.00	*0.86*	0.03

ion, and the rates at which it enters and leaves the soil solution. NO_3^-, for example, diffuses about 3–4 orders of magnitude faster than P_i in moist soil. Therefore, accumulations of NO_3^- in soil, e.g. around a fertiliser pellet, would be expected to disperse faster, on average, than would accumulations of P_i. However, during the rapid production of NO_3^- by soil microbes, e.g. in decomposing patches of fresh organic matter, differences in NO_3^- concentration can persist in soil, leading to patterns such as those shown in Fig. 3.1.

Plants can experience different patterns of nutrient availability from year to year at the same site. For example, Farley and Fitter (1999; Fig. 3.2) showed that as well as considerable month-to-month variations in extractable P_i and NO_3^- concentrations in a woodland soil, P_i availability at that site also varied from one year to the next. The main cause of this inter-annual variability is climate, principally rainfall and temperature. These influence physical features such as soil moisture content which, in turn, modify local microbial activity.

Although plant responses to spatial variations in nutrient supply have received greater attention (see Robinson 1994; Hodge 2004), those to temporal pulses are equally important. The relatively small amount of work which has been done on this has revealed some important interspecific differences in response. Perhaps the best known is the superior specific growth and uptake rates of the inherently slow-growing grass *Festuca ovina* compared with those of the usually faster-growing *Arrhenatherum elatius* when nutrients were supplied as pulses less than 10 h long (Campbell and Grime 1989). This ranking was reversed when the nutrient pulses were longer than 10 h. This suggests that slow-growing species are advantaged in environments where nutrient supply is chronically deficient but punctuated by unpre-

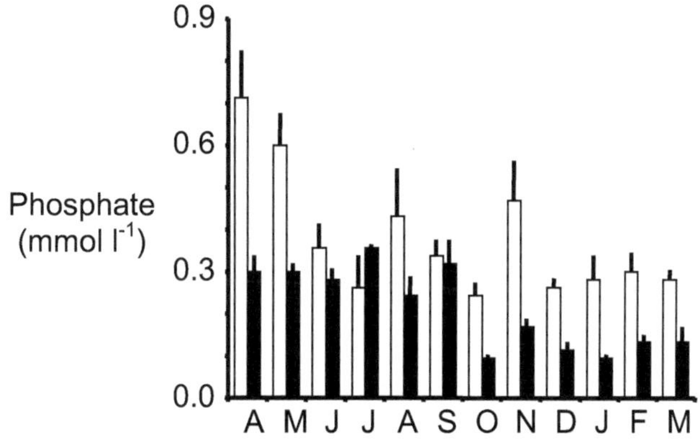

Fig. 3.2. Mean (+SE) extractable phosphate concentrations measured in a deciduous woodland soil in Yorkshire, England, in 1993–1994 (*open bars*) and 1994–1995 (*closed bars*). Data from Farley and Fitter (1999)

dictable episodes of localised nutrient enrichment in patches which are rapidly dispersed by microbial activity or physical processes.

Fitter (1994) introduced a simple 'taxonomy' of soil nutrient patches that recognised their temporal as well as spatial characteristics, classifying them in terms of their size, frequency, and distribution in time and space. Although extremely useful, this scheme has yet to find practical application, not least because of the practical difficulties of measuring and defining patch 'size', 'frequency' and 'distribution' using conventional techniques of soil chemistry, microbiology and geostatistics. It must be acknowledged that no measurements have been made in situ of the kinds of nutrient pulses simulated in the laboratory by Campbell and Grime (1989). Because the results of those experiments conformed to how the responses of inherently slow-growing species were *expected* to differ from those of faster-growing species, it has been assumed subsequently that patterns of nutrient supply in habitats where each type of plant dominates approximate to those which were created experimentally. This assumption requires testing, although the practicalities of doing this (involving destructive, frequent micro-sampling of soil solution components within discrete volumes of the rooting zone) are daunting.

3.3 Single Roots and Root Systems

The responses of plants to spatial and temporal variations in nutrient supply, and to even more extreme laboratory simulations, have been well studied by plant physiologists and, increasingly, ecologists (Robinson 1994; Robinson and van Vuuren 1998). It is of clear practical relevance to understand, first, how crops maximise their uptake of patchily distributed nutrients and, second, the extent to which such nutrient supplies may act as a selective force among wild plants (Grime et al. 1997). In addition, discrete nutrient inputs provide an excellent test-bed to study the mechanisms of the biophysical and developmental responses elicited in plants by such stimuli (Leyser and Fitter 1998).

The most durable and influential model that integrates information about water and nutrient uptake by single roots into resource capture by whole root systems can be traced back to Gardner (1960; see Tinker and Nye 2000, p. 297). The model assumes that each root is cylindrical and positioned at the centre of a cylinder of soil. The radial transport of water and ions through the soil to the root surface is described by diffusion–convection equations. The flux at the root surface is described by a proportionality with soil water content, water potential, or ion concentration, or in terms of Michaelis-Menten kinetics. The root system in a unit volume of soil is imagined to consist of regular arrays of these cylinders-within-cylinders. The greater the root length density, the narrower each soil cylinder, so growth by the root system is accommodated by reducing the volume of soil available to each root between successive time steps. The whole root system is constructed from many such arrays, enough to fill the plant's rooting volume. Uptake by the whole root system is, therefore, a simple summation of the uptake by each single root from within its soil cylinder.

This model has been tremendously influential. It and its variants have been used extensively to explore the constraints on nutrient and water acquisition by crops. Cases where the agreement between model and experiment is good (e.g. Baldwin 1975) give confidence in the model. However, there are cases where agreement is not good. For example, Brewster et al. (1976) discovered that the model underestimated P uptake by *Brassica napus* (rape). The explanation for this was that rape acidifies its rhizosphere, altering the P equilibrium between the soil solution and solid phase, thereby increasing the concentration of soluble P close to the root and, in turn, the actual P uptake rate achieved.

Robinson et al. (1991) found that the model significantly overestimated NO_3^- uptake by wheat (Fig. 3.3). Similar results were obtained for the uptake by wheat of K^+ (Seward et al. 1990), water (Passioura 1980) and Mg^{2+} (Rengel and Robinson 1990). The reason for these discrepancies is probably that the assumption of geometrical uniformity in the root system and soil is not strictly valid. The branching nature of root systems and the fractal-like structure of the soil's pore space (Anderson et al. 1998) mean that a uniform spacing of real roots is impossible. In the case of Robinson et al.'s (1991) experiment, the lack of agreement between experiment and theory (Fig. 3.3) was probably because many roots had clumped around the edges of the pots in which the plants were grown; the same could have applied to Passioura's (1980) pot experiment.

In field experiments, root clumping can occur in highly structured soils, confining many roots to macropores and preventing a more uniform disper-

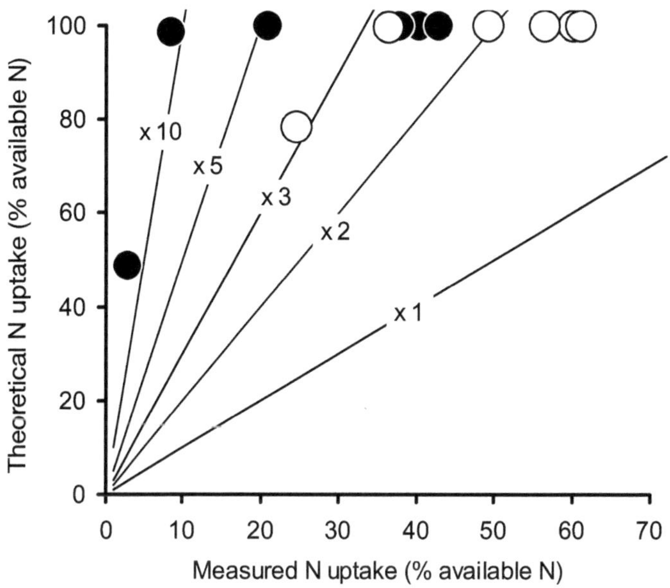

Fig. 3.3. Comparison between measured N uptake by spring wheat (*Triticum aestivum*) grown on soil without N fertiliser (*open symbols*) or with N fertiliser (*closed symbols*), with the uptake predicted using a simulation model assuming uniform root distribution. N uptake was measured by destructively harvesting replicate plants 13, 29, 47, 72, 84 and 97 days after sowing. Data are from Table 1 of Robinson et al. (1991). *Lines* show the extent to which the model overestimated N uptake; perfect estimation is indicated by the *x 1 line*

sion of roots throughout the rooting volume (Passioura 1991). Should soil heterogeneities be even more marked, as in experiments where the distributions of nutrient-rich and nutrient-poor patches are manipulated (e.g. Casper and Cahill 1998), it is unlikely that the basic model would precisely simulate resource capture by root systems without special and extensive modification. Nevertheless, Jackson and Caldwell (1996) did use such a model to demonstrate the theoretical benefits to plants of localised root proliferation and up-regulation of ion transport. Without such responses, plants were predicted to capture significantly less P_i and NO_3^- from patchy supplies compared with plants capable of responding to local increases in the availabilities of these ions.

In principle, there is no reason why spatial and temporal variations in soil conditions and plant response should not be included routinely in such models. However, "... water and nutrient uptake in plant growth and water flow models is mostly described by empirical means, lacking a sound physiological or biophysical basis" (Hopmans and Bristow 2002). That is, we cannot include the extra complexity associated with uptake by non-uniform root systems which respond to internal and external stimuli unless we use available information on a case-by-case basis, and each model is then genotype-, species-, nutrient- or soil-specific.

Introducing extra complexity into models is, in itself, not the problem it once was. We now have sufficient computing power to simulate root growth, soil transport and uptake by plants in crop monocultures in detail sufficient to dwarf the geometrical simplicity of the original single-root model. The real problem (and which is not a new one; Passioura 1996) is in knowing that the output from such complex models is empirically *and theoretically* sound:

"A major drawback of computer models is their apparent insatiable appetite for complexities, thereby providing the computer programmer with the opportunity to increase the number of *a priori* unknown parameters without limitations, and thereby giving the 'false' appearance of mechanistic understanding of the simulated system" (Hopmans and Bristow 2002).

So, whereas computing power no longer limits the detail with which soil–root interactions can be described, the lack of an integrative, robust (and, preferably, simple) theory which could serve as a fundamental reference point for empirically based simulations does limit our understanding and our confidence in simulation outputs. This point is revisited in Section 3.5. Much of the physiological and molecular information emerging about the responses of plants to variable nutrient supplies is species- and nutrient-specific rather than integrative, and this information is discussed in the following section.

3.4 Local and Systemic Response Systems

Potentially, each root of a plant growing in soil could experience a slightly different pattern of nutrient supply. Yet, responses to those supplies must somehow be coordinated within individual plants. Information about the condition of one part of the root system must be generated and transmitted by those cells, and received and processed by others elsewhere in the plant. Current evidence suggests that at least two types of nutrient response system exist in plants: local and systemic (Forde 2002a).

3.4.1 Local Responses

Local responses to nutrients involve the sensing and signalling of information by cells exposed to an external stimulus that elicits responses only within those cells, i.e. the signal is not transmitted beyond the region that senses the stimulus. The sequence of events leading from the sensing of a variable nutrient supply to the expression of a local response is probably as follows. The sensing part of the system responds, in plants, to cellular (probably cytosolic) nutrient status and not, as far as we know, directly to the external ion concentration (unlike in prokaryotes; Forde and Lorenzo 2001). How this works is unknown – does the system detect an absolute concentration, a concentration above or below a certain threshold, or a concentration relative to those of other ions or metabolites in the cell?

Once the nutrient status of the cell(s) has been sensed, signals are generated which activate or modify regulatory genes controlling the rates of ion transport into those cells, and the growth and development of lateral roots around those cells. The best-characterised example of a local, ion-induced developmental response is the NO_3^--induced extension of lateral roots in *Arabidopsis thaliana* by local increases in NO_3^- concentration (Zhang and Forde 1998; Zhang et al. 1999; Forde and Lorenzo 2001; Forde 2002a). A locally elevated NO_3^- concentration induces the expression of a gene, *ANR1*, which increases the extension rate of lateral roots exposed to the NO_3^-. This effect is independent of a nutritional influence of extra N – the effect on lateral extension is the same if mutants deficient in nitrate reductase are treated with NO_3^-.

Before such developmental responses become apparent, changes in the expression of ion transporter genes occur. Modifications to ion transport rates arise from gene-regulated changes in the densities of ion transporters in root plasma membranes (see Chap. 6, this Vol.). Evidence for this comes from increases in transcript expression in root plasma membranes of, e.g. the barley (*Hordeum vulgare*) SO_4^{2-} transporter gene *HVST1* in response to prolonged SO_4^{2-} starvation (Smith et al. 1997). NO_3^- starvation increases the abundance of *NRT2* mRNAs in NO_3^--starved cells (Forde 2000), increasing the capacity of those cells to take up NO_3^- should it subsequently become available.

How many local response systems exist in plants? A conservative estimate would be that every root cell able to transport ions across its membranes would have at least one response system for each ion. These millions of potential responses to variable nutrient supplies do not operate independently of systemic responses. Indeed, it is only because systemic responses exist that local systems can potentially compensate for a shortfall in nutrient demand at the level of the whole plant.

3.4.2 Systemic Responses

If a particular nutrient is in great demand compared with others, this imbalance is sensed and transmitted through the plant. This systemic response 'primes' the relevant local root response systems, should the cells in which they are contained encounter a source of the deficient nutrient. The most thoroughly investigated systemic responses of plants are those to an increased demand of S.

Within 20 h of *Arabidopsis thaliana* plants being deprived of SO_4^{2-}, Lappartient et al. (1999) detected a decrease in the concentration of glutathione

(GSH) in roots. This coincided with increases in the transcription of a SO_4^{2-} transporter gene (*AST68*), in the activity of ATP sulfurylase (which catalyses the first step in SO_4^{2-} assimilation), in the abundance of ATP sulfurylase mRNA, and in short-term SO_4^{2-} influx. These patterns were reversed when SO_4^{2-} was re-supplied to the plants. Such responses are those that a local signal–response system might be expected to express.

Crucially, however, similar patterns were seen in certain *Brassica napus* roots when they were constantly supplied with SO_4^{2-} at the same time that the rest of the root system was deprived of SO_4^{2-} (Lappartient et al. 1999). This is strong evidence that a signal generated in the S-starved roots was transmitted to the shoot (via the xylem), and then to the S-rich roots (via the phloem). Once in the S-rich roots, the signal initiated the suite of responses to the increased plant demand for S – increased SO_4^{2-} transporter production and S assimilation capacity. The best candidate for the phloem-translocated signal is GSH. Ubiquitous in living organisms as a controller of redox potential, GSH is also an end product of S assimilation and, in vascular plants, suppresses the expression of SO_4^{2-} transport and assimilation genes when present in plants at high concentrations, i.e. when S supply is plentiful. When S supply falls, GSH concentrations decrease, and the S-starvation responses are initiated.

Similar systemic systems will exist whereby plants respond to variations in the supplies of nutrients other than S but, obviously, the mechanisms will differ. There is evidence that systemic responses to N starvation or enrichment (e.g. as a result of atmospheric deposition of NH_3 and NO_2) involve the repression of root N uptake by phloem-translocated amino acids which are abundant when plant N demands are small (Imsande and Touraine 1994; Rennenberg and Geßler 1999). Such effects have been found for the repression of NO_3^- uptake by glutamine and asparagine in beech (*Fagus sylvatica*; Geßler et al. 1998), and of NH_4^+ uptake by these and other amino acids in soybean (*Glycine max*; Muller and Touraine 1992; also Chap. 6, this Vol.). Other evidence casts doubt on whether phloem-translocated amino acids are universal shoot-to-root carriers of N demand signals. For example, Tillard et al. (1998) found that the amino acid composition of phloem sap entering N-starved roots was effectively the same as that entering N-rich roots of *Ricinus communis*. The up-regulation of NO_3^- uptake in the N-starved roots could not, therefore, be ascribed to a signal carried by a phloem-translocated amino acid.

Whatever the identity of phloem-transported signals, it remains to be seen whether they act directly on gene expression in roots, or are merely the long-distance carriers of the demand signal. Likewise, whether demand-induced developmental responses in roots are also under the control of phloem-translocated signals is an unexplored problem. The recent discovery of systemic transport of mRNA in plants (see Citovsky and Zambryski 2000) offers the intriguing possibility that such molecules could act as an ‘RNA-based information superhighway' (Jorgensen et al. 1998) to signal and trigger specific developmental events in response to environmental cues.

These examples illustrate the significant advances which are revealing the molecular mechanisms by which plants respond to their internal and external nutritional environments. It must be remembered, however, that these mechanisms are largely nutrient- and species-specific. Not all species respond to S-starvation via changes in phloem GSH concentration, as do the two members of the Cruciferae mentioned above (see Forde 2002b); not all species respond to localised nutrients as strongly as do annual crop species (Robinson and Van Vuuren 1998).

It is unclear precisely where in the plant a demand-induced signal is generated. In meristematic tissues? In mature cells of leaves or other organs? In vascular tissues? It is also unclear if developmental responses to increased S demand occur that are analogous to the effects of increased N demand on root development (Drew et al. 1973; Robinson 1994; Forde 2002a). The apparently close coupling between physiological and developmental responses to NO_3^- starvation (see Sect. 3.4.1) is not shared by all nutrient response systems. For example, variability in K^+ supply does not usually induce strong developmental responses in plants such as barley (Drew 1975). Yet, the expression of at least one high-affinity K^+ transport gene (*HKT1*) does increase in the expected manner in K^+-starved barley and wheat (Wang et al. 1998).

It is difficult to separate the causes and effects of local and systemic response systems at a whole-plant level without recourse to the use of mutants or genetically transformed material and careful experimental design. Gansel et al. (2001) provided a superb example of this approach. When some roots of *Arabidopsis thaliana* were starved of N, the expression of *AtNrt2.1* (which encodes for high-affinity NO_3^- transporters in response to N starvation) increased in roots that were supplied with NO_3^-, implying a systemic cause-and-effect mechanism by which roots are 'informed' of the whole-plant N demand. By contrast, the expression of a corresponding gene for high-affinity NH_4^+ transporters, *AtAmt1.1*, depended more on the local N status of N-starved roots. The net result of these molecular responses was that NO_3^--grown *Arabidopsis* plants achieved full short-term (7 h) compensation for the localisation of the external N supply, whereas their NH_4^+-grown counterparts captured only about one third as much N from the localised supply (Fig. 3.4). Equivalent, long-term responses to the mixed supplies of inorganic N which are available locally and transiently to the roots of soil-grown plants are illustrated in Fig. 3.5.

How many systemic response systems exist in plants? Clearly, the 'at least one per nutrient per cell' answer given for the local response systems cannot apply. As a minimum, there will be one per nutrient per plant; species which seem to be composed of several 'integrated physiological units' (Watson 1986), between which there is limited vascular communication, may have more. The many local nutrient response systems are primed and coordinated

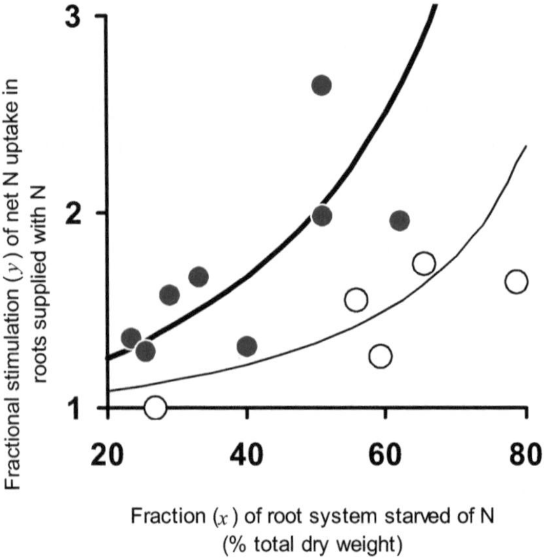

Fig. 3.4. Stimulation (y) of net uptake of NO_3^- (*closed symbols*) and NH_4^+ (*open symbols*) in roots of *Arabidopsis thaliana* plants supplied with these ions when various fractions (x) of the roots were N-starved. Each *data point* represents a separate experiment and is the mean of 4–10 replicates; error bars have been omitted for clarity. The *thick curve* corresponds to full compensation of N uptake for the localisation of the external supply, i.e. $y=100/(100-x)$. The *thin curve* is the corresponding relation for only one third of full compensation. (Modified from Gansel et al. 2001)

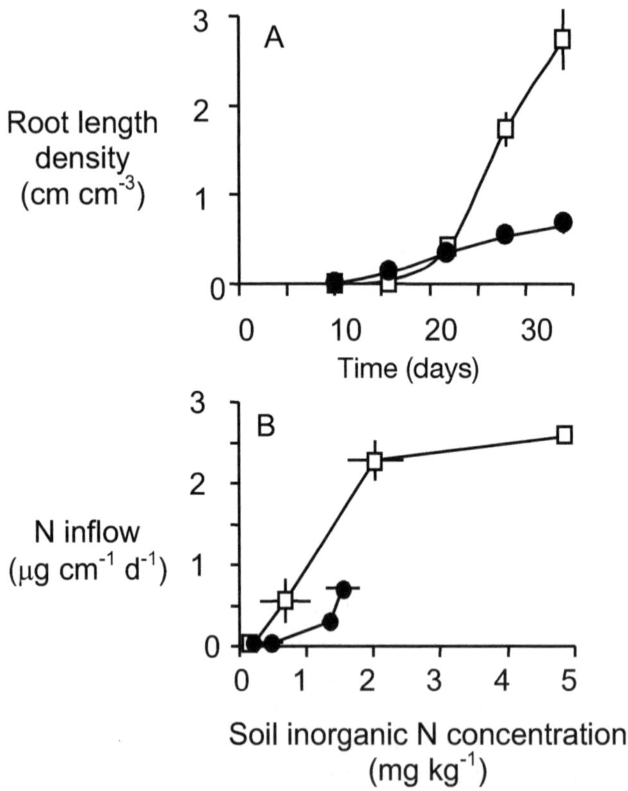

Fig. 3.5. A Temporal changes in the root length density of spring wheat (*Triticum aestivum*) within (*open squares*) and outside (*closed circles*) a patch of N-rich organic matter.'
B Variation in N inflow (net N uptake rate per unit root length) with inorganic N concentration within (*open squares*) and outside (*closed circles*) a patch of N-rich organic matter. Data from van Vuuren et al. (1996)

by the few systemic, demand-induced systems. Precisely how this works in a growing plant remains hazy, but the rate at which new and decisive information is emerging is encouraging.

3.5 Emergent Integration

At some level, the local and systemic responses described in Section 3.4 must integrate with, or form part of, the mechanisms regulating the coordinated growth of root systems, stems and leaf canopies. The usually close allometric relations among root, stem and shoot growth in young, intact plants imply that the balance between the production of these organs is constrained – within limits, only so much root can be produced for a given amount of stem and leaf, and vice versa.

Within these limits, variability or chronic shortages of nutrient supply usually promote root growth at the expense of shoot growth, resulting in an increase in the plant's root fraction (root dry weight as a fraction of the total dry weight). Nutrient availability has, on average, a larger effect on root fraction than do irradiance, water supply or atmospheric CO_2 concentration (Poorter and Nagel 2000; Fig. 3.6). Nutrients and irradiance tend to have

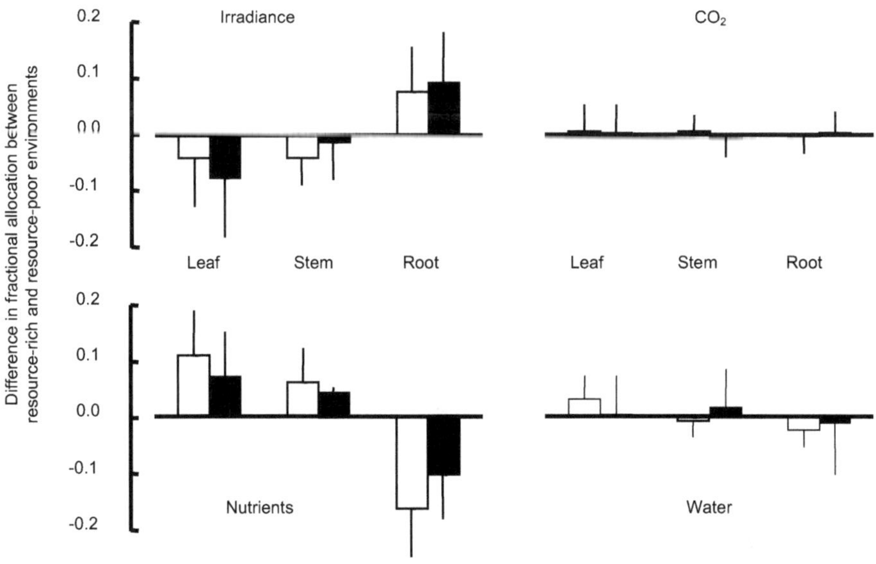

Fig. 3.6. Mean (+SD) difference in biomass allocation among leaves, stems and roots between plants grown under resource-rich and resource-poor conditions. *Open bars* Herbaceous species, *closed bars* woody species. Data collated by Poorter and Nagel (2000) from many sources

opposite effects on allocation to roots. This pattern is consistent with the old idea of a 'functional balance' between root and shoot growth and physiological activities in resource capture (Brouwer 1962; but see Poorter and Nagel 2000). Because plants derive resources simultaneously from above ground and below, it is essential that sufficient absorbing organs with adequate physiological capacity are produced and maintained in those environments to sustain resource capture.

Despite the ease with which allocation responses to nutrients can be demonstrated and their relative consistency among species, we lack a coherent physiological model of how these responses occur, and how they are integrated locally and systemically when challenged by soil heterogeneity. The roles of putative signals such as NO_3^-, cytokinins, auxins, abscisic acid and sucrose in regulating allocation continue to be explored in increasing degrees of detail, from the whole-plant level to the molecular (e.g. Gollan et al. 1986; Tardieu and Davies 1992; Beck 1996; Farrar 1996; van der Werf and Nagel 1996; Munns and Cramer 1996; Scheible et al. 1997; Jeschke and Hartung 2000; Martin et al. 2000; Berleth and Sachs 2001). The picture that has emerged is that such signals are necessary, *but not sufficient*, to account for observed physiological and developmental responses of roots and shoots to variations in nutrient availability (and to other stimuli such as drought; Magnani and Grace 2000). By this I mean that while the signalling mechanisms tell us how such responses occur at the molecular level, they cannot say how responses are integrated to achieve a certain compensation for a variable nutrient supply in terms of growth, resource capture or, ultimately, fitness.

An analogous situation once existed with regard to the stomatal control of gas exchange in leaves. The regulatory roles of ion fluxes, hormones and turgor generation in guard cells were (and remain) the focus of much research. However, only when Cowan (1977) and Cowan and Farquhar (1977) developed a general model (which ignored much of the biochemical information then available) was it clear that leaf stomatal conductance should vary to minimise transpiration for a given amount of carbon assimilation. That breakthrough then provided a macroscopic raison d'être for the molecular processes being studied in guard cells.

No equivalent theory explains why, in response to N deprivation, a plant's root fraction and NO_3^- transport systems should alter by certain amounts in species whose responses are regulated by different molecular mechanisms. What is it that these and similar responses are doing – if anything – for the plants which express them? I suspect that answers will not come from molecular studies of specific signal–response systems. As with stomatal conductance, satisfying explanations are more likely to emerge by taking a more macroscopic view.

For example, the identification of a statistical structure–function convergence in the leaves of species in different biomes (Reich et al. 1997) implies a common set of fundamental constraints under which all land plants operate, irrespective of the particular molecular mechanisms responsible for growing and maintaining the leaves of different species. In a similar vein, allometric scaling theories are providing common biophysical explanations for many structure–function patterns across a wide range of structural scales and taxonomic groups (Enquist et al. 1998; Enquist and Niklas 2002; Gillooly et al. 2002; Enquist 2003). From such apparently universal and robust explanations, a theoretical framework may yet emerge which will find practical use in the simulation of plant structure and function (see Sect. 3.3).

It is important to recognise that the processes we observe in whole plants are not necessarily simple summations of molecular processes. Interactions, including feedbacks, among molecules, between molecules and cells, cells and organs, and plants and their environment, including other plants (Robinson et al. 1999; Gersani et al. 2001), are highly non-linear and hierarchical. One consequence of this is that the emergent behaviour of an organised biological system, such as a plant, at one level (e.g. the whole plant) may ultimately be inexplicable in molecular terms (Cheeseman 1993; Magnani and Grace 2000). If that is true, classical notions of 'regulation', 'coordination' or 'control' based on engineering principles may have to be replaced by newer concepts involving 'stability', 'complexity' and 'self-organisation' if we are to understand how *and* why plants work as they do in the face of environmental variability.

3.6 Summary

This chapter addresses the problem of how a whole plant integrates its physiological functions in response to variations in external nutrient supplies. Plants can respond morphologically to spatial or temporal increases in nutrient availability by, for example, preferentially growing roots in nutrient-rich zones, or by shifting growth allocation to or from roots. Purely physiological responses include the up-regulation of membrane transporters during temporary nutrient starvation. The identities and control mechanisms of many of these transporters are now being revealed by molecular approaches. However, it is unclear how local molecular-level responses coordinate in whole plants with the higher-level systemic responses (which will, themselves, also have molecular bases) to achieve some compensation for nutrient shortages. I argue that to achieve clarity we will need, as well as detailed molecular studies, to develop a suitably macroscopic theory for functional integration in plants. One possibility is the use of metabolic scaling theory as a macro-physiological framework for our ever-expanding micro-physiological knowledge.

Acknowledgements. I am grateful to Prof. A Gallardo for providing Fig. 3.1, and to the numerous people who kindly commented on my equally numerous drafts.

References

Anderson AN, McBratney AB, Crawford JW (1998) Applications of fractals to soil studies. Adv Agron 63:1–76

Baldwin JP (1975) A quantitative analysis of the factors affecting plant nutrient uptake from soils. J Soil Sci 26:195–206

BassiriRad H, Gutschick VP, Lussenhop J (2001) Root system adjustments: regulation of plant nutrient uptake and growth responses to elevated CO_2. Oecologia 126:305–320

Beck E (1996) Regulation of shoot/root ratio by cytokinins from roots in *Urtica dioica*. Plant Soil 185:3–12

Berleth T, Sachs T (2001) Plant morphogenesis: long distance coordination and local patterning. Curr Opin Plant Biol 4:57–62

Brewster JL, Bhat KKS, Nye PH (1976) The possibility of predicting solute uptake and plant growth response from independently measured soil and plant characteristics. V. The growth and phosphorus uptake of rape in soil at a range of phosphorus concentrations and a comparison of the results with the predictions of a simulation model. Plant Soil 44:295–328

Brouwer R (1962) Nutritive influences on the distribution of dry matter in the plant. Neth J Agric Sci 10:361–376

Campbell BD, Grime JP (1989) A comparative-study of plant responsiveness to the duration of episodes of mineral nutrient enrichment. New Phytol 112:261–267

Casper BB, Cahill JF (1998) Population-level responses to nutrient heterogeneity and density by *Abutilon theophrasti* (Malvaceae): an experimental neighborhood approach. Am J Bot 85:1680–1687

Cheeseman JM (1993) Plant-growth modeling without integrating mechanisms. Plant Cell Environ 16:137–147

Citovsky V, Zambryski P (2000) Systemic transport of RNA in plants. Trends Plant Sci 5:52–54

Cowan IR (1977) Stomatal behavior and environment. Adv Bot Res 4:117–228

Cowan IR, Farquhar GD (1977) Stomatal function in relation to leaf metabolism and environment. Symp Soc Exp Biol 31:471–505

Drew MC (1975) Comparison of the effect of a localized supply of phosphate, nitrate, ammonium, and potassium on the growth of the seminal root system and the shoot system of barley. New Phytol 75:479–490

Drew MC, Saker LR, Ashley TW (1973) Nutrient supply and the growth of the seminal root system in barley. I. The effect of nitrate concentration on the growth of axes and laterals. J Exp Bot 24:1189–1202

Elser JJ, Sterner RW, Gorokhova E, Fagan WF, Markow TA, Cotner JB, Harrison JH, Hobbie SE, Odell GM, Weider LJ (2000) Biological stoichiometry from genes to ecosystems. Ecol Lett 3:540–550

Enquist BJ (2003) Scaling the macroecological and evolutionary implications of size and metabolism within and across plant taxa. In: Blackburn TH, Gaston KJ (eds) Macroecology: concepts and consequences. Blackwell, Oxford, pp 321–341

Enquist BJ, Niklas KJ (2002) Global allocation rules for patterns of biomass partitioning in seed plants. Science 295:1517–1520

Enquist BJ, Brown JH, West GB (1998) Allometric scaling of plant energetics and population density. Nature 395:163–165

Farley RA, Fitter AH (1999) Temporal and spatial variation in soil resources in a deciduous woodland. J Ecol 87:688–696

Farrar J (1996) Regulation of root weight ratio is mediated by sucrose. Plant Soil 185:13–19

Fitter AH (1994) Architecture and biomass allocation as components of plastic response of root systems to soil heterogeneity. In: Caldwell MM, Pearcy RW (eds) Exploitation of environmental heterogeneity by plants. Academic Press, San Diego, pp 305–323

Forde BG (2000) Nitrate transporters in plants: structure, function and regulation. Biochim Biophys Acta 1465:219–235

Forde BG (2002a) Local and long-range signalling pathways regulating plant responses to nitrate. Annu Rev Plant Physiol Mol Biol 53:203–224

Forde BG (2002b) The role of long-distance signalling in plant responses to nitrate and other nutrients. J Exp Bot 53:39–43

Forde B, Lorenzo H (2001) The nutritional control of root development. Plant Soil 232:51–68

Gallardo A, Rodríguez-Saucedo JJ, Covelo F, Fernández-Alés R (2000) Soil nitrogen heterogeneity in a Dehesa ecosystem. Plant Soil 222:71–82

Gansel X, Munos S, Tillard P, Gojon A (2001) Differential regulation of the NO_3^- and NH_4^+ transporter genes *AtNrt2.1* and *AtAmt1.1* in *Arabidopsis*: relation with long-distance and local controls by N status of the plant. Plant J 26:143–155

Gardner WR (1960) Dynamic aspects of water availability to plants. Soil Sci 89:63–73

Gersani M, Brown JS, O'Brien EE, Maina GM, Abramsky Z (2001) Tragedy of the commons as a result of root competition. J Ecol 89:660–669

Geßler A, Schultze M, Schrempp S, Rennenberg H (1998) Interaction of phloem-translocated amino compounds with nitrate net uptake by the roots of beech (*Fagus sylvatica*) seedlings. J Exp Bot 49:1529–1537

Gillooly JF, Charnov EL, West GB, Savage VM, Brown JH (2002) Effects of size and temperature on developmental time. Nature 417:70–73

Gollan T, Passioura JB, Munns R (1986) Soil water status affects the stomatal conductance of fully turgid wheat and sunflower leaves. Aust J Plant Physiol 13:459–464

Grime JP, 33 others (1997) Integrated screening validates primary axes of specialisation in plants. Oikos 79:259–281

Hodge A (2004) The plastic plant: responses to heterogeneous supplies of nutrients. New Phytol 162(1):9–24

Hopmans JW, Bristow KL (2002) Current and future needs of root water and nutrient uptake modeling. Adv Agron 77:103–183

Imsande J, Touraine B (1994) N demand and the regulation of nitrate uptake. Plant Physiol 105:3–7

Jackson RB, Caldwell MM (1993) Geostatistical patterns of soil heterogeneity around individual perennial plants. J Ecol 82:683–692

Jackson RB, Caldwell MM (1996) Integrating resource heterogeneity and plant plasticity: modelling nitrate and phosphate uptake in a patchy soil environment. J Ecol 84:891–903

Jeschke WD, Hartung W (2000) Root-shoot interactions in mineral nutrition. Plant Soil 226:57–69

Jorgensen RA, Atkinson RG, Forster RLS, Lucas WJ (1998) An RNA-based information superhighway in plants. Science 279:1486–1487

Lappartient AG, Vidmar JJ, Leutsek T, Glass ADM, Touraine B (1999) Inter-organ signalling in plants: regulation of ATP sulfurylase and sulfate transporter genes expression in roots mediated by phloem-translocated compound. Plant J 18:89–95

Leyser O, Fitter A (1998) Roots are branching out in patches. Trends Plant Sci 3:203–204

MacDonald AJS, Davies WJ (1996) Keeping in touch: responses of whole plants to deficits in water and nitrogen supply. Adv Bot Res 22:229–300

Magnani F, Grace J (2000) Plants as self-organising systems. In: Marshall B, Roberts JA (eds) Leaf development and canopy growth. Sheffield Academic Press, Sheffield, pp 36–58

Marschner H (1986) Mineral nutrition in higher plants. Academic Press, London
Martin A, del Pozo JC, Iglesias J, Rubio V, Solano R, de la Peña A, Leyva A, Paz-Ares J (2000) Influence of cytokinins on the expression of phosphate starvation responsive genes in *Arabidopsis*. Plant J 24:559–567
Muller B, Touraine B (1992) Inhibition of NO_3^- uptake by various phloem translocated amino acids in soybean seedlings. J Exp Bot 43:617–623
Munns R, Cramer GR (1996) Is coordination of leaf and root growth mediated by abscisic acid? Plant Soil 185:33–49
Passioura JB (1980) The transport of water from root to shoot in wheat seedlings. J Exp Bot 31:333–339
Passioura JB (1991) Soil structure and plant growth. Aust J Soil Res 29:717–728
Passioura JB (1996) Simulation models: science, snake oil, or engineering? Agron J 88:690–694
Poorter H, Nagel O (2000) The role of biomass allocation in the growth response of plants to different levels of light, CO_2, nutrients and water: a quantitative review. Aust J Plant Physiol 27:595–607
Reich PB, Walters MB, Ellsworth DS (1997) From tropics to tundra: global convergence in plant functioning. Proc Natl Acad Sci USA 94:13730–13734
Rengel Z, Robinson DL (1990) Modeling magnesium uptake from an acid soil. II. Barber-Cushman model. Soil Sci Soc Am J 54:791–795
Rennenberg H, Geßler A (1999) Consequences of N deposition to forest ecosystems – recent results and future research needs. Water Air Pollut 116:47–64
Robinson D (1994) The responses of plants to non-uniform supplies of nutrients. New Phytol 127:635–674
Robinson D, van Vuuren MMI (1998) Responses of wild plants to nutrient patches in relation to growth rate and life-form. In: Lambers H, Poorter H, van Vuuren MMI (eds) Inherent variation in plant growth: physiological mechanisms and ecological consequences. Backhuys, Leiden, pp 237–257
Robinson D, Linehan DJ, Caul S (1991) What limits nitrate uptake from soil? Plant Cell Environ 14:77–85
Robinson D, Hodge A, Griffiths BS, Fitter AH (1999) Plant root proliferation in nitrogen-rich patches confers competitive advantage. Proc R Soc Lond B 266:431–435
Rorison IH, Robinson D (1984) Calcium as an environmental variable. Plant Cell Environ 7:381–390
Rowell DL (1994) Soil science: methods and applications. Addison Wesley Longman, Harlow
Scheible W-R, Lauerer M, Schulze E-D, Caboche M, Stitt M (1997) Accumulation of nitrate in the shoot acts as a signal to regulate shoot-root allocation in tobacco. Plant J 11:671–691
Schlesinger W (1997) Biogeochemistry, 2nd edn. Academic Press, London
Seward P, Barraclough PB, Gregory PJ (1990) Modelling potassium uptake by wheat (*Triticum aestivum*) crops. Plant Soil 124:303–307
Smith FW, Hawkesford MJ, Ealing PM, Clarkson DT, Vanden Berg PJ, Belcher AR, Warrilow AGS (1997) Regulation of expression of a cDNA from barley roots encoding a high affinity sulphate transporter. Plant J 12:875–884
Tardieu F, Davies WJ (1992) Integration of hydraulic and chemical signalling in the control of stomatal conductance and water status of droughted plants. Plant Cell Environ 9:540–545
Tillard P, Passama L, Gojon A (1998) Are phloem amino acids involved in the shoot to root control of NO_3^- uptake in *Ricinus communis*? J Exp Bot 49:1371–1379
Tinker PB, Nye PB (2000) Solute movement in the rhizosphere. Oxford University Press, Oxford

van der Werf A, Nagel O (1996) Carbon allocation to shoot and roots in relation to nitrogen supply is mediated by cytokinins and sucrose. Plant Soil 185:21–32

van Vuuren MMI, Robinson D, Griffiths BS (1996) Nutrient inflow and root proliferation during the exploitation of a temporally and spatially discrete source of nitrogen in soil. Plant Soil 178:185–192

Wang T-B, Gassmann W, Rubio F, Schroeder JI, Glass ADM (1998) Rapid up-regulation of *HKT1*, a high-affinity potassium transporter gene, in roots of barley and wheat following withdrawal of potassium. Plant Physiol 118:651–659

Watson MA (1986) Integrated physiological units in plants. Trends Ecol Evol 1:119–123

Yanai J, Linehan DJ, Robinson D, Young IM, Hackett CA, Kyuma K, Kosaki T (1996) Effects of inorganic nitrogen application on the dynamics of the soil solution composition in the root zone of maize. Plant Soil 180:1–9

Zhang H, Forde BG (1998) An *Arabidopsis* MADS box gene that controls nutrient-induced changes in root architecture. Science 279:407–409

Zhang H, Jennings A, Barlow PW, Forde BG (1999) Dual pathways for regulation of root branching by nitrate. Proc Natl Acad Sci USA 96:6529–6534

4 Internal Regulation of Nutrient Uptake by Relative Growth Rate and Nutrient-Use Efficiency

V.P. Gutschick and J.C. Pushnik

4.1 Introduction

Plants are observed to regulate their uptake of nutrients in elaborate patterns according to their environment of growth and their stage of development. Among the common patterns is that plants reduce their nutrient uptake capacities (per mass of root, v, or of the whole plant, v_{plant}) as nutrient concentrations increase (Clements et al. 1979; Godwin and Blair 1991; Youssefi et al. 1999; see Chap. 6, this Vol.). Similarly, changes in the shoot's environment, such as in CO_2 partial pressure, also induce changes in nutrient uptake rates (v). A number of questions arise – for one, why should a 'good' such as nutrient acquisition ever be curtailed, or not expressed at a maximal rate? One must infer that downregulating the acquisition of a beneficial resource confers a net benefit in Darwinian fitness, for which most plants or their immediate ancestors have been heavily selected. Admittedly, long-domesticated plants may diverge from the fitness functions of wild plants (Gutschick 1987, 1997, 1999; Jackson and Koch 1997). In some cases, the explanation lies at the immediate physiological level, in that some nutrients in excess are toxic, such as boron (Nable et al. 1990) and even phosphate (e.g., Romera et al. 1992). Nonetheless, such downregulation occurs even for nutrients that show no apparent toxicity in luxury consumption, such as nitrogen. Plant performance does not follow the guidelines espoused by the actress Mae West, "Too much of a good thing is wonderful". Some experimental evidence shows that overexpression of nontoxic nutrient acquisition is deleterious to plant growth and fitness – witness the stunting of supernodulating legumes (Carroll et al. 1985), which can perhaps be attributed to excess diversion of photosynthate to N_2 fixation. Nonetheless, downregulation occurs even at modest, physiological nutrient content, most markedly in woody plants (Gessler et al. 1998).

Prediction of uptake capacity in changing environments, whether for crops or wild plants, is highly desirable for studies of global change. We, as a

Ecological Studies, Vol. 181
H. BassiriRad (Ed.) Nutrient Acquisition by Plants
An Ecological Perspective

research community, might (and must) achieve description suited to wide ranges of plants and environments. This is a worthwhile task, but an unending one, given the infinite continuum of possible environments and of possible landscapes of Darwinian fitness or agronomic value. Far more useful in the long term would be a predictive capability based on demonstrably shared mechanisms (biochemical and genetic), or on a knowledge of the overarching selection pressures for regulation of uptake mechanisms. A comprehensive theoretical framework for uptake capacity should be sought simultaneously on two levels, physiological and ecological/evolutionary.

In the next few pages, we will put forth plausible mass-action forms that fit observations of how plants respond – in nutrient uptake rate, photosynthetic rate, tissue nutrient content, and root allocation – particularly to changes in N availability, or to elevated CO_2. The individual rate processes are formulated to respond properly (e.g., the fraction of reduced nitrogen (*RN*) loaded into the xylem should increase if the root *RN* increases). The forms we propose are heuristic (mass actions based on gross pools, not on pools of intermediate metabolites). These are meant to be guides to the origins of the plant responses, specifically to promote considerations of how processes must change relative to each other in order to give observed plant responses. The resultant model extends simple *functional balance* models, which only resolve gross root and shoot capture of resources (nutrients, CO_2, light) but do not explain why resource-capture capabilities attain the values observed or how these capabilities might be regulated (Gutschick and Kay 1995).

Foremost, we attempt here to provide a semi-mechanistic 'explanation' of how the uptake capacity (V_{max}) and root:shoot ratio (r) should acclimate to the growth environment. The functional balance model derived from the experiments of Gutschick and Kay (1995) sought to identify optimal V_{max} and r values by fixing either V_{max} or r values and varying the other. One intriguing result was that there should be an optimal root:shoot ratio (r=1), independent of environment (N only, not considering water). Second, the optimal V_{max} should be infinity: incremental gains in relative growth rate (*RGR*) continue, if at smaller rates, for any increase in V_{max}. A mechanistic model obviously would disallow such an extreme; it incorporates responses that evolved despite not maximizing the relative growth rate. If the predictions are realistic, then one might seek an explanation as to why these mechanisms evolved.

We will start our discussion with the question of 'what sets nutrient demand?'. Demand is surprisingly difficult to formulate, at least in terms of the external environmental variables and the basic growth attributes of the plant, including physiological capacities for nutrient uptake and photosynthesis, and growth patterns of root and shoot. Most of the literature on plant function defines demand as a single point value, the current uptake rate that one might calculate from current growth rate and current tissue nutrient content, f_n, of the plant. This definition is what economists would call the 'quantity demanded'. In contrast, 'demand' is a mathematical function – the quantity demanded as a function of price (cost of nutrient acquisition). With this view of demand, we must inquire why the rate of uptake has its current magnitude, and what will be the rate if we change environmental conditions, say, CO_2 levels.

4.2 Phenomenology of Uptake Rate Responding to Nutrient-Use Efficiency and Growth Rate

4.2.1 The Evolutionary and Ecological Perspective of Physiological Demand

Economists propose that demand is determined by a rational consumer who knows the utility of a resource, and seeks to maximize net marginal benefit (the utility less the cost, as derivative with respect to quantity consumed) of a resource. For a human consumer, this is monetary value, other values being converted to this as a medium of exchange. For a plant, demand could be defined as the continuous function that relates the mass of nutrient acquired as a function of the cost of acquiring and the benefit of using that nutrient. Cost is a function of, above all, external nutrient concentration – for example, it requires more energy in root growth to obtain dilute nutrients. It is necessary that this mass of nutrient optimize the plant's function, say, its growth or, more importantly, its Darwinian fitness.

This is where the difficulty arises – in what sense(s) and to what extent do plants optimize their performance? Fitness can never be fully maximized (Stenseth and Maynard Smith 1984). Even if fitness is nearly optimized, the relation to readily measured physiology or even simple growth in mass is complicated. Greatest vegetative biomass is not equivalent to greatest fitness. For example, the dispersal dynamics of seeds does matter, and for perennials, reproduction in any one year involves tradeoffs with future reproductive output (Boutin and Harper 1991; Edwards and Crawley 1999). Consider further the relation of total biomass growth to nutrient acquisition. Fast early growth may deplete the soil of nutrients for later reproductive growth. For nutrients that are poorly mobile in plants, this may more than negate benefits of early growth, and may explain why early *RGR* is manifestly held below maximal *RGR* in some plants. Moreover, any prediction of optimal uptake capacity for a specified environment requires that we possess information on aspects such as the availability of nutrients at all future times, especially at times of reproduction. Yet there are uncertain, risky, or stochastic elements in the availability of a nutrient, which may depend on, e.g., weather (precipitation, temperature). The same risk and uncertainty are associated with the utility of a nutrient. Changes in the concurrent availability of water modify the photosynthetic utility of nitrogen, as discussed in the following sections. On an ecological level, high N content brings stochastic risks of herbivory; these risks may be the most potent limiters of high nutrient content, particularly for N.

Even for deterministic costs and benefits, the plant cannot accurately sense current availability in a distributed soil volume. Future availability depends as well on other organisms for nutrient mobilization, such as by microbial mineralization. Of course, there are broad, predictable patterns of future availability, such as seasonal flushes, and plants have been naturally selected to respond effectively to these patterns, but the risk lies in the details, especially in competitive growth. Many of the limitations and promises of optimization theory applied to plants have been discussed by, e.g., Bloom et al. (1985) and Gutschick (1987). We shall provisionally accept a number of limitations, and explore some details of what is involved in optimizing *RGR* alone with a known external nutrient concentration, known cost of acquiring and metabolizing the nutrient, and known benefit (current photosynthetic utility) as a function of tissue nutrient content (f_n).

4.2.2 A View from Physiology: Response of Growth Rate to Nutrient Availability, Cost, and Utility

An understanding of the regulation of nutrient acquisition and use requires a physiological perspective. Considerable detail at the levels of physiology and gene expression has been achieved in model systems, such as *Arabidopsis*. However, these model systems are somewhat limiting, in that they do not adequately represent the range of selection pressures that have molded diverse physiological patterns among different species and environments. Ecologists and whole-plant physiologists are positioned to broaden this view, generating clues as to the nature of regulatory signals and regulatory actions. One example is in elucidating the roles of nutrient costs (for acquisition and metabolism), and of saturation of benefits in making it beneficial to limit nutrient uptake. Nutrients are costly, in several ways – for example, in the construction of roots, and in metabolism for uptake, reduction, and maintenance (Gutschick 1987; Zerihun et al. 1998). Moreover, as nutrients become more abundant, there is a decline in their marginal benefit, as measured by photosynthetic utility, p^* (mass of photosynthate made per gram of nutrient per day; Gutschick and Kay 1995). The photosynthetic utility of nitrogen, in particular, is much increased when the shoot is exposed to high CO_2.

As a first accounting of such costs and benefits, we present here a model applicable primarily to growth in a steady environment, and omitting the complication of transitions from vegetative to reproductive growth. In this model (elaborated from Gutschick and Kay 1995, Appendix III therein), nutrient uptake and nutrient use for photosynthesis achieve a functional balance – that is, root and shoot functions attain a balance. Such models do appear to correctly predict the trends in f_n and in *RGR* at high CO_2 (Zerihun et al. 2000; BassiriRad et al. 2000). Can such models also predict trends in V_{max}? They can do so only on the basis that a given V_{max} may maximize growth rate, rather than giving a mechanistic explanation. However, this view must be accommodated, if only to explain why the physiological mechanisms exist and act as they do. We show that yet more physiological and developmental traits must be considered.

Consider the rate of dry-matter gain of a plant limited by its nutrient uptake. Uptake occurs at velocity v per mass of root. Total root mass is m_r, so that the uptake rate of the whole plant is $m_r v$. New plant tissue has a fractional nutrient content, $\tilde{f}_n$, where the tilde indicates increment in new tissue. The uptake-limited growth rate is then

$$\dot{m}^{ul} = m_r v / \tilde{f}_n \tag{4.1}$$

and the relative growth rate of the whole-plant is then divided by whole-plant mass:

$$RGR^{ul} = \frac{m_r v}{(m_r + m_s)\tilde{f}_n} = \frac{rv}{(1+r)\tilde{f}_n} \tag{4.2}$$

Here, r is the root:shoot ratio, as usual.

The photosynthesis-limited growth rate is simply the whole-plant photosynthetic rate, A_{plant}, multiplied by the conversion efficiency from raw photosynthate to dry matter, β. Now, A_{plant} equals the average photosynthetic rate of leaves per mass, $A_{L,m}$, multiplied by leaf mass, m_L. Much photosynthesis is done at light saturation, where the rate per leaf area is proportional to the nitrogen mass per leaf area. Equivalently, then, the rate per leaf mass is proportional to the mass fraction of nitrogen (and perhaps also of phosphorus; see Gutschick 1993), that is, $A_{L,m}=p^* f_n$. The leaf mass may be expressed as a fraction of shoot mass, $\alpha_L=m_L/m_s$. Thus, the growth rate is

$$\dot{m}^{pl} = \beta p^* f_n \alpha_L m_s \tag{4.3}$$

and the photosynthesis-limited *relative* growth rate is the above divided by total plant mass,

$$RGR^{pl} = \beta p^* \alpha_L \frac{m_s}{m_s + m_r} f_n = \beta p^* \alpha_L f_n / (1+r) \tag{4.4}$$

If root and shoot are in functional balance, the two *RGR* expressions above are equal. Dividing out a common factor of $1/(1+r)$, we obtain

$$rv / f_n = \beta p^* \alpha_L f_n \tag{4.5}$$

In functional balance, the incremental and average nutrient contents are equal, so that we can solve for f_n as $\sqrt{rv/(\beta p^* \alpha_L)}$. Note that it is determined by functional balance between the root and shoot, and is not freely adjustable as a plant response. Substituting this expression for f_n into either *RGR* expression gives an expression for *RGR* in terms of physiology, allocation, and environment:

$$RGR = \frac{\sqrt{rv}}{1+r}\sqrt{\beta p^* \alpha_L} \tag{4.6}$$

This growth rate is realistic, except that it increases indefinitely as v or p^* increases. In order for v, hence, V_{max}, to have an optimum (that is, for *RGR* to decline beyond that), it is necessary that there be both a cost to nutrient acquisition and a saturation of the benefit of the nutrient in photosynthesis. The costs have been discussed at length for N by, e.g., Gutschick (1981) and Zerihun et al. (1998). We incorporate them as changing the whole-plant photosynthetic rate in Eq. (4.4) above, introducing a factor $(1-c_N f_N)$, with c_N being the energetic cost of acquiring and metabolizing N. The saturation of photosynthetic carbon assimilation with f_N is also observed empirically (Sinclair and Horie 1989). It is also interpretable in theory. Consider leaves of a given mass per area – roughly, their thickness; specifying f_N then determines the carboxylation capacity $V_{C,max}$. For a specified leaf irradiance I_L, one can derive estimates of leaf (and also whole-plant) assimilation per unit area and per unit mass of N. Both of these functions begin as linear in f_N at low f_N, and then approach an asymptote as light, rather than N, begins to be limiting the photosynthetic rate *per mass of leaf*. We thus introduce another factor that decreases the whole-plant photosynthetic rate in Eq. (4.4); one of the simplest mathematical forms with a single saturation parameter q is $\exp(-qf_N)$. Here, q is the inverse of a 'saturating' N content in the shoot, $f_{N,s,sat}$. (Other forms, such as Michaelis-Menten, give very similar results.)

4.2.3 Does a Simple Model Predict an Optimal Uptake Capacity and Root Allocation?

We now use this model, with the modified costs and benefits of nutrients, to predict the optimal value of V_{max}, for nitrogen in particular. We estimate the other physiological and environmental variables, and then vary V_{max} to find the value that maximizes *RGR*. We consider various values of the N-reduction cost, c_N, as well. The solution is obtained numerically (we use a binary search in f_N; note that our Fortran programs are freely available). The other variables we must set are: the nutrient concentration at the root surface, c_e (a nominal 100 μM here); the Michaelis constant, K_m (taken as 30 μM), in order to compute v as $V_{max}c_e/(c_e+K_m)$; the root:shoot ratio, r (0.4 is taken as a typical value); the biosynthetic conversion efficiency, β (almost invariant among plants, so that we fix it at 0.77, excluding the N-reduction cost; BassiriRad et al. 2000); the fraction of shoot mass as leaves, α_L (fixed at 0.55, appropriate for a young plant); and p^*, the photosynthetic utility of the nutrient (fixed at 15.26 g carbohydrate g^{-1} dry matter day^{-1}, to represent a modest competitor).

Figure 4.1 presents the results. Note first that *RGR* responds more weakly than linearly to changes in uptake capacity. Following the elementary model of Eq. (4.6), it rises initially as the square root of v, and hence as the square root of V_{max}. It then flattens out as N uptake increases, mostly because the photosynthetic capacity saturates in f_N. Increasing costs of N reduction also exact a penalty in *RGR*, as expected. For all cases with relatively high costs (>6 g glucose per g nutrient, there is a clear optimum value of V_{max}, above which *RGR* declines. At uptake capacities significantly higher than the optimum and at high nutrient cost, however, no functional balance can be attained. For such cases, but also in general, one needs a model with realistic repression of uptake by the accumulation of N in tissue, as we develop below.

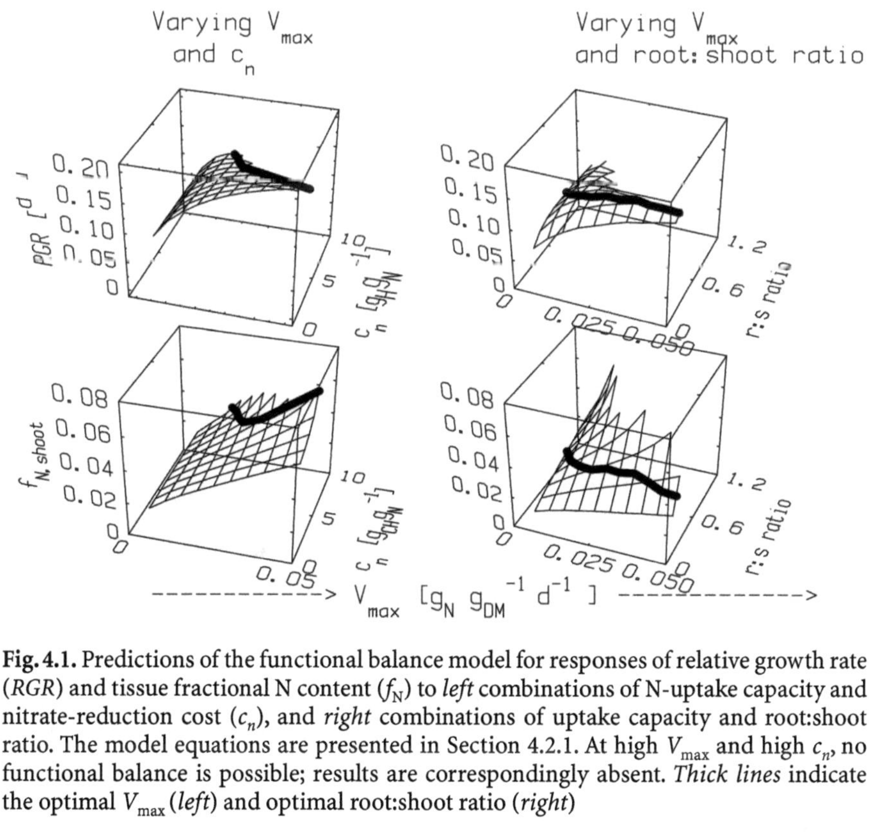

Fig. 4.1. Predictions of the functional balance model for responses of relative growth rate (*RGR*) and tissue fractional N content (f_N) to *left* combinations of N-uptake capacity and nitrate-reduction cost (c_n), and *right* combinations of uptake capacity and root:shoot ratio. The model equations are presented in Section 4.2.1. At high V_{max} and high c_n, no functional balance is possible; results are correspondingly absent. *Thick lines* indicate the optimal V_{max} (*left*) and optimal root:shoot ratio (*right*)

Predicted tissue N content is only realistic in some regions of the simulations presented here. When V_{max} is low or moderate, the tissue content, f_n, is realistic (below 5% N), but not for high V_{max}. Surprisingly, unrealistically high f_N occurs when there is a high cost of N that might be assumed to suppress N accumulation. However, this cost penalizes carbon accumulation, not uptake. Feedback to uptake is clearly not represented, nor is it warranted if increased N were really to confer a growth benefit, even a small one. Also unrealistic is that at high V_{max}, the limited diffusibility of N in soil (even as nitrate) draws down the concentration of N at the root. In essence, V_{max} is partly ineffective. This phenomenon can be modeled readily, for a specified root geometry and diffusivity (Gutschick and Kay 1995). Figure 4.2 shows such a calculation.

On the right-hand side of Fig. 4.1 are results in which the root:shoot ratio is varied as well as is V_{max}. This simulation is aimed at determining if there is an optimal value of r (as real plant behavior implies), when realistic costs and realistic saturation of benefits occur. Indeed, for any V_{max} there is an optimal

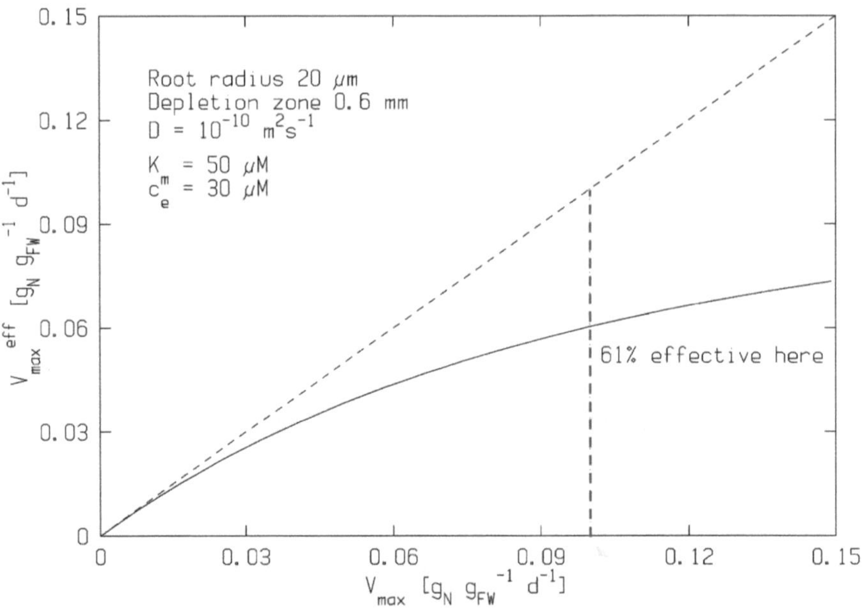

Fig. 4.2. Diminishing returns in actual uptake rate at high N-uptake capacity, when diffusibility of N in soil is limited at the stated realistic value. The steady-state uptake rate was computed for cylindrical roots of 20-μm radius with a depletion zone 0.6 mm wide, using standard equations (see Gutschick and Kay 1995). The achieved uptake rate, v, equals $V_{max}c_a/(c_a+K_m)$, where c_a is the concentration at the root surface. This rate is equated to an effective V_{max} ($V_{max}{}^{eff}$) operating at the N concentration in bulk soil, c_e, as $V_{max}{}^{eff}c_e/(c_e+K_m)$

value of r, and it declines at high uptake rates, following one's intuition that N demand is more readily satisfied in such cases.

Another interesting perspective from physiology is how the benefit of a nutrient is affected by other nutrients, as well as by water and other environmental factors. Consider how a decrease in water availability induces stomatal closure. The ensuing decrease in leaf-interior partial pressure of CO_2 (C_i) reduces the carboxylation rate per Rubisco enzyme by well-known formulae (Farquhar et al. 1980), thus reducing the photosynthetic gain per mass of N per unit time. In the long term (weeks), plants can approach an optimal mixed usage of water and N (Wright et al. 2003; Buckley et al. 2002). We have not entered this consideration in our model.

For greater realism and more insight, we must propose that the uptake capacity V_{max} and root allocation are not free parameters, but are both tied to internal indicators of plant status, to achieve optimal values (maximal *RGR*), or to meet other physiological, developmental, and environmental constraints (including risks) that we have not yet considered. Uptake then should respond to N status, although not to that alone, since the optimal *RGR* is attained at different magnitudes of f_N when the photosynthetic utility, p^*, changes: the magnitude of f_N is predicted to decrease with increasing p^*, as is seen in elevated-CO_2 experiments almost uniformly. Evaluating if this is optimal requires more detail. We have presented some detail, but have not answered the question of how optimal uptake shifts as p^* changes, in other publications (BassiriRad et al. 2000). With the realization that internal N and carbohydrate (*CH*) statuses both represent signals of sub- or supra-optimality, we must proceed to a model that resolves internal pools of N and *CH*. Such a model is presented below. Before proceeding to this larger model, we analyze another, little appreciated limit on the utility of nutrients, namely, internal developmental limits on growth rate. This surely must be a strong modifier of growth rate, and of any near-optimizing responses in V_{max} and in root allocation.

4.3 Toward a Model of Uptake Regulation in Response to Nutrient Utility

4.3.1 Predicted Response to an Intrinsic, Physiological or Developmental Limit on Relative Growth Rate

A phenomenon that is insufficiently appreciated as determining *RGR* and nutrient content or uptake rates is the maximal relative growth rate, RGR_{max}. This has been measured by, e.g., Poorter and Remkes (1990), and has been incorporated into a model of *RGR* by BassiriRad et al. (2000). Presumably, it arises from the existence of limits on the number of growing points (meristems) in a plant, and of limits on organ expansion rates (e.g., Tardieu and

Granier 2000). This may be taken as a quantitative expression of the qualitative term ‘vigor’. As the environment becomes more favorable, *RGR* is approaches RGR_{max}. As the utilization of growth substrates nears saturation, presumably the use of growth substrates stagnates and their internal pools rise markedly. These pools should then function as feedback signals. In our simple model of functional balance with finite RGR_{max} (BassiriRad et al. 2000; see Fig. 4 therein), the bulk nutrient content, f_n, does rise sharply, although we do not explain this rise mechanistically.

To explore how the existence of RGR_{max} is predicted to affect optimization (and thus regulation) of V_{max}, we use an expanded functional balance model (BassiriRad et al. 2000). To express the origin of *RGR* limitation, we resolve a new variable, f_C, the nominal (molar) concentration of nonstructural carbohydrates internal to the plant. *RGR* initially increases with f_C but reaches an asymptote because growing points saturate in their ability to use carbohydrates. We express *RGR*, then, as $RGR_{max} f_C\ /(f_C+K_e)$, with k_C as a parameter. Photosynthesis, too, is affected (repressed) by high f_C, being decreased by a factor $1/(1+k_C f_C)$. As in BassiriRad et al. (2000), we choose values of $k_C=1$, and $K_e=0.25$; both are unitless. The value of f_C itself is set by the balance between photosynthetic production and consumption in growth – that is, functional balance occurs in carbohydrate generation and use, analogous to the balance for N. As in the model without this developmental limitation, we then explored predicted performance with varied uptake capacity and cost of N reduction. Figure 4.3 shows that *RGR* is curtailed even when it is significantly

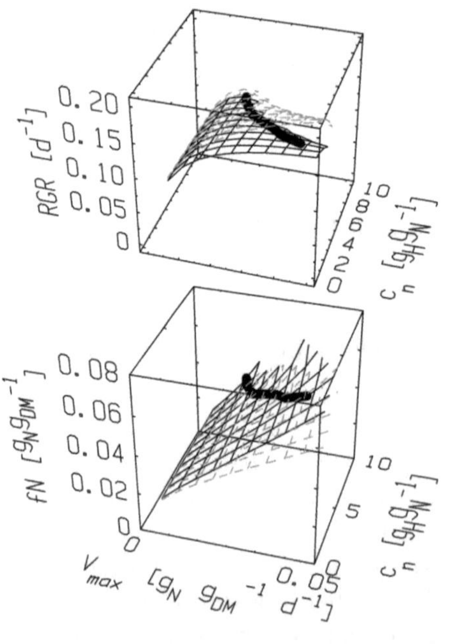

Fig. 4.3. Predictions of growth rate and tissue N content for the functional balance model with a developmental limit (RGR_{max}=0.20 day^{-1}) imposed. Physiological parameters are as in Fig. 4.1, with the addition of growth-limiting parameters described in the text. *Thick lines* indicate the optimal V_{max} values. *Shaded lines* represent the simulations of Fig. 4.1 for comparison

below RGR_{max}, while tissue N content is higher because use of N for growth is increasingly restricted. Optimal values of V_{max} remain apparent, and at closely the same values as for unlimited growth in Fig. 4.1. More signatures of developmental limitation may be discerned if internal pools of metabolites are resolved, as we now proceed to do.

4.3.1.1 Do Internal N and Carbohydrate Pools Explain Responses of Vmax to Photosynthetic N Utility and to Developmental Limitations on Relative Growth Rate?

A model with an account of developmental limitations must resolve an internal pool of nonstructural carbohydrates. Further insight might be gained by resolving additional internal pools. The predicted plant performance might thus be more realistic. Conversely, such a model might be useful in predicting the 'signatures' (patterns of changes in internal pools) that indicate that a plant is responding either to developmental limitations (RGR_{max}) or to changes in photosynthetic N utility (p^*), or to additional internal factors.

There is substantial qualitative evidence, and significant quantitative evidence, that uptake (as V_{max}, and K_m, etc.) responds to internal pools, which are reflected as well in tissue nutrient concentration, f_n (see Chap. 6, this Vol.). At the root, one observes downregulation of v by high concentrations of the nutrient itself, or of some key pool involving the nutrient. High $[K^+]$ appears to downregulate V_{max} of K^+ (Glass and Dunlop 1979; Siddiqi and Glass 1987); soluble K^+ is, of course, the only significant component of total tissue K^+. For nitrogen, the situation is more complex; v is regulated by a small pool of reduced-N compounds (King et al. 1992), which broadly track total tissue content. However, rapid transients occur in the regulatory pool in response to sudden (and normal) changes in shoot photosynthesis with time of day, for example (Morcuende et al. 1998), and these are important manifestations of regulation that must occur faster than changes in bulk tissue content. Similar small pools appear to regulate iron uptake, for example (Schmidke et al. 1999). In steady growth, bulk tissue content is related to the regulatory metabolite pools, and thus is correlated with V_{max} (see Chap. 6, this Vol.).

Also relevant to the role of internal pools in regulation and any associated optimization are the changes in f_n induced by changes in the shoot environment, such as light (I_L=irradiance in the photosynthetically active radiation region or PAR) or CO_2 (expressed as C_a=partial pressure in ambient air) or temperature. Increases in C_a cause increases in carboxylation rate per Rubisco molecule (Farquhar et al. 1980), and thus in carbohydrate pools. These commonly cause significant decreases in nitrogen and sulfur contents (Penuelas and Matamala 1990; Peterson et al. 1999). In previous papers (Gutschick 1993; Gutschick and Kay 1995), we expressed this effect solely with the physiological parameter p^*, the nutrient-use efficiency (or efficacy) in photosynthesis.

In functional balance, *RGR* is predicted (and observed; ibid., and Zerihun et al. 2000) to respond directly to $\sqrt{p^*}$, while f_n responds inversely as $\sqrt{p^*}$. Here, to explain optimal performance, we must resolve more detailed internal pools of metabolites.

The most parsimonious set of pools is composed of the following: nitrate in the root (N_r^o, '*o*' representing the oxidized form), reduced N in the root and shoot (N_r^r and N_s^r, respectively), and nonstructural carbohydrate in the root and shoot (C_r, and C_s, respectively). Correspondingly, we must resolve the processes of: N uptake; N reduction (taken to be all in the root; this may be generalized readily); growth of root tissue – with attendant incorporation of nitrate, reduced N, and carbohydrate; xylem loading of reduced N; shoot growth with incorporation of reduced N and *CH*; phloem loading of *CH*; and phloem loading (recycling) of reduced N back to the root. We derive a set of kinetic equations for all these processes in terms of the metabolite pools and rate constants for the net metabolic reactions. The rate constants are estimated numerically from 'gross' observables: growth rates under baseline conditions; uptake rate per mass of root and its response to reduced-N accumulation in the root; tissue N contents in the root and shoot (N_r^o, N_r^r, N_s^r) under baseline conditions; the fraction of reduced N recycled to the root; the response of xylem and phloem loading to reduced-N pools; and photosynthetic CO_2 assimilation per mass of N, and its response to tissue N (f_N) and to *CH* accumulation. These data are not trivial to obtain, but they are much simpler than a more detailed biochemical dataset. Once the rate constants are computed, it is assumed that all acclimations of the plant (to altered N availability, shoot CO_2 or temperature, etc.) are represented by the kinetic scheme with these rate constants fixed – that is, the kinetic scheme captures all the acclimation of the plant.

Many kinetic schemes have been proposed for the various facets of root N uptake, its feedback regulation, root and shoot growth, photosynthesis, and its regulation. (Le Bot et al. 1998). We propose here a similarly parsimonious synthesis of these ideas, and some new ideas. First, we take the evidence that internal or cytoplasmic pools of reduced N, as they increase, can downregulate N uptake from soil (King et al. 1992; Kronzucker et al. 1995; Sivasankar et al. 1997), and propose the simplest functional form for whole-plant N uptake, *U*, to be:

$$U = m_r v_0 / \left(1 + a_U \mathrm{N}_r\right) \tag{4.7}$$

Here, v_0 is the Michaelis-Menten form, $V_{max}{}^0 c_e/(c_e+K_m)$, where $V_{max}{}^0$ is the fully unrepressed value. The denominator is a downregulating factor, with a_U being a physiological constant (parameter). Next, we assume that our plant reduces all its nitrate in the root. (The model can readily be generalized to include nitrate reduction partially in the shoot, and also uptake of some N as reduced N. Roughly, the partial uptake of N as reduced N can be simulated by decreasing the cost of N reduction, ignoring the difference in nitrate and ammonium uptake kinetics.) The reduction rate, R_{NR}, is taken as simple mass action, responsive to both nitrate and carbohydrate in the root:

$$R_{NR} = m_r k_{NR} \mathrm{N}_r^o C_r \tag{4.8}$$

Root growth, $G_r = dm_r/dt$, similarly follows mass action in both substrates,

$$G_r = m_r k_{RG} C_r \mathrm{N}_r^r \tag{4.9}$$

The individual rates of incorporation of reduced N and of carbohydrates must be specified, in a form that generates consistent fractions of N in tissue. For reduced-N incorporation into root tissue, $I_{\mathrm{N},r}$, many empirical forms may be proposed, in advance of detailed experimental evidence. Having tested the behavior of many polynomial forms, we propose a Michaelis-Menten form that realistically generates increased incorporation of N into tissue (increased f_{N}) as the availability of reduced N increases (that is, as N_r increases):

$$I_{\mathrm{N},r} = R_{NR} \mathrm{N}_r^r / \left(\mathrm{N}_r + K_{sel} C_r\right) \tag{4.10a}$$

$$= k_{NRG} C_r \left(\mathrm{N}_r\right)^2 / \left(\mathrm{N}_r + K_{sel} C_r\right) \tag{4.10b}$$

Here, the rate constant for incorporating reduced N into growing tissue, k_{NRG}, is proportional to the rate constant k_{RG} for total growth, with a proportionality that produces the proper f_{N} at low N availability. A selectivity coefficient, K_{sel}, prevents excessive incorporation of N relative to C. Similarly, the incorporation of carbohydrate into new tissue, $I_{C,r}$, must be consistent with the overall growth rate and with the fraction of biomass deriving from reduced N. We derived the form

$$I_{C,r} = k_{RG} C_r \mathrm{N}_r{}' - k_{NRG} \frac{C_r \left(\mathrm{N}_r{}'\right)^2}{\left(\mathrm{N}_r + K_{sel} C_r\right)} + \left[\frac{1}{f_{\mathrm{N},RN}} - \frac{1}{f_{\mathrm{N},NO}}\right] k_{RG} C_r \mathrm{N}_r{}' F_{DM} \mathrm{N}_r{}^o \tag{4.11}$$

The third term enforces mass balance of reduced and nitrate N in new tissue. Here, $f_{\mathrm{N},RN}$ and $f_{\mathrm{N},NO}$ are the mass fractions of N in the form of reduced N (about 0.18 for amino acids) and of nitrate (18/62=0.29), respectively; F_{DM} is that fraction of fresh mass represented by dry matter.

Proceeding toward the shoot, we describe xylem loading of reduced N, X_{N}, as an attenuating quadratic in N_r,

$$L_{\mathrm{N}} = k_X \left(\mathrm{N}_r^r\right)^2 / \left(1 + a_X \mathrm{N}_r^r\right) \tag{4.12}$$

This is a modification of simple linear kinetics by a Michaelis-Menten factor, $N_r^r/(1+a_X N_r^r)$, which increases the export fraction at high N_r^r. Shoot growth in mass, $G_s=dm_s/dt$, is described as by simple mass action,

$$G_s = m_s k_{SG} C_s N_s^r \tag{4.13}$$

Note that the ratio of shoot growth to root growth remains constant (root:shoot fraction is constant) if the C and N pools (which are concentrations) are stable. The incorporation of reduced N into shoot biomass has a form analogous to that in roots, excluding the nitrate term. The shoot as source of carbohydrates is described as in our basic model, with total carbohydrate production rate A given as

$$A = p^* \, f_{N,s} m_L = p^* f_{N,s} \alpha_L m_s \tag{4.14}$$

To account for repression by accumulated carbohydrate, we add a simple factor,

$$A \rightarrow A^0 \left(1 + a_{PS} C_s^0\right) / \left(1 + a_{PS} C_s\right) \tag{4.15}$$

The factor in the numerator simply normalizes A to its value at a chosen reference environment in which A attains a known rate A^0 and C_s has its base value, C_s^0. The parameter a_{PS} describes the sensitivity of photosynthesis to repression by free carbohydrate. To describe phloem loading of carbohydrate to circulate to roots, P_C, we use a model based on the pressure-flow model (admittedly of modest accuracy: Farrar and Jones 2000),

$$L_C = m_s k_{PC} \left[\Delta\psi + RT\left(C_s - C_r\right)\right] \tag{4.16}$$

Here, $\Delta\psi$ is the water-potential difference from shoot to root (<0), R is the gas constant, and T the absolute temperature. Partly as a signal for regulating N uptake, reduced N is also loaded into phloem to return to roots. We take the rate, P_N, as

$$L_N = L_C N_s^r / \left(N_s^r + K_{PN} C_s\right) \tag{4.17}$$

This form makes the fractional return of reduced N increase as reduced N increases – that is, when N supply exceeds growth demand.

These equations were developed by some trial and error in order to get realistic responses of growth, f_N, and uptake to changes in external N availability, in p^* (hence, in CO_2 or irradiance), and in growth potential. It is important to note that a maximal RGR is implicit in the rate constants k_{RG}, k_{SG}. Most of the rate constants can be fitted from observed values of RGR, f_N, and the pool sizes for a given environment. Other parameters may be estimated to explore the sensitivity of growth and f_N to partitioning and feedback parameters such as K_{sel} and a_X. The model is coded in Fortran 77 and will be provided to anyone interested (please contact the authors). The code is voluminously commented; the output is pseudo-graphic, providing a schematic printout of pools and flows to a plain ASCII terminal.

4.3.2 Applying the Model: Differences from Simple Functional Balance

This explicit carbon–nitrogen pool model (CN model, or CN) has been applied to a few illustrative cases (Fig. 4.4). These are (1) a base case with the same physiology as that used in the FB model above; (2) the same, but with a developmental limit, RGR_{max}=0.15 per day; (3) doubled p^*, as would be the case if atmospheric CO_2 approximately doubled. The other parameters are similar to the other simulations here: shoot leaf fraction α_L=0.55; root:shoot ratio r=0.4; external solution concentration of nitrate c_e=100 mmol m^{-3} (100 μM); uptake Michaelis constant K_m=30 mmol m^{-3}; biosynthetic efficiency is 0.77; cost of nitrate reduction c_N=6 g carbohydrate g^{-1} N; saturation of photosynthetic N utility given with $f_{N,sat}$=0.03 g N g^{-1} dry matter; p^*=15.3 g carbohydrate g^{-1} N day^{-1}; fresh mass/dry mass ratio=4; free nitrate concentration in roots in base case=12.5 mM; repression parameter for uptake a_U=100 g reduced N g^{-1} fresh mass; repression parameter for xylem loading a_{XL}=500 g N g^{-1} fresh mass; recycling ratio in base case=0.33; $\Delta\psi$=0.1 MPa; a_{PS}=5.19 in the CN model, to give a fractional expression of photosynthesis equal to 1/1.3=0.77 in the base case. For N incorporation into new root growth, the parameter K_{sel} is set to 0.01, making the fraction N in new tissue fairly sensitive to the reduced-N:*CH* balance in the root.

Several features immediately distinguish the CN model from the simple functional balance model. Most notably, there is no optimal V_{max} when expressed as unrepressed value, V_{max}^0 : *RGR* increases smoothly toward an asymptote, and never decreases (Fig. 4.4A, top row of graphs). Even as V_{max}^0 grows very large, the buildup of reduced N (*RN*) in the root represses V_{max} (Eq. 4.7) and forces its decline. The effect is shown in Fig. 4.4C for all three physiological constitutions modeled. The modest repression of root growth at high N availability (Fig. 4.4C, bottom row) acts to further limit N uptake. This consequently limits the associated reduction costs and tissue N buildup. In the shoot, the benefit of increased N is progressively but not inordinately cut by the inefficient partitioning of N (the saturation of photosynthetic rate, as $f_{N,s}$ increases; Fig. 4.4B, middle row of graphs). Thus, increased N availability – achieved either by high external concentration or by high uptake capacity as V_{max}^0 – again gives *no* optimal value of uptake capacity as V_{max}^0. The observed control of uptake capacity, particularly to low values, must originate in selection on some traits coupled to nutrition, such as the cation–anion balance that might be adversely affected at high N uptake. Even more likely is the rising

A

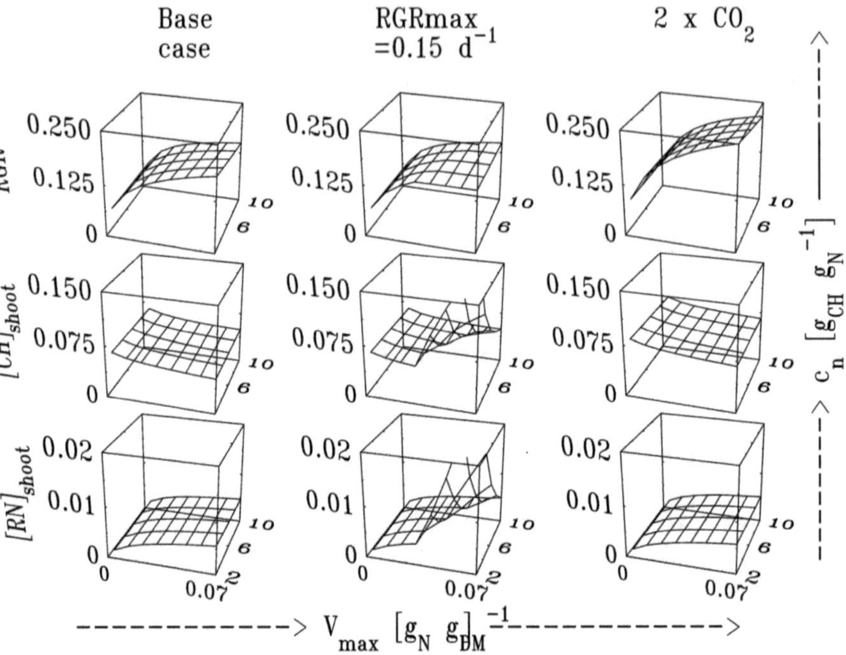

Base case
RGRmax =0.15 d^-1
2 x CO2
RGR
0.250
0.125
0
[CH]shoot
0.150
0.075
0
[RN]shoot
0.02
0.01
0
0.07
10
6
2
Vmax [gN gDM]^-1
cn [gCH gN^-1]

B

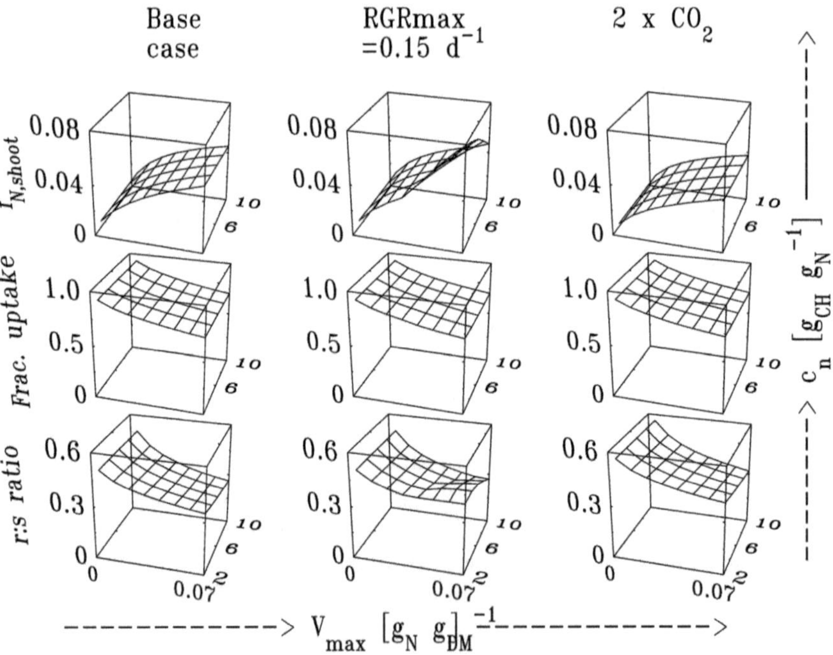

Base case
RGRmax =0.15 d^-1
2 x CO2
fN,shoot
0.08
0.04
0
Frac. uptake
1.0
0.5
0
r:s ratio
0.6
0.3
0
0.07
10
6
2
Vmax [gN gDM]^-1
cn [gCH gN^-1]

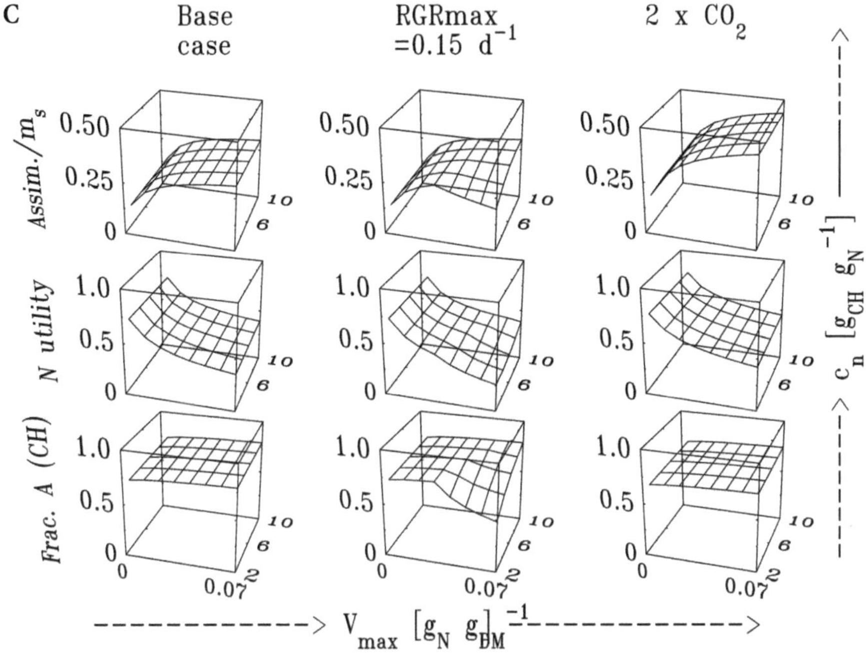

Fig. 4.4A–C. Predictions of the CN model that resolves internal metabolite pools, for a base-case physiology (described in the text); the same physiology with a developmental limitation on *RGR*; and increased photosynthetic N-use efficiency. Responses to combinations of fundamental (unrepressed) uptake capacity ($V_{max}{}^0$) and N-reduction cost are shown as **A** relative growth rate (*RGR*, per day) and in-shoot concentrations of free carbohydrate ($[CH]_{shoot}$, g g^{-1} dry matter) and free reduced-N compounds ($[RN]_{shoot}$, predominantly amino acids); **B** shoot total-N content ($f_{N,shoot}$, mass fraction), fraction of maximal N uptake (*Frac. uptake*, relative to fully expressed capacity), and root:shoot ratio; **C** photosynthetic carbon assimilation per unit mass of shoot (*Assim./m$_s$*, g carbohydrate g^{-1} dry mass day^{-1}), fraction of maximal photosynthetic N utility (N *utility*, unitless, cf. relative to assimilation linearly proportional to f_N), and fraction of maximal photosynthesis not repressed by carbohydrate accumulation [*Frac. A* (*CH*), unitless]. *CH* Carbohydrates, *DM* dry matter

risk of herbivory at high N and associated high water content (Sect. 4.3.3, below). In our view, the problem of explaining uptake regulation remains open. Second, there is a clear signal of developmental limitation (finite RGR_{max}; middle column of graphs), as a sharp rise in both tissue *CH* and reduced-N contents as N availability rises. With such growth limitation, the root:shoot ratio is predicted to become insensitive to N at high N availability; this may or may not be realistic.

Other behavior simulated in the model often parallels that in the simple FB model. In all three cases, *RGR* initially rises in proportion to the square root of N availability (expressed as $V_{max}{}^0$). It also increases closely as the square root of photosynthetic utility – that is, modeled *RGR* is about 1.4-fold greater when p^* is doubled. This increase is maintained at all values of $V_{max}{}^0$ and c_n. Shoot *CH* increases notably because of higher photosynthetic rates, while reduced N is depleted by faster growth; thus, the repression of photosynthesis by *CH* increases, while the saturation of photosynthesis by *RN* is relieved. The effect of high N-reduction cost, c_N, is modest in all cases, with no more than a 10 % reduction of *RGR* at c_N=10 relative to c_N=0; N reduction is only a moderate fraction of the energy budget.

Interestingly, the simulated shoot *CH* content (Fig. 4.4A, middle row) remains fairly stable as V_{max} rises, and so therefore does the fractional repression of photosynthesis by *CH* buildup (Fig. 4.4B, bottom row). Simply, high N enables more *CH* production, but in almost the same proportion it increases the consumption of *CH* in tissue growth. Shoot *RN* content, $N_s{}^r$, rises initially as the square root of V_{max}, as in the FB model. It becomes flatter yet at high $V_{max}{}^0$, as 'excess' uptake is repressed.

4.4 Additional Questions: Stress, Root Turnover, Low Set Points of Woody Plants

4.4.1 Role of Root Lifetime

Chapter 8(this Vol.) treats a number of the many interesting ways in which root turnover affects growth and nutrient content. Root turnover patterns vary among species and among broad functional groups (Gill and Jackson 2000). The selection pressures favoring faster turnover may include the danger of being parasitized (Aiken and Smucker 1996). Relevant to the current topic is how root turnover affects responses of nutrient uptake to photosynthetic nutrient utility p^* and to achieved growth rate. In the simplest accounting, turnover represents a capital cost over and above the initial construction costs of the root system (Gutschick 1987; Eissenstat 1992; Comas et al. 2000; Eissenstat et al. 2000; Chap. 8, this Vol.). It therefore inflates the metabolic cost of operating roots. This cost is not incurred per mass of nutrient, so that it is not appropriate to incorporate it into the cost c_n used above. The cost of turnover resembles the cost of supporting mycorrhizal tissue that is more metabolically active than the bulk of root mass. We have published (Bassiri-Rad et al. 2000) an expanded version of the functional balance model that accounts for mycorrhizal activity, and this version is appropriate for exploring how turnover may affect uptake responsiveness to p^* and growth rate.

High metabolic activity, such as from high turnover, 'burns off' carbohydrates and makes f_N rise (ibid.). It thus can accentuate sensitivity of uptake to the p^* and growth rate. However, dilution of this high metabolic activity into a base of low-metabolism, long-lived roots would correspondingly dilute these changes in uptake sensitivity. Thus, we have not provided simulations of turnover, which can occur in a wide range of rates, fractions of the root system, and costs. We estimate that the effects of turnover may be related more strongly to reducing the local depletion of nutrients in the rhizosphere. Improved access to undepleted soil regions then makes the root kinetic parameters, especially V_{max}, more important determinants of uptake rate, as seen in our model (BassiriRad et al. 2000, Fig. 3 therein, using mathematical approximations developed by Gutschick and Kay 1995).

4.4.2 Effects of Water Stress and Salinity

Water stress has at least three effects that are very relevant to the topic of this chapter. First, it reduces growth rates, particularly leaf expansion rates (Granier and Tardieu 1999). It thus acts as a reduction in RGR_{max}, and is expected to magnify the downregulation of uptake. It is not yet clear if nutrient uptake at the roots responds directly to stress signals such as ABA (Goldbach et al. 1975; Roberts and Snowman 2000); it may be that downregulation is a response to growth repression with accumulation of reduced N, as in our model. Second, it reduces nutrient availability in the root zone, and thus decreases uptake rates. Third, water stress reduces stomatal conductance. This induces a lowering of leaf-internal CO_2 (C_i) and thus of p^*, as discussed in the first section of this chapter. Our models in any form predict that lower p^* will increase the tissue nutrient content, f_N, and consequently induce additional downregulation of uptake as V_{max}. The two effects of stress are not readily separated, except with a simultaneous change in CO_2 that can restore p^* under water stress while leaf and whole plant growth remains repressed. We do not yet have experimental data for experiments in which CO_2 partial pressure was regulated to control growth repression independently of p^*, nor in which all the relevant variables, including p^*, C_i, f_N, and V_{max}, were measured. We look forward to the challenge of such results to interpret with our model. A distinct challenge is interpreting or predicting patterns of recovery from stress. BassiriRad et al. (1999) found contrasting patterns among desert species in N uptake following pulses of water availability. Our model, currently a steady-state model, is not directly amenable to predicting regulation under these conditions, but its extension with resolution of internal N and C pools might be useful.

Salinity stress, like water stress, initially has but minor effects on leaf photosynthetic capacity (Munns and Termaat 1986). Its major effect is on leaf life span and on leaf expansion rate, which are decreasing functions of soil salinity. Thus, an increasing fraction of leaves are senescent or dying. In our model, this acts as a reduction in p^* averaged over the plant, and one expects a rise in f_n to accompany a decrease in growth rate as RGR. This is seen in some experiments in which N fertilization effects are not confounding (e.g., Zidan 1992). Direct effects of anions in salts upon uptake, as by competitive uptake, have the opposite effect of reducing uptake and f_n, of course.

4.4.3 Why Is Uptake so Low in Some Species?

Some species, particularly woody species, have significantly lower uptake rates than herbaceous species (mostly ruderals) in the same physical location. One must assume that the divergence in uptake rates represents evolutionarily stable strategies of the co-occurring species (Maynard Smith and Price 1973; example of application to plant growth: Vincent and Vincent 1996). Nonetheless, for the low-uptake species, one must ask why the evolved set point of uptake is so low. More significant for the current inquiry in this chapter and for global-change studies is whether a low set point indicates the existence of internal controls that differ fundamentally from those we have considered here. That is, under changes in atmospheric CO_2, will woody plants alter their V_{max} very differently from herbaceous plants? Some surveys have been made by BassiriRad et al. (2000), Nowak et al. (2004) on this topic.

Our basic model of functional balance does not predict optimal V_{max} directly. Nonetheless, our model with V_{max} adjusted to find its optimal value would likely predict a nearly universal curve among species, if basic physiological factors such as p^* are shared among species. Woody plants do have moderately lower p^* than herbs (lower photosynthetic rates per unit leaf area, and not much lower N content per unit leaf area). However, the lower value is insufficient to explain the much lower values of V_{max}. We must seek explanations in the model that is expanded to resolve internal N and C pools. Several contributing factors are plausible, each of which would generate selection pressure for altered kinetic parameters. First, risks of herbivory rise with higher N content (Letourneau 1997; see also Hartley and Jones 1997 for some qualifications about this relationship). Second, high water content may be needed for high soluble protein content and high photosynthetic capacity (Roderick et al. 1999), and this may also correlate with higher risk of herbivory. Third, woody plants may often exist at climax with low N-mineralization rates relative to demand, and with poorly diffusible ammonium as the dominant form of N. Thus, they operate near the limit of zero concentration at the root surface, which gives minimal effect of V_{max} on uptake as well as low f_N. The first two contributing factors can be expressed in the CN model naturally by setting f_N, etc., in a base case. One obtains predicted responses, but one does lose some universality, in that base values of f_N (thus, of V_{max} and other physiological parameters) are not optimized, but rather species-specific suboptimal values. The third factor can be expressed in a soil+plant model that includes diffusion in soil. The patterns of response become specific at least to the levels of functional group of plant and of soil type. This may be inescapable.

4.5 Conclusions

The trends in V_{max} that our models predict are realistic and might guide research into plant responses to new environments that develop with global change. However, the predictions are not universal; different species and ecotypes show variation in their responses. We propose that the variation arises from selection pressures unique to different seasonal patterns of nutrient availability, or to herbivory risks, etc. Some of these formative patterns, such as seasonal nutrient flushes, might be incorporated into the model if it were integrated over time for the whole season. We might then explain a significant fraction of the variation among plants of different functional types and successional status. Other patterns such as herbivory risk are likely to be notably more difficult to account for.

Internal N and carbohydrate pools are being resolved in current experimental research on the biochemical and genetic regulatory basis of uptake regulation. These studies are intensive, and of necessity are focused on a few model species such as *Arabidopsis*. Our models, expanded to resolve a limited number of these internal pools, may offer simpler if somewhat less accurate prediction of uptake responses from a few biochemical assays. These predictions would thus be based on mechanisms specific to species and genotypes. Nonetheless, they would contain implicitly the complexities in physiology, ecology, and evolution for the species, without the need to understand patterns among all species at once. The models thus might be useful to predict responses of some key species that dominate biomass in some ecosystems, or that act as indicators for global change.

4.6 Summary

Nitrogen availability is a well-known controlling factor for plant growth. An earlier model of functional balance between the root and shoot offered an explanation of plant relative growth rate in terms of nutrient uptake rate, photosynthetic nitrogen-use efficiency (a notable function of CO_2 availability), biosynthetic efficiency, and allocation patterns among roots, stems, and leaves. Nitrogen content was determined by functional balance and was not independently adjustable; its responses to environmental factors (concentration of N compounds in soil solution, etc.) were clearly fixed. However, this model left the uptake capacity for nitrogen itself undetermined, having a virtually unconstrained set point, determined genetically. It also presented a dilemma in defining optimal allocation patterns. In this chapter, we offer an expansion of the functional balance model, aimed at explaining how controls over N uptake, allocation, and use in photosynthesis – hence, plant growth –

are exerted via intermediate pools of reduced N and carbohydrates, as well as more detailed bioenergetic costs of their transformations. New physiological parameters are defined for the controls over pool sizes and fluxes; these parameters should be more closely related to identifiable biochemical processes and their genetic bases than are the broader allocation ratios in the original model. We also consider that growth rate may be internally limited by meristem number and developmental rates; this limit affects N uptake and allocation in unique quantitative patterns. New questions are then posed about the variable patterns of performance among diverse plant species, particularly about the role of root lifetime and water stress, and about the occurrence of very low N-uptake capacities in many species.

Appendix: Definitions of Terms and Symbols

Terms and symbols in the Appendix are given in the text and are detailed in a computer model available on the website: http://biology-web.nmsu.edu/vince.

α_L	fraction of shoot mass as leaves
β	conversion efficiency of raw photosynthate into biomass
A	carbon assimilation rate
A^0	maximal (unrepressed) assimilation rate
A_{plant}	whole-plant photosynthetic rate
a_U	parameter for nitrate uptake repression by accumulated reduced N in root
a_{XL}	repression parameter for xylem loading
a_{PS}	repression parameter for photosynthesis (CN model only)
CH	free carbohydrate
CN	carbon–nitrogen pool model
c_e	concentration of nutrient external to root
c_n	cost of nutrient acquired per unit mass
C_r	nonstructural carbohydrate in roots
C_s	nonstructural carbohydrate in shoots
DM	dry matter
FB	functional balance model
f_C	nominal whole-plant molar concentration of nonstructural carbohydrates
f_n	tissue nutrient content, as mass fraction
$\tilde{f}_n$	fractional nutrient concentration in new tissue
f_N ($f_{N,s}$, $f_{N,r}$)	fractional nitrogen content (in shoot, in root)
$f_{N,s,sat}$	63 % saturation point for photosynthesis as f_N in shoot increases

G_r	root growth rate
G_s	shoot growth rate
$I_{C,r}$	incorporation, carbohydrates into roots
$I_{C,s}$	incorporation, carbohydrates into shoots
$I_{N,r}$	incorporation, nitrogen into roots
$I_{N,s}$	incorporation, nitrogen into shoots
K_e	Michaelis parameter in growth response to whole-plant non-structural carbohydrate
$K_{sel,RG}$	selectivity parameter for carbohydrate over reduced-N incorporation in root growth
K_{PN}	selectivity parameter for loading *CH* vs. reduced N into phloem in the shoot
k_C	parameter for repression of photosynthesis by carbohydrate buildup
k_{RG}	mass-action rate constant for root growth
k_{SG}	mass-action rate constant for shoot growth
L_C	loading of phloem to circulate carbohydrates to roots
L_N	loading of phloem to recycle N
m_r	mass of root
m_s	mass of shoot
$\dot{m}^{pl}$	photosynthesis-limited growth rate
$\dot{m}^{ul}$	uptake-limited plant growth rate
N_r^o	oxidized nitrogen (nitrate) in root
N_r^r	reduced nitrogen in root
N_s^r	reduced nitrogen in shoot
p^*	nutrient-use efficiency (efficacy) in photosynthesis; photosynthetic utility
P_{LM}	photosynthetic rate per unit mass tissue
R_{NR}	rate of nitrate reduction in roots
RGR	relative growth rate (increase in biomass per unit time)
RGR_{max}	maximum RGR
U	uptake rate of the whole plant
V_{max}	maximal uptake velocity per mass of roots
$V_{max}{}^0$	fully unrepressed V_{max}
$V_{C,max}$	maximum carboxylation capacity in photosynthesis
v	achieved rate of uptake per mass of roots

References

Aikens RM, Smucker AJM (1996) Root system regulation of whole plant growth. Annu Rev Phytopathol 34:325–346

BassiriRad H, Tremmel DC, Virginia RA, Reynolds JF, deSoyza AG, Brunell MH (1999) Short-term patterns in water and nitrogen acquisition by two desert shrubs following a simulated summer rain. Plant Ecol 145:27–36

BassiriRad H, Gutschick VP, Lussenhop J (2000) Root system adjustments: regulation of plant nutrient uptake and growth responses to elevated CO_2. Oecologia 126:305–320

Bloom AJ, Chapin FS III, Mooney HA (1985) Resource limitation in plants – an economic analogy. Annu Rev Ecol Syst 16:363–392

Boutin C, Harper JL (1991) A comparative study of the population dynamics of five species of *Veronica* in natural habitats. J Ecol 79:199–221

Carroll BJ, McNeil DL, Gresshoff PM (1985) A supernodulation and nitrate-tolerant symbiotic soybean mutant. Plant Physiol 78:34–40

Clements CR, Jones LHP, Hopper MJ (1979) Uptake of nitrogen from free-flowing nutrient solution: effect of termination and intermittent nitrate supplies. In: Hewitt EJ, Cutting CV (eds) Nitrogen assimilation by plants. Academic Press, London, pp 123–133

Comas LH, Eissenstat DM, Lasko AN (2000) Assessing root death and root system dynamics in a study of grape canopy pruning. New Phytol 147:171–178

Edwards GR, Crawley MJ (1999) Herbivores, seed banks and seedling recruitment in mesic grasslands. J Ecol 87:423–435

Eissenstat DM (1992) Cost and benefit of constructing roots of small diameter. J Plant Nutr 15:763–782

Eissenstat DM, Wells CE, Yanai RD, Whitbeck JL (2000) Building roots in a changing environment: implications for root longevity. New Phytol 147:33–42

Farquhar GD, von Caemmerer S, Berry JA (1980) A biochemical model of photosynthetic CO_2 assimilation in leaves of C_3 species. Planta 149:78–90

Farrar JF, Jones DL (2000) The control of carbon acquisition by roots. New Phytol 147:43–53

Gessler AM, Schultze S, Schrempp S, Rennenberg H (1998) Interaction of phloem-translocated amino compounds with nitrate uptake by roots of beech (*Fagus sylvatica*) seedlings. J Exp Bot 49:1529–1537

Gill RA, Jackson RB (2000) Global patterns of root turnover for terrestrial ecosystems. New Phytol 147:13–31

Glass ADM, Dunlop J (1979) The regulation of K^+ influx in excised barley roots. The relationships between K^+ influx and electrochemical potential differences. Planta 145:395–397

Godwin DC, Blair GJ (1991) Phosphorous efficiency in pasture species. V. A comparison of white clover accessions. Aust J Agric Res 42:531–540

Goldbach E, Goldbach H, Wagner H, Michael J (1975) Influence of N-deficiency on abscisic acid content of sunflowers. Physiol Plant 34:138–140

Granier C, Tardieu F (1999) Water deficit and spatial pattern of leaf development. Variability of responses can be simulated using a simple model of leaf development. Plant Physiol 119:609–619

Gutschick VP (1981) Evolved strategies of nitrogen acquisition in plants. Am Nat 118:607–637

Gutschick VP (1987) A functional biology of crop plants. Croom Helm, London

Gutschick VP (1997) Nutrient-limited growth rates: roles of nutrient-use efficiency and of adaptations to increased uptake rate. J Exp Bot 44:41–51

Gutschick VP (1997b) Photosynthesis, growth rate, and biomass allocation. In: Jackson LE (ed) Ecology in agriculture. Academic Press, San Diego, pp 39–78
Gutschick VP (1999) Biotic and abiotic consequences of differences in leaf structure. New Phytol 143:3–18
Gutschick VP, Kay LE (1995) Nutrient-limited growth rates: quantitative benefits of stress responses and some aspects of regulation. J Exp Bot 46:995–1009
Hartley S, Jones CG (1997) Plant chemistry and herbivory, or why the world is green. In: Crawley MJ (ed) Plant ecology, 2nd edn. Blackwell, Oxford, pp 284–324
Jackson LE, Koch GW (1997) The ecophysiology of crops and their wild relatives. In: Jackson LE (ed) Ecology in agriculture. Academic Press, San Diego, pp 3–37
King BJ, Siddiqi MY, Glass ADM (1992) Studies of the uptake of nitrate in barley. V. Estimates of root cytoplasmic nitrate concentration using nitrate reductase activity – implications for nitrate influx. Plant Physiol 99:1582–1589
Kronzucker HJ, Siddiqi MY, Glass ADM (1995) Compartmentation and flux characteristics of ammonium in spruce. Planta 196:691–698
Le Bot J, Adamowicz S, Robin P (1998) Modelling plant nutrition of horticultural crops: a review. Sci Hortic 74:47–82
Letourneau DK (1997) Plant-arthropod interactions in agroecosystems. In: Jackson LE (ed) Ecology in agriculture. Academic Press, San Diego, pp 239–290
Maynard Smith J, Price GR (1973) The logic of animal conflict. Nature 246:15–18
Morcuende R, Krapp A, Hurry V, Stitt M (1998) Sucrose feeding leads to changes in the rates of alpha-oxoglutarate synthesis and increased synthesis of a wide spectrum of amino acids in tobacco leaves. Planta 206:394–409
Munns R, Termaat A (1986) Whole-plant responses to salinity. Aust J Plant Physiol 13:143–160
Nable RO, Lance RCM, Cartwright B (1990) Uptake of boron and silicon by barley genotypes with differing susceptibilities to boron toxicity. Ann Bot 66:83–90
Nowak RS, Ellsworth DS, Smith SD (2004) Functional responses of plants to elevated atmospheric CO_2 – do photosynthetic and productivity data from FACE experiments support early predictions? New Phytol 162:253–280
Penuelas J, Matamala R (1990) Changes in N and S leaf content, stomatal density and specific leaf area of 14 plant species during the last three centuries of CO_2 increase. J Exp Bot 41:1119–1124
Peterson AG, Ball JT, Luo Y, Field CB, Reich PB, Curtis PS, Griffin KL, Gunderson CA, Norby RJ, Tissue DT, Forstreuter M, Rey A, Vogel CS, CMEAL Participants (1999) The photosynthesis–leaf nitrogen relationship at ambient and elevated atmospheric carbon dioxide: a meta-analysis. Global Change Biol 5:331–346
Poorter H, Remkes C (1990) Leaf area ratio and net assimilation ratio of 24 wild species differing in relative growth rate. Oecologia 83:553–559
Roberts SK, Snowman BN (2000) The effects of ABA on channel-mediated K+ transport across higher plants roots. J Exp Bot 51:1585–1594
Roderick ML, Berry SL, Noble IR (1999) The relationship between leaf composition and morphology at elevated CO_2 concentrations. New Phytol 143:63–72
Romera FJ, Alcantara E, de la Guardia MD (1992) Effects of bicarbonate, phosphate and high pH on the reducing capacity of Fe-deficient sunflower and cucumber plants. J Plant Nutr 15:1519–1530
Schmidke I, Kruger C, Frommichen R, Scholz G, Stephan UW (1999) Phloem loading and transport characteristics of iron in interaction with plant-endogenous ligands in castor bean seedlings. Physiol Plant 106:82–89
Siddiqi MY, Glass ADM (1987) Regulation of K^+ influx in barley: evidence for a direct control of influx by K^+ concentration in root cells. J Exp Bot 38:935–947
Sinclair TR, Horie T (1989) Leaf nitrogen, photosynthesis, and crop radiation use efficiency: a review. Crop Sci 29:90–98

Sivasankar S, Rothstein S, Oaks A (1997) Regulation of the accumulation and reduction of nitrate by nitrogen and carbon metabolites in maize seedlings. Plant Physiol 114:583–589

Stenseth NC, Maynard Smith J (1984) Coevolution in ecosystems: Red Queen or stasis? Evolution 38:870–880

Tardieu F, Granier C (2000) Quantitative analysis of cell division in leaves: methods, developmental patterns and effects of environmental conditions. Plant Mol Biol 43:555–567

Vincent TLS, Vincent TL (1996) Using the ESS maximum principle to explore root-shoot allocation, competition and coexistence. J Theor Biol 180:111–120

Youssefi F, Brown PH, Weinbaum SA (1999) Regulation of nitrogen uptake at the whole plant level: a study in almond trees. Am Soc Hortic Sci 9:598–600

Zerihun A, McKenzie BA, Morton JD (1998) Photosynthate costs associated with the utilization of different nitrogen-forms: influence on the carbon balance of plants and shoot-root biomass partitioning. New Phytol 138:1–11

Zerihun A, Gutschick VP, BassiriRad H (2000) Compensatory roles of nitrogen uptake and photosynthetic N-use efficiency in determining plant growth response to elevated CO_2: evaluation using a functional balance model. Ann Bot 86:723–730

Zidan I, Shaviv A, Ravina I, Neumann PM (1992) Does salinity inhibit maize leaf growth by reducing tissue concentrations of essential mineral nutrients? J Plant Nutr 15:1407–1419

5 Biological Nitrogen Fixation Associated with Angiosperms in Terrestrial Ecosystems

J.I. Sprent

5.1 Introduction

Roots are the main source of nutrients for plants, although both roots and foliage can take up various forms of combined nitrogen. Uptake of combined nitrogen can also be enhanced by symbiosis with fungi, forming various types of mycorrhiza (see Chap. 9, this Vol.). Where nitrogen is the major limiting element, symbiotic associations with nitrogen-fixing organisms enable plants to access the normally unreactive dinitrogen molecule. This has enabled plants from all major groups to invade, and sometimes persist in all major habitats of the world. In natural situations, nitrogen-fixing organisms often form part of succession, being out-competed when the system becomes nitrogen-sufficient. However, there are cases, for example, some tropical rain forests (nodulated legume trees) and boreal forests (various lichens), where they are prominent in the mature system. As will be seen below, there is a wide range of symbiotic systems, each of which has evolved in its own particular type of habitat within which it makes its contribution. Even among nodulated legumes, there are plants of widely different sizes and special adaptations. The South African genus *Aspalathus*, for example, co-exists with other plants adapted to the nutrient-poor soils of the fynbos vegetation. Amounts of nitrogen fixed may be small (although measurements in natural ecosystems are few), but are adequate for sustaining the natural biodiversity. It is when exotic legumes are introduced that problems arise, as discussed in Section 5.5.2.6.

Biological nitrogen fixation is an ancient process, which may have been universal in primitive organisms (Young 1992). It is confined to prokaryotic organisms, both eubacteria and archaea, and current evidence suggests that it was present before these two lines of microorganisms diverged (Leigh 2000). Some eubacteria (a term which includes cyanobacteria) live in closer or looser association with plants of all major taxa (Sprent and Sprent 1990).

Ecological Studies, Vol. 181
H. BassiriRad (Ed.) Nutrient Acquisition by Plants
An Ecological Perspective

Although the biochemistry of nitrogen fixation is described in most standard texts (e.g. Sprent and Sprent 1990), the central reaction is given here:

$$N_2 + 8H^+ + 8e^- \rightarrow 2NH_3 + H_2$$

An inevitable consequence of the reduction of nitrogen is the production not only of reduced N but also of hydrogen gas, a process using a significant amount of the total energy of the reaction. Although many organisms have the ability to recover some of the energy lost in hydrogen production, by oxidising it to water, this is not always true, particularly of some legume/rhizobial systems. In the 1970s, attempts were made to assess whether hydrogen-generating symbioses were wasteful (in terms of carbon used) compared with those which recycled hydrogen. In parallel, selection of host–rhizobial combinations that did not evolve hydrogen was carried out. Overall, these very logical approaches did not result in significantly improved nitrogen fixation or plant growth. Based on the observation that in field soil, little hydrogen is evolved even when expected from the type of plant grown, Layzell and others (e.g. Dong and Layzell 2001) considered the possibility that hydrogen evolution might actually be beneficial! Experimental testing of this hypothesis showed that 60% of electrons from hydrogen oxidation were passed to oxygen and used for carbon dioxide fixation (Dong and Layzell 2001). It appears that hydrogen evolution by nodules results in more and more active soil organisms that can oxidise hydrogen and enhance carbon deposition in soils – an added bonus in times of global concern over carbon dioxide emissions?

In order to be of general benefit, nitrogen fixed by microorganisms must be in a suitable form to be assimilated by other organisms. In the case of free-living soil microbes and those in loose associations with plants, this usually requires microbial death and decay. The advantage of more advanced symbiotic systems is that they obtain reduced N directly from the microorganism, in exchange for carbon compounds used to fuel the nitrogen reduction process. In the case of both legume and non-legume nodules, the endophyte usually fixes only nitrogen when inside the nodule; there is a more-or-less simultaneous repression of the ability to assimilate ammonium and of the induction of nitrogenase synthesis (Sprent and Sprent 1990).

With rapid advances in molecular methods, there has been great progress in understanding the genetics and molecular biology of nitrogen fixation in a few free-living microorganisms and some symbioses. However, the ecological significance of these systems and of the vast majority of those which have been less intensively studied is poorly understood.

One of the biggest frustrations for those working on nitrogen fixation and nitrogen cycling at a field level is the lack of good techniques for quantifying these processes. Neither nitrogen fixation nor denitrification can be accurately quantified in any but a few types of ecosystem. The situation is particularly difficult in the case of woody perennials (Boddey et al. 2000). For accounts of methods available for use with plant symbioses, the reader is referred to Peoples et al. (1989). Life is much simpler for those who work with free-living organisms, where the acetylene reduction assay (the nitrogenase enzyme also reduces acetylene to ethylene, easily measured by gas chromatography) can usually be applied.

In this chapter, most of the different types of nitrogen-fixing organisms/ symbioses found in terrestrial ecosystems will be considered, but with an emphasis on those involving angiosperms.

5.2 Cyanobacteria and Their Symbioses

Not all cyanobacteria can fix nitrogen, although all those known to form symbioses are able to do so. Unlike some of their free-living relatives, symbiotic cyanobacteria carry out nitrogen fixation in specialised cells called heterocysts, and many belong to the genus *Nostoc* (Sprent and Sprent 1990; Rai et al. 2000). All are potentially photosynthetic, although this activity is frequently suppressed when they are symbiotic.

5.2.1 Free-Living Nitrogen-Fixing Cyanobacteria

These occur in all ecosystems studied, but may be especially important in freshwater and marine environments, which are not dealt with in this volume. In terrestrial habitats, coastal and desert sands have been well studied. Cyanobacteria in coastal sands often occur as a distinct layer of a microbial mat in intertidal zones. The interactions between the layers of such mats are complex, and they appear to act as a functional unit (Herbert et al. 1988). Both on coasts and in deserts, one important attribute of cyanobacteria is that they bind sand particles and hence stabilise the systems. Another important attribute is that cyanobacteria found in these habitats are very resistant to desiccation, surviving long periods (even years) in a desiccated state. They are rapidly rehydrated during a short rainfall event, and are therefore able to fix nitrogen when there would be insufficient water to support activity in larger, higher organisms. Factors affecting nitrogen fixation include particle texture, mineral content and mobility, salinity, and interactions thereof. Within their ecosystems, they fill an important niche (Scherer et al. 1984; Skujins et al. 1987; Sprent 1993). Another harsh environment where cyanobacteria are found is Antarctica, where one of the habitats they colonise is within translucent rocks, where they may also be part of complex microbial mats (Wynne-Williams et al. 1999). Comparisons have been drawn between these environments and the surface of Mars (Wynne-Willams and Edwards 2000), and there are plans to look for cyanobacteria in future space programmes.

5.2.2 Cyanobacterial Symbioses

These occur in all main groups of plants and with fungi (Sprent and Sprent 1990; Rai et al. 2000). Those with bryophytes are of considerable scientific interest, but only of restricted ecological significance. The one symbiosis with a fern, *Azolla*, is aquatic and therefore not considered here, although it may be important both ecologically (for fixing nitrogen or blocking waterways, depending on the area) and agriculturally, as a nitrogen fertiliser for wetland rice (Rai et al. 2000). Symbioses with fungi (lichens), gymnosperms (cycads) and angiosperms (*Gunnera* spp.) all have important ecological roles, and will each be considered briefly in turn.

Nitrogen-fixing lichens are of two main types – two-membered, consisting only of a cyanobacterium and fungus, and three-membered, in which there is also a green algal partner (Sprent and Sprent 1990). These types are found in similar habitats, which are often harsh, such as surfaces of rocks (where they may be exposed to very high and very low temperatures, as well as being often desiccated or exposed to saline water), and in deserts and boreal forests. Like some free-living cyanobacteria, lichens are often very desiccation-tolerant and, in relevant environments, fulfil a similar role (Sect. 5.2.1, above). They also play a significant part in the nitrogen nutrition of some forests. Two types have been well studied. First, those of the Pacific Northwest of the USA and Canada, where lichens dominate the epiphytic flora and may fix enough nitrogen to replace that which is lost in climax forests (Denison 1979; Sprent and Sprent 1990). Second, in boreal forests and arctic tundra, where they are found on the soil surface and, sometimes together with free-living cyanobacteria, may be the major source of nitrogen for the ecosystem (Nash 1996).

5.2.3 Cycads

These are unusual in that all known extant and fossil species are symbiotic with cyanobacteria, which are located in modified, highly branched 'coralloid' roots. Most cycads are native to warm parts of the southern hemisphere, although they extend northwards to parts of southern Asia, islands such as Cuba and to Florida (Jones 1993). Most are desiccation-tolerant and often form a dominant part of the vegetation. Parts of southern Africa, for example, have extensive open stands of *Encephalartos* and, in some eucalypt forests of Australia, cycads form the dominant understorey, taking the position occupied by *Acacia* and other legume genera elsewhere in the country. Like some legumes, cycads may regenerate well after fire. *Macrozamia reidlei* in western Australia is one such fire-resistant species, and may fix over 18 kg N ha^{-1} $year^{-1}$, a significant quantity in its ecological context (Halliday and Pate 1976). Cycads evolved over 200 Ma ago when continents were far removed from their present distribution, and when the concentration of CO_2 in the atmosphere was considerably greater than even the most pessimistic forecasts for the next 50 years. Perhaps they will stage a comeback!

5.2.4 *Gunnera*

This is the only genus of angiosperm which enters into a nitrogen-fixing symbiosis with a cyanobacterium. *Gunnera* has been placed in a family of its own, with about 40 species largely confined to the southern hemisphere. Species vary in size from small, creeping herbs to giant species that are virtually stemless, with leaves up to 1.5 m across. The latter are widely cultivated in parks and gardens, where they thrive near water. In evolutionary terms, *Gunnera* is towards the base of the angiosperms, and its experiments with cyanobacteria do not appear to have been repeated. However, all species are symbiotic with *Nostoc*, which appears able to fix sufficient nitrogen for the needs of the host (Osborne et al. 1992). Like all nitrogen-fixing symbioses with angiosperms (and unlike most of those in lower plants), the microsymbiont is confined within the boundaries of the host cell walls. Since it is bounded by host plasmalemma-derived membrane material, however, it is not strictly intracellular. Symbiotic cells are found in modified glands at the base of petioles. As in other cyanobacterial symbioses, the *Nostoc* in *Gunnera* has a high proportion of its cells in the nitrogen-fixing form – heterocysts. Its vegetative cells also have a full complement of photosynthetic pigments, even when buried so deeply inside host tissue that they may not have access to light. However, isolated glands of *Gunnera magellanica* show light-stimulated nitrogenase activity, indicating that the photosynthetic machinery remains functional (Silvester et al. 1996)

The natural distribution of *Gunnera* is interesting because it is found in many places lacking (or only sparsely populated by) both legumes and actinorhizal plants. In some cases, such as the Falkland Islands (Malvinas), *G. magellanica* is the only endemic symbiotic nitrogen-fixing plant (inferred from Moore 1968). There, it is widely distributed, from the shoreline to the highest parts of the islands. In all areas sampled, it contains abundant colonies of *Nostoc* (pers. observ.). Apparently, extant *Gunnera* species are derived from ancient Gondwanan stock (Hill and Scriven 1995). However, based on pollen analysis, it appears that in late Cretaceous times, *Gunnera* may have had a wider distribution, ranging from high northern to high southern latitudes (Jarzen and Dettman 1989). Bergman et al. (1992) suggest that the retreat of *Gunnera* from some of its earlier locations was caused by climate changes, in particular to drier conditions. Physiologically, *Gunnera* appears to have a requirement for high humidity, although it can grow in a wide range of temperatures (Bergmann et al. 1992). In common with other angiosperm symbioses, the microsymbiont passes fixed nitrogen in the form of ammonium to its host, where it is incorporated into organic compounds, especially asparagine (Silvester et al. 1996). Unlike the situation in root nodules, these organic compounds are translocated out of the nitrogen-fixing glands (sometimes called internal nodules) in the phloem (Stock and Silvester 1994).

There is insufficient information to assess the role of nitrogen fixation by *Gunnera* in global terms, but at least in areas where it is the sole or major type of nitrogen-fixing symbiosis, it is likely to be important.

5.3 Non-nodulating Associations Between Angiosperms and Non-photosynthetic Nitrogen-Fixing Microorganisms

5.3.1 Loose Associations

Most work has been carried out with crop and pasture species from the grass family Gramineae. In some areas of the world, most notably in South America, C_4 forage grasses such as *Paspalum notatum* and C_4 crops such as sugar cane have been grown for many years without adding nitrogen fertiliser. Clearly, these must obtain nitrogen from somewhere, and much work on the topic, most notably by the late Johanna Döbereiner and colleagues in Brazil (Döbereiner et al. 1988), suggested that the source of this nitrogen was bacterial. Several new genera and species of bacteria were isolated and grown in special media (in particular, semi-solid agar) which enabled their physiology to be studied. These bacteria in themselves are of considerable scientific interest, but their interaction with plants has proved more controversial. The problem is that soil bacteria (including rhizobia) synthesise a variety of substances, such as auxins, which may affect plant growth. Disentangling the effects of bacteria on plant growth from those of nitrogen fixation has generated at least as much heat as light amongst scientists concerned. There is little doubt that the genus *Azospirillum*, which has been widely studied in different parts of the world, can stimulate root growth and thus enable plants to access soil nitrogen and other nutrients more efficiently. How much, if any, of the raised nitrogen content in plants inoculated with *Azospirillum* comes from N_2 gas is less clear. The arguments have been debated many times (e.g. Okon and Labandera-Gonzales 1994; Vande Broek et al. 2000) and will not be pursued here. The effects of *Azospirillum* on root growth can also be seen in legumes, and co-inoculation with rhizobia may lead to enhanced nodulation (Burdman et al. 1996). Both rhizobia and *Frankia* (see below) can also enhance plant growth other than by nitrogen fixation (Wall 2000).

5.3.2 Endophytic Associations

These have been the subject of a number of recent reviews, for example, James (2000). Plant scientists have known for many years that non-pathogenic fungi can grow endophytically inside plants, but it is only comparatively recently that endophytic bacteria have aroused interest outside of their role as plant growth promoters. The major systems which have been studied include the graminaceous plants Kallar grass (*Leptochloa fusca*), rice (*Oryza sativa*) and sugar cane (*Saccharum* spp.), other plants such as maize (*Zea mays*), wheat (*Triticum* spp.), *Sorghum* spp., cassava (*Manihot esculenta*) and *Coffea* spp. having been studied to a lesser extent.

5.3.2.1 Kallar Grass

This is a halophytic grass found in Pakistan and neighbouring countries. In its intercellular spaces, it may house large populations of *Azoarcus* sp. which can fix nitrogen. Some of this N, presumably following cell death, may pass to the host plant. Indeed, this can account for up to 26 % of the nitrogen pool in the grass. A crucial experiment in this work was the demonstration that plants inoculated with a *nif*⁻ mutant (i.e. one which cannot synthesise nitrogenase) of the wild-type strain incorporate less ^{15}N than plants inoculated with the parent strain (Hurek et al. 1998).

5.3.2.2 Rice

As the world's major cereal grain, and one whose productivity is often limited by available N, nitrogen-fixing rice has been the subject of much research. One aspect of this work has involved the study of endophytic bacteria. There is evidence that, in some cases, N fixed by endophytes may be incorporated into rice (Shrestha and Ladha 1996); in others, no such evidence was obtained (Urquiaga et al. 1992). These studies were conducted in the Philippines and Brazil respectively, so rather than assume that one group is wrong and the other right, it is better to consider factors such as host genotype and soil conditions as contributing to the results obtained (see Sect. 5.3.2.3, below). Whereas in Kallar grass the endophyte always appears to be *Azoarcus* sp., in rice there may be several possible genera, including *Azoarcus*, *Alcaligenes*, *Bacillus*, *Enterobacter*, *Herbaspirillum*, *Klebsiella* and *Pseudomonas* (James 2000). Some of these are strict anaerobes, which is particularly relevant to rice grown in flooded fields.

5.3.2.3 Sugarcane

It was in sugar cane that the first generally accepted evidence for nitrogen fixation in grasses was obtained, after some very careful experimentation by Boddey and coworkers in Döbereiner's group (e.g. Boddey 1995). However, workers in some other countries have been unable to reproduce these findings (cf. references in James 2000). One important factor is the host genotype. In Brazil, these have been selected for more than a century on soils low in N, encouraging them to associate with nitrogen-fixing organisms, whereas in countries such as Australia germplasm has been selected and raised on fertilised fields. Additional factors necessary are sufficient supplies of water, K, P and Mo. Bacteria involved so far have been classified as *Herbaspirillum* and *Gluconacetobacter diazotrophicus* (originally called *Acetobacter diazotrophicus*). The latter grows well on high sugar, is acid-tolerant and lacks nitrate reductase. All these properties have advantages for living in the vascular system of plants, in particular the xylem. Lack of nitrate reductase removes one of the obstacles to nitrogen fixation found in many systems, namely, that it is inhibited by nitrite. This means that the host plant can benefit both from soil nitrate transported from the root in the xylem and assimilated in the shoot (in sugar cane, but not all plants), and from nitrogen fixed by the bacteria. However, McCully and coworkers (Dong et al. 1994) have argued that bacteria cannot invade the xylem of healthy plants, and published evidence that these were present rather in intercellular spaces, as in the *Azoarcus*/Kallar grass system. As there have been several reviews of this subject recently (e.g. James 2000), all the evidence will not be reconsidered here. The main gap lies between the undoubted fact that many genotypes of sugarcane can incorporate N_2 from the atmosphere into their tissues (Boddey 1995), and the equally undoubted fact that sugarcane (and other grasses) can house a number of potentially nitrogen-fixing bacteria (James 2000). The critical questions to be answered are (1) where are the bacteria located and how do they get there, and (2) are there enough of them to fix a significant amount for plant nitrogen demand?

5.4 Nodulated Plants: Symbioses with Frankia and rhizobia

For many decades, legume and non-legume root nodules were considered quite distinct for various reasons, including the structure and type of microsymbiont. Legumes nodulate with a group of eubacteria collectively known as rhizobia, whereas non-legumes nodulate with the filamentous actinomycete genus *Frankia* – the latter are now usually referred to as actinorhizal plants. Legume nodule structure has the stem-like feature of a peripheral vascular system; actinorhizal nodules have a root-like central vascular system. However, recent molecular studies of the relations between different major groups of plants have shown that all root-nodulated plants belong to a single grouping, known as the Rosid I clade (Soltis et al. 2000). Within this clade, there are six major branches: two have no known nodulated taxa, the other four do. Table 5.1 outlines the families in these four groups. This and other evidence strongly suggests that there is a group of plants which have some predisposition to form nitrogen-fixing nodules, which may or may not be expressed and, if expressed, may lead either to actinorhizal or to legume nodules. Full expression of nodulation may have evolved at least three to four times in both groups of nodulated plants (Wall 2000; Sprent 2001). Workers interested in the evolution of nodulation are attempting to explain how these different types of nodule may have arisen from a common ancestor. An important paper in this area is that of Gualtieri and Bisseling (2000), who distinguished two key, independent processes in nodule development – the formation of the initial infection (pre-nodule), and the formation of vascular connections with the subtending root. Sprent (2001) developed these arguments to include various morphological types of legume nodule, in particular some with determinate and others with indeterminate growth. Underlying these distinctions is a delicate balance of hormones, produced by the host plant and possibly also the microsymbiont. This balance may change with time and cause, for example, older legume nodules to produce rootlets, as is normal for some types of actinorhiza (Table 5.1). Environmental factors

Table 5.1. Parts of the Rosid I clade of Soltis et al. (2000), indicating the four (out of six) branches with nodulating families (branches are numbered arbitrarily). Where known, the mode of infection by the microsymbiont is shown, and also whether the nodules can be associated with nodule rootlets

Family	Mode of infection	Nodule rootlets
Branch 1		
Rhamnaceae	Intercellular	Absent
Rosaceae	Intercellular	Absent
Ulmaceae (*Parasponia*)	Wound (crack)	Absent
Branch 2		
Leguminosae (Fabaceae)	Root hair, epidermal or wound (crack)	Occasional, usually in older nodules
Branch 3		
Betulaceae	Root hair	Absent
Myricaceae	Root hair	Present
Casuarinaceae	Root hair	Present
Branch 4		
Coriariaceae	Not known	Absent
Datiscaceae	Not known	Present

Table 5.2. The main types of legume root nodule and their taxonomic position (for further details, see Sprent 2000, 2001), and for the dalbergioid group (Lavin et al. 2001)

Morphological/growth type	Occurrence
Indeterminate; may be branched; may be perennial	All Mimosoideae, Caesalpinioideae and most Papilionoideae
Determinate; ureide exporting; amide exporting	Papilionoid tribes Desmodeae, Phaseoleae and Psoraleae (?) Papilionoid tribe Loteae
Aeschynomenoid; infection via root junctions; sometimes found on stems	The Dalbergioid group of legumes which includes tribes Aeschynomeneae, most Dalbergieae and some others
Primitive nodules, where bacteroids are not released from infection threads	All Caesalpinioideae, except some species of *Chamaecrista*. A few papilionoid genera

such as ethylene production in the rhizosphere may also lead to changes in nodule morphology (see discussion and references in Sprent 2001). It has been known for some time that both actinorhizal and legume nodules have more than one mode of infection by their symbionts. These can be broadly divided into two types, i.e. those which do or do not involve root hairs. In legumes at least, there are variants of the latter type (Sprent 2001). In both types of nodule, the type of infection is related to the host plant family (actinorhizas) or tribe (legumes; Tables 5.1 and 5.2; Sprent 2000). Also found in both groups is the pigment haemoglobin which, as in blood, acts as an oxygen carrier. This pigment is part of the solution to the oxygen problem faced by all aerobic nitrogen-fixing systems, namely, how to satisfy the need for oxygen for oxidative phosphorylation to fuel the nitrogenase reaction when the nitrogenase enzyme is inactivated by molecular oxygen.

5.4.1 Actinorhizal Plants

Frankia-induced nodules are found scattered among angiosperm families, but all are now known to be members of the Rosid I clade. Table 5.1 lists the families/genera known to form nodules, together with some of their individual characters. Further details can be found in the proceedings of several recent meetings on actinorhizal plants, published in the peer-reviewed journals Physiologia Plantarum vol. 99 (1997), and Canadian Journal of Botany vol. 77 (1999); the reader is also referred to the reviews of Huss-Danell (1997) and Wall (2000) as well as the book edited by Schwintzer and Tjepkema (1990).

Research on actinorhizas lagged behind that on legumes for a variety of reasons, including a lack of funding (these being non-agricultural species), the difficulty of growing *Frankia* in pure culture (a problem shared with arbuscular mycorrhizas) and the fact that nodules are generally woody, with a high tannin content, making it difficult to extract nitrogenase and other key enzymes. However, there have been rapid advances in the last few years, stimulated partly by the recognition of the common ancestry and characters of all nodulated plants, and partly by the advent of new molecular tools for studying plant–microbial signalling and other aspects.

Frankia has several morphological forms, including filamentous hyphae, vesicles (which are often the site of nitrogen fixation in the nodule) and complex sporangia. These structures enable the organism to live both inside the nodule and in the soil (Benson and Silvester 1993). However, the extent to which living in soil is an active process, rather than mere survival is not yet clear and may vary with genotype. Although some strains cultured under low oxygen concentrations produce vesicles and nitrogenase (Parsons et al. 1987), there is no evidence to suggest that nitrogen fixation in soil occurs in the free-living form. The thickness of the wall of the vesicles is inversely related to oxygen concentration, as the wall forms a barrier to protect nitrogenase from inactivation. Inside the nodules, the vesicle walls form part of the oxygen protection mechanisms in some genera but in others there are alternative mechanisms, and vesicles are not formed. Similarly, there are wide variations in the production of haemoglobin, depending on factors controlling/facilitating oxygen diffusion. Where present, as in legume nodules, the role of haemoglobin is to maintain a high flux of oxygen at the low concentrations established by other anatomical features. In nodules of *Myrica*, which are often found under water, negatively gravitropic nodule roots, with well-developed aerenchyma, grow to the water surface, allowing oxygen to reach the nodules.

At present, *Frankia* is the only genus recognised as inducing actinorhizal nodules. Classification at the subgeneric level is currently being undertaken but no overall scheme has as yet been accepted. Some commonality in signal exchange systems between host and microsymbiont in actinorhizas and legumes is now emerging (Wall 2000). Exchange of signals between host and microsymbiont in both actinorhizal plants and legumes implies a degree of specificity. There is evidence for this in both types of symbiosis, but it has not proved possible in either case to draw a close parallel at the host genus/family/tribe level with particular genotypes of microsymbiont (Wall 2000; Sprent 2001).

5.4.2 *Parasponia*

This member of the Ulmaceae, originally identified as *Trema*, is the only reported case of nodulation with rhizobia outside the legume family. It is a native of Papua New Guinea and nearby areas where it may fix significant amounts of nitrogen. Nodule structure is similar to that of actinorhizas, in that the vascular tissue is central and the infected tissue cortical. Within the infected tissue, bacteria are retained within threads bounded by host cell wall material (fixation threads). At the time, this was thought to be a reflection of the fact that *Parasponia* was an unusual host for rhizobia, not fully recognised as a symbiont but rather as a pathogen. In legumes, release from fixation threads into membrane-bound vesicles (peribacteroid units or symbiosomes) was thought to be a necessary condition for the induction of nitrogenase, and this holds true for the more advanced genera. However, a number of primitive woody legumes also have fixation threads, very similar to those of *Parasponia* (Sprent 1989). For a review of the *Parasponia* symbiosis, the reader is referred to Becking (1992).

5.4.3 Legumes

Leguminosae, also called Fabaceae, is the third largest family of flowering plants and it also contains the genus with the largest number of species (*Astragalus*, 2,500–3,000 species, although the genus may be subdivided in the future). One of the properties most often associated with legumes is the ability to form nitrogen-fixing root nodules with rhizobia. However, not all legumes *can* fix nitrogen and, of these, not all *do* fix nitrogen under all conditions. Twenty years after the classic text of Allen and Allen (1981), which chronicled all information about legume nodules and their rhizobia known at the time, I attempted an updated assessment of our knowledge in this field of research (Sprent 2001). Most people divide legumes into three subfamilies, but some still prefer to allocate a family status to these subfamilies. Table 5.2 gives the main morphological/structural types of nodules known to date, and indicates which are confined to certain legume tribes. The taxonomic utility of nodulation in legumes was pioneered by Corby (1988), and has since been extended by others (Sprent 2000). In some sections of the Papilionoideae at least, nodule features are the most discriminating of the non-molecular systematic characters (Lavin et al. 2001). However, in spite of this variation within the Leguminosae, very little recent attention has been given to any aspect of the structure and function of legume nodules, outside a few agronomically important species such as the soybean (*Glycine max*), common bean (*Phaseolus sativus*), alfalfa (lucerne, *Medicago sativa*) and pea (*Pisum sativum*). These four species all have root hair infection pathways; the former two species have determinate, and the latter two indeterminate nodules. Nodulation and nodule development in these types of nodule have been extensively described in the book edited by Spaink et al. (1998), which is strongly recommended. Unfortunately, there has been a regrettable tendency to generalise from this small set of species. For example, all known ureide- (rather than amide) exporting legumes have determinate nodules and are of tropical/subtropical origin, facts which have given rise to the commonly held misconception that most tropical legumes have these characters. At present, they appear to be confined to the tribes Desmodieae and Phaseoleae (which are likely to be merged in future), and possibly Psoraleae. The type of nodule (known as aeschynomenoid) found in the peanut (groundnut, *Arachis hypogaea*), which forms following rhizobial entry at the junction of lateral roots and which never has infection threads, appears common to a number of tribes which, on the basis of molecular and morphological characters, especially those of nodules, are now grouped into a dalbergioid clade (Lavin et al. 2001). Some genera in these tribes form nodules on stems; these are always associated with adventitious root initials. All are tropical/subtropical, and none is known to export ureides. Many tropical genera (all members of the subfamily Mimosoideae, and several papilionoid tribes) have indeterminate nodules and, where known, are amide exporters. Some have hair infections, others infections via root junctions (but later forming infection threads), and yet others have rhizobial infection directly between epidermal cells (Sprent 2001). These distinctions are of more than only taxonomic interest. Indeterminate nodules have the potential advantage of being able to survive periods of adverse conditions, such as drought and freezing, and can resume growth very rapidly when conditions improve. It takes longer to produce a new crop of determinate nodules. Aeschynomenoid nodules can be formed at any site on a root (or, in some cases, stem) which has the potential to form lateral roots. Thus, they can be produced on relatively old roots, where old and new nodules may be interspersed, a feature which may be of interest in some important timber legumes such as species of *Dalbergia* and *Pterocarpus*. As with actinorhizal plants, these various types of nodule probably evolved under different environmental conditions.

Earlier work divided rhizobia into fast- and slow-growing forms, and this distinction is still useful today. Colony morphology, motility, acid or alkali production, salt and antibiotic sensitivity, and substrate utilization are other factors used in this context. Recently, various molecular techniques have been added to this suite of characters and, with increasing interest in tropical rhizobia, the number of genera has proliferated rapidly over the last 10 years and now includes members of the β as well as the α sub-groups of the Proteobacteria (for example Chen et al. 2003). It has further become apparent that some rhizobia are more related to animal pathogens and other bacteria than they are to other rhizobia. The old concept of a family Rhizobiaceae is no longer acceptable. An additional complication is the fact that rhizobia may live

Table 5.3. Genera of rhizobia currently recognised[a]

Genus	Species	Comments
Agrobacterium[b]	5+[c]	Plant-pathogenic, widely used as vector for genetic engineering of plants
Allorhizobium	1	Fast-growing, closely related to *Agrobacterium*
Azorhizobium	1(+)	Closely related to *Xanthobacter*. Only known from stem nodules on *Sesbania rostrata*. Able to grow on N_2 ex planta
Bradyrhizobium	3+	Symbiotic genes chromosomal. Alkali-producing
Mesorhizobium	4+	Symbiotic genes chromosomal. Acid-producing. Polar flagella
Rhizobium	8+	Symbiotic genes on plasmids. Acid-producing. Peritrichous flagella
Sinorhizobium	4+	Symbiotic genes on plasmids. Acid-producing. Peritrichous flagella

[a] In addition to these 'classical' rhizobia, all of which belong the α-proteobacteria, other members of this branch (e.g. *Methylobacterium nodulans*, nodulating some species of *Crotalaria*, Sy et al. 2001) and from a separate branch, the β-proteobacteria (e.g. *Ralstonia taiwanensis*, now renamed *Wautersia taiwanensis*, Vaneechoutte et al. 2004 nodulating some species of *Mimosa*, Chen et al. 2003) are now known to fix nitrogen effectively in legume nodules

[b] It has been suggested that *Agrobacterium* be incorporated into *Rhizobium* for various reasons, including the fact that a number of 'rhizobia' isolated from nodules are virtually indistinguishable from it

[c] The exact number of species for many of these genera is not yet clear

saprophytically in soil, and may lose their nodulation genes to sometimes gain them again, not always from the same source. Table 5.3 lists the rhizobial genera recognised (by most, but not all) at the time of writing (cf. Sprent 2001), but more recently a possible alternative classification has been suggested (Young et al. 2001).

Another area which has progressed rapidly in the last few years is that of signal exchange between rhizobia and their host legumes, a prerequisite for nodulation and nitrogen fixation. These are also well covered in the volume edited by Spaink et al. (1998), are summarised in Sprent (2001), and have been the subject of a number of recent reviews, of which Broughton et al. (2000) is particularly recommended as being both informative and entertaining. A breakdown in the signalling process may account for nodulation failure in some soils. For example, low pH, aluminium and calcium affect the expression of nodulation genes in clover rhizobia (Richardson et al. 1988). More such studies are needed if we are to understand why, apart from the presence of suitable rhizobia, potentially nodulating plants do not do so in certain environments.

5.5 Ecological Role and Uses of Nodulated Plants

5.5.1 Actinorhizal Plants

Geographically, actinorhizal plants are distributed rather differently from legumes. The latter are thought to have evolved in the tropics in the late Cretaceous (Sprent 2001) and to have radiated thereafter, whereas there is considerable evidence that nodulated non-legumes emerged following the retreat of the Pleistocene glaciations (Silvester 1977), although some may have evolved much earlier (Benson and Clawson 2000). Today, and with the exception of the Casuarinaceae, they occur only in temperate regions and, if at lower latitudes, only at higher altitudes. In the Himalayas, for example, *Alnus nepalensis* is a successful coloniser of rocky outcrops and of areas affected by landslides (Sharma 1993).

Most field studies of actinorhizal plants have been concerned with forestry. Some actinorhizal genera, such as *Alnus*, may be used directly as timber (Gordon and Wheeler 1983). Some may contribute fixed nitrogen as an understorey component (similar to legumes in other habitats) – for example, *Cercocarpus* in some pine forests (Lepper and Fleschner 1977). Others, such as *Ceanothus*, may invade forests after logging or burning activities. Although *Ceanothus* may be important in contributing nitrogen to subsequent replanting, it is often also regarded as a weed (Gordon and Wheeler 1983; Sprent and Sprent 1990).

When introduced into new areas, actinorhizal plants can quickly become dominant and then substantially modify the successional sequence. This has been shown clearly for *Myrica faya* which was introduced to the Hawaiian islands by Portuguese settlers, and its seed further distributed by birds. It nodulates well in this region, presumably with *Frankia* from local soil, even though there are no native actinorhizal plants. It has now colonised large areas and, by increasing soil nitrogen levels, has had a major impact on the natural ecosystem (Vitousek and Walker 1989). Actinorhizal plants are often found slightly later in the succession – for example, *Hippophäe* behind dune slacks (Bond 1976) and *Coriaria* on alluvial flood terraces (Silvester 1975). The ability of some species of *Casuarina* to tolerate drought and salinity reflects the environment in which they normally occur, but has also enabled them to be introduced for purposes such as wind protection and as a source of firewood in parts of Africa and Asia (Anonymous 1984; Diem and Dommergues 1990).

5.5.2 Legumes

5.5.2.1 Nodulated and Non-nodulated Plants

There are many legumes which lack the ability to nodulate. Almost all of these are found in tropical/subtropical areas, none at high latitudes. This may simply reflect the fact that legumes first evolved in the tropics, and that nodulation was acquired later. However, many of the non-nodulating legumes, especially those in the Caesalpinioideae, are dominant in their ecosystems. Often they have leaf nitrogen contents similar to those of nodulating legumes (McKey 1994), indicating that they have very efficient ways of obtaining nitrogen from the soil. These probably include good root systems and various types of mycorrhiza, but I know of no detailed studies on the physiology of nitrogen uptake by non-nodulating legumes. On the other hand, a great deal is known about the acquisition of combined nitrogen by potentially nodulating plants. Most of this knowledge stems from agriculturally important species, where nodulation is often suppressed in the presence of soil N, especially nitrate. It is argued that this is because nitrogen fixation requires more energy than does the uptake and assimilation of combined nitrogen. Apart from the costs of making nodules, nitrogenase is large, slow and expensive to produce (Sprent and Raven 1985). However, many legumes grow in natural environments where the predominant form of nitrogen is ammonium, and this may be far less inhibitory to nodulation and nitrogen fixation (Sprent 1999).

5.5.2.2 Geographical Trends

The most pronounced of these is latitudinal. Legumes occurring at higher latitudes are all from the Papilionoideae, and all are nodulated. Most are herbs or small shrubs and are commonly perennial. As mentioned above, they usually have indeterminate nodules. They also generally exhibit a higher degree of host–rhizobial specificity than at lower latitudes (Sprent 2001). Overall, this suggests that the more advanced legumes are more dependent on nitrogen fixation than their more primitive counterparts. The experimental evidence published to date supports this interpretation (Schulman et al. 1988). Interestingly, this is not necessarily true of actinorhizas. *Dryas octopetala*, which has a circumpolar distribution, apparently nodulates only in Alaska and Canada (Silvester 1977), although more work is needed on the nodulation status of this species. Temperature optima for nitrogen fixation in cold-adapted legumes are lower than for legumes adapted to warmer climates. This adaptation lies in part in the rhizobia, since inoculation of plants adapted to warmer environments confers cold tolerance to the nodules (Cloutier et al. 1996).

Legumes occurring at high altitude also experience cold, but day length variations are usually less extreme than at high latitude. However, high-altitude legumes are also papilionoid and nodulated, sharing many of the characters of high-latitude ones. They also may fix up to 100% of their nitrogen, even at low soil pHs of 4.1 (Jacot et al. 2000). Atmospheric pressure drops with altitude by about 1 millibar per 10 m, so that plants growing above 4,000 m experience pressures about half those at sea level. Bearing in mind that the control of oxygen diffusion is an important feature of nodule physiology, it is rather surprising that there has been no interest in the effects of reduced pressure on nodule functioning.

5.5.2.3 Dry, Flooded and Saline Environments

Many workers (including the present author) have published evidence that drought, flooding and salinity inhibit nodulation and nitrogen fixation and, furthermore, that nitrogen acquisition via fixation is more sensitive to these stresses than acquisition from combined nitrogen. However, these studies were conducted on plants of agricultural importance. Examination of the floras from areas where these conditions are common shows them often to have major nodulated legume components (Sprent 2001). In dry environments, nitrogen fixation may be confined to periods of rainfall, when corky nodules may resume growth (see above). The physiology of nodules from plants native to saline soils has, to my knowledge, not been studied to date, although some plants, such as species of the mimosoid genus *Prosopis*, are very tolerant of high salinity. *Prosopis tamarugo* grows in the extremely dry/saline environment of central Chile. It was one of the first plants in which hydraulic lift was considered to be important for water relations. Surprisingly, although several species of *Prosopis* are known to nodulate, I have found no record of this for *P. tamarugo*. The potential of the species has been studied by an FAO programme (Habit et al. 1981), but no mention of nitrogen nutrition was made. If nodulated, this species would make an excellent experimental plant for studies on both drought and salinity tolerance.

Legumes native to flooded environments have received some attention in recent years and, in surveying this literature, I found that all are apparently nodulated, regardless of site or subfamily (Sprent 1999, 2001). Some, such as *Discolobium pulchellum* Benth., a stem-nodulating species, nodulate *only* when flooded (Loureiro et al. 1994). Nodulated legumes appear to be essential components of some wetland areas of the world, such as the Pantanal of Brazil and neighbouring countries, although the amounts of nitrogen fixed have yet to be quantified (James et al. 2001).

5.5.2.4 Possible Chemical Constraints

Studies with agricultural species have shown that nodulation and nitrogen fixation are inhibited by pH extremes and the frequently co-occurring high levels of aluminium (Hartwig 1998). However, wild plants may be adapted to one or both of these constraints (Sprent 1999). Available phosphorus is perhaps the most widely recognised limitation to both nodulation and nitrogen fixation, and there is vast literature on this topic and on the role of arbuscular mycorrhizas in aiding phosphorus uptake (see review by Hartwig 1998, and Chap. 9, this Vol.). However, there are both legumes and actinorhizal plants that do not appear to have a component of their phosphorus requirement specifically associated with nodules (Sprent 1999, and Sect. 5.5.2.5, below).

5.5.2.5 Rhizobial Constraints

There is wide variation in the rhizobial content of soils. In some cases, compatible rhizobia seem to be present for plants that are non-native – for example, *Lupinus arboreus* nodulates freely in sand dunes of the North Island of New Zealand (Gadgil 1983). On the other hand, successful introduction of new legume crops can be very dependent on inoculation with suitable rhizobia, and this has given rise to inoculant industries found in many countries. They were first extensively developed in Australia and the USA when the economies of these countries were largely dependent on agriculture and, in particular, exotic nitrogen-fixing crops such as subterranean clover and soybean respectively. More recently, introduction of soybean into Europe has required inoculation, as native rhizobia are not compatible with this species. These facts are entirely consistent with the early concepts of host/rhizobial specificity. However, recent work in the tropics and elsewhere has shown that life is not always that simple. Genetic exchange can occur between inoculated rhizobia and saprophytic soil rhizobia. This was elegantly shown in New Zealand where *Lotus corniculatus* plants were grown in a soil in which there were no compatible rhizobia, so they were inoculated with a strain of *Mesorhizobium meliloti*. Re-isolation and analysis of rhizobia from the resultant nodules showed them to be genetically different from the inoculated form (Sullivan et al. 1995). In stressed environments, the ability to fix nitrogen (or even to nodulate plants) may be less important for rhizobia than the ability to survive in soils. Under such conditions, for example, in hot, dry environments, a wide range of rhizobia may be found, with varying abilities to nodulate plants growing there and producing nodules of very varied effectiveness in nitrogen fixation (discussed in Sprent 2001). In such situations, inoculation may be difficult because of the need (as in all environments) for the inoculant strain to be competitive with native strains. In such cases, the type of rhizobia found may be more related to soil conditions than to host plants. This may be one of the reasons why the level of host specificity shown by legumes in tropical/subtropical regions is often much lower than that at higher latitudes. The tropics also support a far greater range of legumes, multiplying the possibilities for nodulation of any one species by a range of rhizobia. Although it has been recognised for some time that broad host-range rhizobia occur, the extent of this has been impressively demonstrated by the work of Pueppke and Broughton (1999) who showed that the fast-growing rhizobial strain NGR234, originally isolated in Papua New Guinea by Trinick, can nodulate (not always effectively) 375 species (at least) of legumes, from all three subfamilies. A good example of the variation in effectiveness of rhizobia nodulating the same species is the work of Burdon et al. (1999) on Australian species of *Acacia*. Associations varied from being close to parasitic to highly effective.

5.5.2.6 Problems with Nodulated Legumes

It is often assumed by environmentalists and organic farmers that use of legumes is preferable to mineral fertilizers. This is an oversimplification – it depends on conditions and, in particular, the type of farming practiced. Like the nitrogen in any other organic material, legume residues are mineralised and, if ploughed in at the wrong time of year, this mineralisation can result in the formation of nitrate which behaves in exactly the same way as excess inorganic fertiliser (Sprent 1987).

A different problem may arise when nodulated legumes are introduced into new areas and get out of control. There are numerous cases of this in the literature, but I shall cite only two. First, seed production. Fast-growing woody nodulated legumes are being introduced as multipurpose plants in agroforestry. Many, such as *Leucaena leucocephala*, grow well and fix considerable quantities of nitrogen. Unfortunately, many are also prolific producers of seed and, therefore, may become weeds, a topic discussed in Hughes and Styles (1989). Second, many such plants reproduce vegetatively (often producing many seeds, too), so when cut regenerate and spread rapidly. They can invade nutrient-poor areas and destroy fragile ecosystems. In the unique fynbos vegetation of the South African Cape Province, there is a major problem with Australian acacias. These can be very aggressive and, when invading new areas, may fix much nitrogen, thus destroying the very delicate balance of plants adapted to low-nutrient soils (Stock et al. 1995; Holmes and Cowling 1997).

5.6 Mycorrhizas, Cluster (Proteoid) Roots and Other Nutrient Acquisition Systems

Until recently, it was believed that nodulated legumes are arbuscular mycorrhizal, and non-nodulating legumes either arbuscular or ectomycorrhizal. Actinorhizal plants may form one or both kinds. Further, based largely on work with members of the Proteaceae but also consistent with the situation in the legume *Lupinus*, and possibly actinorhizal genera such as *Gymnostoma* and *Myrica* (Huss-Danell 1997; Skene et al. 2000), cluster roots (originally termed proteoid roots because of their almost universal presence in the Proteaceae) were thought to be an alternative to mycorrhizas as a way of accessing poorly available soil nutrients such as phosphorus. However, now that more legumes have been studied under natural conditions, these generalisations are breaking down. For example, a number of Australian acacias can form both arbuscular and ectomycorrhizas, even when grown outside Australia, for example, in Brazil and Senegal (cf. references in Sprent 1994). In these exotic habitats they associate with local ectomycorrhizal fungi, which endemic acacias are unable to do. This ability to form dual mycorrhizas appears confined to certain members of the Australian subgenus *Phyllodineae* (also known as *Heterophyllum*), and may have evolved to cope with particular soil conditions. The family Proteaceae is largely confined to parts of South Africa and Australia, which have similar types of soil. Plants of the endemic South African legume genera *Aspalathus* and *Cyclopia* may form extensive cluster roots as well as being both nodulated and arbuscular mycorrhizal (Sprent 2001). Although cluster roots, like arbuscular mycorrhizas, are most usually associated with low-phosphorus conditions, they have recently been shown to form on the actinorhizal species *Casuarina glauca* under iron-deficiency (Arahou and Diem 1997). It appears that plants adapted to particularly low nutrient conditions can exploit all of the known ways of obtaining nutrients. These are the very qualities which make plants such as the Australian *Acacia auriculiformis* excellent for land reclamation, for example, after mining activities (Franco and Faria 1997). In addition to the structural modifications described above, nodulated plants may have physiological adaptations, such as the ability of pigeon pea (*Cajanus cajan*) to secrete acids which enable it to obtain phosphorus from insoluble iron complexes in soil (Ae et al. 1990).

Although parallels are often drawn between infection processes and signalling among the various types of nodule and mycorrhiza, there is no known case of healthy nodules being infected with mycorrhizal fungi. On the other hand, there are several reported cases of mycorrhizal 'nodules' on plants with and without the ability to form nitrogen-fixing root nodules. A recent, beautifully illustrated paper by Duhoux et al. (2001) describes the formation of such nodules on the two species of the actinorhizal genus *Gymnostoma* with arbuscular mycorrhizal fungi related to *Glomus*. These structures were shown to be modified lateral roots, and it was suggested that their formation may be related to growth of the hosts in exposed and rocky habitats, although their physiology remains to be examined. Other 'nodules' on roots of both legumes and non-legumes have been described which appear to be neither mycorrhizal nor nitrogen-fixing in nature (Sprent 2001). Some of those described from Western Australia by Pate and coworkers (e.g. Pate and Dixon 1981) superficially look like nodules but are rather used for nutrient storage. Since both nodulated legumes and actinorhizal plants are often found in nutrient-poor habitats, we should not be surprised at their versatility of modes of nutrient acquisition.

5.7 How Much Nitrogen Is Fixed Biologically?

This is a question to which many would like an answer. In a very wide-ranging study in the literature, Cleveland et al. (1999) concluded that globally (including aquatic habitats) the annual range was between 100 and 290 Tg N. However, they acknowledged that this was a very approximate value, distorted by many factors including the lack of methods for measuring nitrogen fixation in most ecosystems. Vitousek and others have approached the problem from another aspect, trying to identify the major factors limiting nitrogen fixation in different ecosystems, and incorporating these factors into a testable model (Vitousek et al 2002). It is likely to be some years before we have a better estimate than that of Cleveland et al. (1999).

5.8 What Does the Future Hold for Nitrogen-Fixing Plants?

Global change is a subject of much current debate. Whatever its origin, natural or anthropogenic, it is likely to affect biodiversity in the broad sense and may specifically affect nitrogen-fixing systems. Two aspects are relevant – rising CO_2 levels and nitrogen deposition from the atmosphere. Of these, most work has been done on CO_2, although there is no real consensus as to the likely outcome. Generally, N_2-fixing species respond more positively than their non-fixing counterparts (Soussan and Hartwig 1996; Hungate et al. 1999), but the effect is not universal (Niclaus et al. 1998). There are several reasons for this; for one, the wide range of experimental conditions (particularly CO_2 concentrations and time of exposure) and species used. For nodulated plants, enriched CO_2 may have a direct effect, because it can affect the supply of C-skeleton necessary to assimilate newly generated ammonium from fixation (Sprent and Sprent 1990). Far more likely is that increased atmospheric CO_2 may have an effect via plant growth. If plants photosynthesise more and grow faster, then there may be increased nodulation in proportion to this growth. This in turn may affect biodiversity. For example, after four years atmosppheric CO_2 enrichment of semi-natural grassland growing in a nutrient poor region of Germany, the nodulated legume *Lotus corniculatus* showed increased competitiveness when compared with grasses and forbs (Winkler and Herbst 2004).

As nitrogen fixation is usually reduced when plants have access to combined nitrogen, is this atmospheric nitrogen deposition likely to lead to a suppression of nitrogen-fixing species? This will depend on the particular form of nitrogen being deposited, and the geographical area of interest. Apart from nitrogen compounds which may be formed following forest fires, the principal source of increased atmospheric nitrogen is the transport sector (motor vehicles and jet aircraft) and intensive animal production systems (Fowler et al. 1998) – it is thus currently less of a problem in developing than in developed countries. Deposition is normally divided into two types – dry and wet and, within each, there is a range of nitrogen species, both oxidised (NO_x) and reduced (NH_x). Some forms of gaseous nitrogen, for example, NO_2, may be absorbed and assimilated by leaves, but it is the combined nitrogen entering the soil which is most likely to have an effect on nitrogen fixation. In symbiotic systems, nitrate is generally more inhibitory to nitrogen fixation than ammonium (Sprent 1999). In Europe, it has been estimated that 15–33 kg N ha^{-1} may be deposited, over half of which is in the reduced form (Fowler et al. 1998). This is unlikely to have a major direct effect upon the nodulated plants native to these areas, although by altering species composition, there may be indirect effects. On the other hand, downwind of intensive animal farms, up to 80 kg N ha^{-1} may be deposited, principally in the form of ammonium (Pitcairn et al 1998). This can have a profound effect on N-sensitive species, and could certainly inhibit nodulation and nitrogen fixation in some nodulated plants (Hartwig 1998).

5.9 Summary

Symbiotic nitrogen fixation has enabled plants from all functional groups to invade major habitats of the world. Nitrogen-fixing plants generally have an advantage over their non-fixing counterparts, and tend to be more competitive when N is limiting. Despite this advantage, in many undisturbed systems the relative abundance of fixers to non-fixing species remains low due to limitations in other resources such as phosphorus. This relationship can, however, break down in more disturbed habitats when invasive nitrogen-fixing species become dominant, thereby modifying the biodiversity of the system. Biological nitrogen fixation can also directly impact the N cycle. This is particularly true in systems where internal cycling does not generate a significant amount of inorganic N, e.g. in extremely dry and hot deserts. In such systems, nitrogen fixation by free-living microorganisms could be the dominant source of N input into the soil.

Major gaps still remain in our understanding of symbiotic nitrogen fixation. Some of these challenges arise from technical difficulties in accurately quantifying nitrogen fixation under field conditions. In other instances, it is still problematical unambiguously to assess the mechanism of action of N_2-fixing bacteria. For example, in addition to affecting plant growth through nitrogen supply, some N_2-fixing bacteria can produce growth-enhancing hormones such as auxin. Therefore, it is presently difficult to ascertain the extent to which the faster growth rate of plants inoculated with these bacteria is caused by higher N_2 supply. More studies are also needed to resolve why low soil pH, aluminium and calcium prevent nodulation in some legumes which might otherwise nodulate if suitable rhizobia were present. Future studies of nitrogen-fixing plants and their interactions with non-fixing plants must also consider changes in global climate. In particular, the projected increases in atmospheric CO_2 concentration and N deposition are likely to alter the balance of fixing and non-fixing species in significant ways.

References

Ae N, Arihara J, Okada K, Yoshihara T, Johansen C (1990) Phosphorus uptake by pigeon pea and its role in cropping systems of the Indian subcontinent. Science 248:477–480

Allen ON, Allen EK (1981) The Leguminosae: a source book of characteristics, uses and nodulation. University of Wisconsin Press, Madison

Anonymous (1984) Casuarinas: nitrogen-fixing trees for adverse sites. National Academy Press, Washington

Arahou M, Diem HG (1997) Iron deficiency induces cluster (proteoid) root formation in *Casuarina glauca.* Plant Soil 196:71–79

Becking J-H (1992) The *Rhizobium* symbiosis of the nonlegume *Parasponia.* In: Stacey G, Burris RH, Rvans HJ (eds) Biological nitrogen fixation. Chapman and Hall, New York, pp 497–559

Benson DR, Clawson ML (2000) Evolution of actinorhizal plant symbiosis . In: Triplett EW (ed) Prokaryotic nitrogen fixation. Horizon, Wymondham, Norfolk, pp 207–224

Benson DR, Silvester W (1993) Biology of *Frankia* strains, actinomycete symbionts of actinorhizal plants. Microbiol Rev 57:293–319

Bergman B, Johansson C, Söderbäck E (1992) The *Gunnera-Nostoc* symbiosis. New Phytol 122:379–400

Boddey RM (1995) Biological nitrogen fixation in sugar cane: a key to energetically viable biofuel production. Crit Rev Plant Sci 6:209–266

Boddey RM, Peoples MB, Palmer B, Dart PJ (2000) Use of ^{15}N natural abundance technique to quantify biological nitrogen fixation by woody perennials. Nutr Cycling Agroecosyst 57:235–270

Bond G (1976) The results of the IBP survey of root-nodule formation in non-leguminous angiosperms. In: Nutman PS (ed) Symbiotic nitrogen fixation in plants. Cambridge University Press, Cambridge, pp 443–474

Broughton WJ, Jabbouri S, Perret X (2000) Keys to symbiotic harmony. J Bacteriol 182:5641–5652

Burdman S, Volpin H, Kigel J, Kapulnik Y, Okon Y (1996) Promotion of nod gene inducers and nodulation in common bean (*Phaseolus vulgaris*) roots inoculated with *Azospirillum brasilense* Cd. Appl Environ Microbiol 62:3030–3033

Burdon JJ, Gibson AH, Searle SD, Woods MJ, Brockwell J (1999) Variation in the effectiveness of symbiotic associations between native rhizobia and temperate Australian *Acacia*: within-species interactions. J Appl Ecol 36:398–408

Chen WM, James EK, Prescott AR, Kierans M, Sprent JI (2003) Nodulation of *Mimosa* spp. by the β-proteobacterium *Ralstonia taiwanensis*. Mol Plant-Microbe Interact 16:1051–1061

Cleveland CC, Townsend AR, Schimel DS, Fisher H, Howarth RW, Hedin LO, Perakis S, Latty EF, von Fischer JC, Elseroad A, Wasson MF (1999) Global patterns of terrestrial biological nitrogen (N_2) fixation in natural ecosystems. Global Biogeochem Cycles 13:623–645

Cloutier J, Laberge S, Prévost D, Antoun H (1996) Sequence and mutational analysis of the common *nod*BCIJ region of *Rhizobium* sp. (*Oxytropis arctobia*) strain N33, a nitrogen-fixing microsymbiont of both arctic and temperate legumes. Mol Plant Microbe Interact 9:523–531

Corby HDL (1988) Types of rhizobial nodule and their distribution among the Leguminosae. Kirkia 13: 53–123

Denison WC (1979) *Lobaria oregana*, a nitrogen-fixing lichen in old-growth Douglas fir forests. In: Gordon JC, Wheeler CT, Perry CT (eds) Symbiotic nitrogen fixation in the management of temperate forests. Forest Research Laboratory, Oregon State University, Corvallis, pp 266–275

Diem HG, Dommergue YR (1990) Current and potential uses and management of *Casuarinaceae* in the tropics and sub-tropics. In: Schwintzer CR, Tjepkema JD (eds) The biology of *Frankia* and actinorhizal plants. Academic Press, New York, pp 317–342

Döbereiner J, Reis MV, Lazarina AC (1988) New N_2 fixing bacteria in association with certain cereals and sugar cene. In: H Bothe, FJ de Bruijn, WE Newton (eds) Nitrogen fixation: hundred years after. Gustav Fischer, Stuttgart, pp 717–722

Dong Z, Layzell DB (2001) H_2 oxidation, O_2 uptake and CO_2 fixation in hydrogen treated soils. Plant Soil 229:1–12

Dong Z, Canny MJ, McCully ME, Roboredo MR, Cabadilla CF, Ortega E, Rodes R (1994) A nitrogen-fixing endophyte from sugarcane stems. Plant Physiol 105:1139–1147

Duhoux E, Rinaudo G, Diem HG, Auguy F, Fernandez D, Bogusz D, Franche C, Dommergues Y, Huguerin B (2001) Angiosperm *Gymnostoma* trees produce root nodules colonized by arbuscular mycorrhizal fungi related to *Glomus*. New Phytol 149:115–125

Fowler D, Sutton MA, Smith RI, Pitcairn CER, Coyle M, Campbell G, Stedman J (1998) Regional budgets of oxidised and reduced nitrogen and their relative contribution to the nitrogen inputs of sensitive ecosystems. Environ Pollut 102:337–342

Franco AA, Faria SM de (1997) The contribution of N fixing tree legumes to land reclamation and sustainability in the tropics. Soil Biol Biochem 29:897–903

Gadgil RL (1983) Biological nitrogen fixation in forestry – research and practice in Australia and New Zealand. In: Gordon JC, Wheeler CT (eds) Biological nitrogen fixation in forest ecosystems: foundations and applications. Dr. W. Junk, The Hague, pp 317–332

Gordon JC, Wheeler CT (eds) (1983) Biological nitrogen fixation in forest ecosystems: foundations and applications. Dr. W. Junk, The Hague

Gualtieri G, Bisseling T (2000) The evolution of nodulation. Plant Mol Biol 42:181–194

Habit MA, Contreras D, González RH (1981) *Prosopis tamarugo*: fodder tree for arid zones. FAO, Rome

Halliday J, Pate JS (1976) Symbiotic nitrogen fixation by coralloid roots of the cycad *Macrozamia reidlei*: physiological characteristics and ecological significance. Aust J Plant Physiol 3:349–358

Hartwig UA (1998) The regulation of symbiotic N_2 fixation: a conceptual model of N feedback from the ecosystem to the gene expression level. Perspect Plant Ecol Evol Syst 1:92–120

Herbert RA, Tughan CS, de Wit R, van Gemerden H (1988) Development of laminated mat microbial ecosystems in temperate marine environments. In: Megusar F, Gantar M (eds) Perspectives in microbial ecology. Slovene Society for Microbiology, Ljubljana, pp 105–111

Hill RS, Scriven LJ (1995) The angiosperm-dominated woody vegetation of Antarctica: a review. Rev Palaeobot Palynol 86:175–198

Holmes PM, Cowling RM (1997) The effects of invasion by *Acacia saligna* on the guild structure and regeneration capabilities of South African fynbos. J Appl Ecol 34:317–332

Hughes CE, Styles BT (1989) The benefits and risks of woody legume introductions. In: Stirton CH, Zarucchi JL (eds) Advances in legume biology. Monogr Syst Bot Mo Bot Gard 29:505–531

Hurek T, Handley L, Reinhold-Hurek B, Piche Y (1998) Does *Azoarcus* sp fix nitrogen with monocots? In: Elmerich C, Kondorosi A, Newton WE (eds) Biological nitrogen fixation for the 21st century. Kluwer, Dordrecht, p 407

Huss-Danell K (1997) Actinorhizal symbioses and their N_2 fixation. New Phytol 136:375–405

Jacot KA, Lüscher A, Nösberger J, Hartwig UA (2000) Symbiotic N fixation of various legume species along an altitudinal gradient in the Swiss alps. Soil Biol Biochem 32:1043–1052

James EK (2000) Nitrogen fixation in endophytic and associative symbiosis. Field Crops Res 65:197–209

James EK, Loureiro MF, Pott A, Pott VJ, Martins CM, Franco AA, Sprent JI (2001) Flooding-tolerant legume symbioses from the Brazilian Pantanal. New Phytol 150:723–738

Jarzen DM, Dettman ME (1989) Taxonomic revision of *Tricolpites reticulates* Cookson ex Cooper 1953 with notes on the biogeography of *Gunnera*. Pollen Spores 31:97–112

Jones DL (1993) Cycads of the world. Smithsonian Institute Press, Washington

Lavin M, Pennington RT, Klitgaard BB, Sprent JI, de Lima HC, Gasson PE (2001) The dalbergioid legume (Fabaceae): delimitation of a pantropical monophyletic clade. Am J Bot 88:503–533

Leigh JA (2000) Nitrogen fixation in methanogens – the archaeal perspective. In: Triplett EW (ed) Prokaryotic nitrogen fixation. Horizon, Wymondham, Norfolk, pp 657–669

Lepper MG, Fleschner M (1977) Nitrogen fixation by *Cercocarpus ledifolius* (Rosaceae) in pioneer habitats. Oecologia 27:333–338

Loureiro MF, Faria SM de, James EK, Pott AA, Franco AA (1994) Nitrogen-fixing stem nodules of the legume, *Discolobium pulchellum* Benth. New Phytol 128:283–295

McKey D (1994) Legumes and nitrogen: the evolutionary ecology of a nitrogen-demanding lifestyle. In: Sprent JI, McKey D (eds) Advances in legume systematics. Part 5. The nitrogen factor. Royal Botanic Gardens, Kew, pp 221–228

Moore DM (1968) The vascular flora of the Falkland Islands. Sci Rep no 60, British Antarctic Survey, London

Nash TH (1996) Nitrogen, its metabolism and potential contribution to ecosystems. In: Nash TH (ed) Lichen biology. Cambridge University Press, Cambridge, pp 121–135

Okon Y, Labandera-Gonzales CA (1994) Agronomic applications of *Azospirillum*: an evaluation of 20 years worldwide field inoculation. Soil Biol Biochem 26:1591–1601

Osborne BA, Cullen A, Jones PW, Campbell GJ (1992) Use of nitrogen by the *Nostoc-Gunnera tinctoria* (Molina) Mirbel symbiosis. New Phytol 120:481–487

Parsons R, Silvester WB, Harris S, Gruijters WTM, Bullivant S (1987) *Frankia* vesicles provide inducible and absolute protection for nitrogenase. Plant Physiol 83:728–731

Pate JS, Dixon KW (1981) Plants with fleshy underground storage organs – a Western Australian survey. In: Pate JS, McComb AJ (eds) The biology of Australian plants. University of Western Australia Press, Nedlands, pp 181–215

Peoples MB, Faizah AW, Rerkasen B Herridge DF (eds) (1989) Methods for evaluating nitrogen fixation by nodulated legumes in the field. ACIAR, Canberra

Pitcairn CER, Leith ID, Sheppard LJ, Sutton MA, Fowler D, Munro RC, Tang S, Wilson D (1998) The relationship between nitrogen deposition, species composition and foliar nitrogen concentrations in woodland flora in the vicinity of livestock farms. Environ Poll 102:41–48

Pueppke SG, Broughton WJ (1999) *Rhizobium* sp. strain NGR234 and *R. fredii* USDA257 share exceptionally broad, nested host-ranges. MPMI 12:293–318

Rai AN, Söderbäck E, Bergman B (2000) Cyanobacterium-plant symbioses. New Phytol 147:449–481

Richardson AE, Simpson RJ, Djordjevic MA Rolfe BG. (1988) Expression of nodulation genes in *Rhizobium leguminosarum* bv. *trifolii* is affected by low pH, Ca and Al. Appl Environ Microbiol 54:2541–2548

Scherer S, Ernst SA, Chen T-W, Böger P (1984) Rewetting of drought-resistant blue-green algae: time course of water uptake and reappearance of respiration, photosynthesis, and nitrogen fixation. Oecologia 62:418–423

Schulman HM, Lewis MC, Tipping EM, Bordeleau LM (1988) Nitrogen fixation by three species of Leguminosae in the Canadian high Arctic tundra. Plant Cell Environ 11:721–728

Schwintzer CR, Tjepkema JD (eds) (1990) The biology of *Frankia* and actinorhizal plants. Academic Press, New York

Sharma E (1993) Nutrient dynamics in Himalayan alder plantations. Ann Bot 72:329–336

Shrestha RK, Ladha JK (1996) Genotypic variation in promotion of rice nitrogen fixation as determined by ^{15}N dilution. Soil Sci Am Soc J 60:1815–1821

Silvester WB (1975) Ecological and economic significance of the non-legume symbioses. In: Newton WE, Nyman CJ (eds) Proc 1st Int Symp Nitrogen Fixation. Wiley, New York, pp 489–506

Silvester WB (1977) Dinitrogen fixation by plant associations excluding legumes. In: Hardy RWF, Gibson AH (eds) A treatise on dinitrogen fixation, section IV. Wiley, New York, pp 141–189

Silvester WB, Parsons R, Watt PW (1996) Direct measurement of release of ammonia in the *Gunnera-Nostoc* symbiosis. New Phytol 132: 617–625

Skene KR, Sprent JI, Raven JA, Herdman L (2000) Biological flora of the British Isles: *Myrica gale*. J Ecol 88:1079–1094

Skujins J, Nikkanen T, Hernriksson E (1987) Research note: responses of cyanobacterial growth and N_2 fixation to textural components of soils. Arid Soil Restor Rehab 1:195–198

Soltis DE, Soltis PS, Chase ME, Mort ME, Albach DC, Zanis M, Savolainen V, Hahn WH, Hoot SB, Fay MF, Axtell M, Swensen SM, Prince LM, Kress WJ, Nixon KC, Farris JS (2000) Angiosperm phylogeny inferred from 18S rDNA, *rbc*L, and *atp*B sequences. Bot J Linn Soc 133:381–461

Spaink HP, Kondorosi A, Hooykaas PJJ (eds) (1998) The *Rhizobiaceae* – molecular biology of the model plant-associated bacteria. Kluwer, Dordrecht

Sprent JI (1987) The ecology of the nitrogen cycle. Cambridge University Press, Cambridge

Sprent JI (1989) Which steps are essential to the formation of functional legume nodules? New Phytol 111:129–153

Sprent JI (1993) The role of nitrogen fixation in primary succession on land. In: Miles J, Walton DWH (eds) Primary succession on land. British Ecological Society, Spec Publ 12, Blackwell, Oxford, pp 209–219
Sprent JI (1994) Nitrogen acquisition systems in the Leguminosae. In: Sprent JI, McKey D (eds) Advances in legume systematics. Part 5. The nitrogen factor. Royal Botanic Gardens, Kew, pp 1–23
Sprent JI (1999) Nitrogen fixation and growth of non-crop species in diverse environments. Perspect Plant Ecol Evol Syst 2:149–162
Sprent JI (2000) Nodulation as a taxonomic tool. In: Herendeen PS, Bruneau A (eds) Advances in legume systematics, part 9. Royal Botanic Gardens, Kew, pp 21–44
Sprent JI (2001) Nodulation in legumes. Royal Botanic gardens, Kew
Sprent JI, Raven JA (1985) Evolution of nitrogen fixing symbioses. Proc R Soc Edinb B 85:215–237
Sprent JI, Sprent P (1990) Nitrogen fixing organisms: pure and applied aspects. Chapman and Hall, London
Stock P, Silvester WB (1994) Phloem transport of recently fixed nitrogen in the *Gunerra-Nostoc* symbiosis. New Phytol 126:259–266
Stock WD, Wienand KT, Baker AC (1995) Impacts of invading N_2-fixing *Acacia* species on patterns of nutrient cycling in 2 Cape ecosystems – evidence from soil incubation studies and ^{15}N natural abundance values. Oecologia 101:375–382
Sullivan JT, Patrick HN, Lowther WL, Scott DB, Ronson CW (1995) Nodulating strains of *Rhizobium loti* arise through chromosomal symbiotic gene transfer in the environment Proc Natl Acad Sci USA 92:8985–8989
Sy A, Giraud E, Jourand P, Garcia N, Willems A, de Lajudie P, Prin Y, Neyra M, Gillis M, Boivin-Masson C, Dreyfus B (2001) Methylotrophic *Methylobacterium* nodulate and fix nitrogen in symbiosis with legumes. J Bacteriol 183:214–220
Urquiaga S, Cruz KHS, Boddey RM (1992) Contribution of nitrogen fixation to sugar cane: nitrogen-15 and nitrogen balance estimates. Soil Sci Soc Am J 56:105–114
Vande Broek A, Dobbelaere S, Vanderleyden J, Vandommelen A (2000) *Azospirillum*-plant root interactions: signalling and metabolic interactions. In: Triplett EW (ed) Prokaryotic nitrogen fixation. Horizon, Wymondham, Norfolk, pp 761–777
Vaneechoutte M, Kampfer P, De Baere T, Falsen E, Verschraegen G (2004) *Wautersia* gen. nov., a novel genus accommodating the phylogenetic lineage including *Ralstonia eutropha* and related species, and proposal of *Ralstonia* [*Pseudomonas*] *syzygii* (Roberts et al. 1990) comb. nov. Int J Syst Evol Microbiol 54:317–327
Vitousek PM, Walker LR (1989) Biological invasion by *Myrica faya*in Hawaii: plant demography, nitrogen fixation, and ecosystem effects. Ecol Monogr 59:247–265
Vitousek PM, Cassman K, Cleveland C, Crews T, Field CB, Grimm NB, Howarth RW, Marino R, Martinelli L, Rastetter EB, Sprent JI (2002) Towards an ecological understanding of biological nitrogen fixation. Biogeochemistry 57/58:1–45
Wall LG (2000) The actinorhizal symbiosis. J Plant Growth Regul 19:167–182
Winkler JB, Herbst M (2004) Do plants of a semi-natural grassland community benefit from long-term CO_2 enrichment? Basic and Applied Ecology 5:131–143
Wynne-Williams DD, Edwards HGM (2000) Antarctic ecosystems as models for extraterrestrial surface habitats. Planet Space Sci 48:1065–1075
Wynne-Williams DD, Edwards HGM, Garcia-Pichel F (1999) Functional biomolecules of Antarctic stromatolite and endolithic cyanobacterial communities. Eur J Phycol 34:381–391
Young JPW (1992) Phylogenetic classification of nitrogen-fixing organisms. In: Stacey G, Evans HJ, Burris RH (eds) Biological nitrogen fixation. Chapman and Hall, New York, pp 43–86

Young JM, Kuykendall LD, Martínez-Romero E, Kerr A, Sawada H (2001) A revision of *Rhizobium* Frank 1889, with an emended description of the genus, and the inclusion of all species of *Agrobacterium* Conn 1942 and *Allorhizobium undicola* de Lajudie *et al.* 1998, *R. rhizogenes*, *R. rubri*, *R. undicola* and *R. vitis*. Int J Syst Evol Microbiol 51:89–103

6 Homeostatic Processes for the Optimization of Nutrient Absorption: Physiology and Molecular Biology

A.D. Glass

6.1 Introduction

As early as 1906, laboratory experiments using hydroponically grown wheat plants demonstrated that when a particular inorganic nutrient was withheld for a period of hours, the root absorption capacity for that nutrient was increased several-fold (Brezeale 1906). Brezeale's pioneering experiments demonstrated that this was the case for NO_3^-, Ca^{2+}, K^+ and P_i, whose rates of uptake had increased by 3.3-fold (NO_3^-), 2.2-fold (Ca^{2+}), 4-fold (K^+) and 2.5-fold (P_i), respectively, after 18 h without each of these ions. In the intervening years, this phenomenon has been rediscovered, confirmed and studied extensively, with a view to understanding its adaptive significance and the underlying mechanism(s) responsible for both the sensing and activation of these responses. Biologists might consider that the adaptive significance of this phenomenon was obvious, namely, to allow plants to maintain internal levels of required nutrients at some ideal prescribed values. Nevertheless, soil scientists such as Barber (1995) have concluded, on the basis of mechanistic models of plant nutrient uptake in soils, that this is not necessarily the case. Barber's model predicted that altering the V_{max} (maximum velocity; see below) for ion influx had little effect on ion uptake at low external ion concentration because the delivery of ions to the root surface is diffusion-limited under such conditions. However, the Barber model failed to take into account inter-plant competition; competitive advantage is increased if V_{max} is increased, despite the predictions of the Barber model (Robinson 2001). Moreover, the up-regulation of ion uptake when the supply of that nutrient is inadequate appears to be a universal response, well documented in bacteria, fungi, algae and higher plants. It would therefore be surprising if it were of no relevance to ion acquisition. Even if nutrient supply depends upon a diffusive flux from soil to root, the increased capacity for influx under conditions of deprivation should

Ecological Studies, Vol. 181
H. BassiriRad (Ed.) Nutrient Acquisition by Plants
An Ecological Perspective

steepen the gradient for diffusion, and correspondingly increase the flux to the root surface. Furthermore, under conditions wherein nutrient supply is restored (e.g., following re-hydration of dry soils), the high capacity for influx in nutrient-deprived plants might allow plants to optimize ion capture in competition with other species.

Among the many observations reported since Brezeale's pioneering experiments, it has been established that following re-supply of the withheld nutrient, there is a rapid reduction of the expanded capacity for uptake of that nutrient. As a result of this potential to rapidly up-regulate and down-regulate uptake, nutrient uptake and internal nutrient concentrations may be rendered (to a large extent) independent of external concentration. For example, in a study of ryegrass, maintained under conditions of constant external NO_3^- concentration ($[NO_3^-]_o$), NO_3^- uptake, plant N status and biomass generation were found to be independent of $[NO_3^-]_o$ from 14 μM to 14 mM (Clement et al. 1978). Above and below these concentrations, plant growth and tissue [N] were affected by $[NO_3^-]_o$.

In contrast to laboratory conditions, the soil solutions from which plants absorb required nutrients is extremely heterogeneous with respect to ion concentration. In studies of agricultural soils, concentrations of the major nutrients vary across orders of magnitude (Reisenauer 1966; Wolt 1994). In natural soils, the same broad ranges of concentration may be found in even quite small areas (Jackson and Caldwell 1993). Soil nutrient heterogeneity and its consequences for plant nutrient acquisition were covered in Chapter 6 (this Vol.). It is noteworthy, however, that there are characteristic differences among different habitats with respect to the abundance of different forms of particular nutrients. For example, Stark and Hart (1997) reported that in various forest ecosystems in North America the NH_4^+ to NO_3^- ratio varied from 2.6 to 56.6. In various arctic tundra soils, NO_3^- was undetectable and combined amino acid concentrations were several-fold higher than those of NH_4^+ (Kielland 1995). Species adapted to such habitats may demonstrate reduced capacity to absorb NO_3^- compared to NH_4^+ or amino acids (Kielland 1994, 1997; Kronzucker et al. 1997). By contrast, early successional species are more opportunistic, capable of satisfying their N demands equally well from NO_3^- or reduced N (Min et al. 1999). In addition to the above-described heterogeneity, seasonal activities of soil organisms produce substantial changes in the availabilities of required nutrients.

In the context of such variability, plants appear to have evolved a wide variety of adaptations that enable them to optimize nutrient acquisition. These include life history adaptations (e.g., mycorrhizal associations), morphological adaptations (e.g., root hairs and plastic root:shoot ratios), and physiological adaptations (e.g., the plasticity of ion flux in response to changing external ion concentration). The latter adaptation is the focus of this chapter. Much of the earlier work in this area is summarized in reviews by Cram (1976), Glass (1983), and Clarkson (1985). The review by Cram (1976) emphasized not only the nutritional role of absorbed ions, but also their osmotic functions, citing examples of ion fluxes that were regulated in response to turgor and/or volume perturbations. Thus, the rapid down-regulation of the expanded ion uptake capacity following re-supply of the withheld nutrient may serve to prevent osmotic dysfunction or problems of specific ion toxicity associated with excessive accumulation. While the potential for toxicity is relatively easy to comprehend in the cases of ions such as P_i or NH_4^+, the apparent universality of the response as it extends to less problematic ions (e.g., SO_4^{2-} and K^+) weakens this argument. Thus, although dramatic advances have been realized in recent years, involving the cloning of a large number of genes encoding nutrient transporters of higher plants, conjectures still abound with respect to the adaptive and functional aspects of these phenomena.

While the mechanisms involved in up-regulating or down-regulating ion fluxes in unicellular organisms might be essentially similar to those utilized by higher plants at the cellular level, the spatial separation of the region of uptake from that involved in utilization might be as great as 100 m in forest trees. Thus, in such plants, there is an additional need for long-distance communication of plant "demand" for inorganic nutrients. The cycling and recycling of absorbed ions or their metabolic products, from root to shoot and back, has been proposed as the means of signaling whole-plant requirements to the root, so that uptake may match demand (Cooper and Clarkson 1989; Marschner et al. 1997).

It has commonly been asserted that the mechanisms responsible for activating the up- or down-regulation of ion uptake involve feedback from internal pools of ions or their metabolic products (Cram 1976; Clarkson 1985; Glass and Siddiqi 1984, 1995; Lappartient and Touraine 1996). An alternative hypothesis, namely, the detection of changes of external ion concentration via external (perhaps membrane-localized) sensors seems improbable based upon two sorts of experiments. Split-root experiments have commonly demonstrated that when only one region of the root is supplied with a particular nutrient, rates of ion uptake in the fed root increase substantially, despite a constant external ion concentration (Drew and Saker 1984). Another approach that leads to the same conclusion has made use of root pruning experiments. Borkert and Barber (1983), for example, demonstrated that reducing the proportion of soybean roots exposed to nutrient solution from 100 to 12.5% caused the V_{max} for P_i influx to increase >sixfold, despite a constant external P_i concentration.

6.2 Kinetic Constants for Ion Influx

6.2.1 Definitions

Prior to current developments in the cloning of genes encoding plant ion transporters, one approach to probing the underlying mechanisms responsible for up-regulating and down-regulating influx was to determine effects of nutrient deprivation and/or re-supply on V_{max} and K_m values for uptake. These terms (V_{max} and K_m) are derived from classical enzyme kinetics (Epstein 1976). V_{max} is the predicted maximum velocity (the Y asymptote of a typically hyperbolic concentration versus substrate plot). The Michaelis constant is commonly interpreted as the substrate concentration required to achieve 50 % of V_{max} and/or as a measure of the enzyme affinity for its substrate (low values of K_m being equivalent to higher affinity). The K_m is more formally defined as the dissociation constant of the enzyme substrate complex.

6.2.2 Methodology

Ideally, the experimental determination of K_m and V_{max} values involves measuring ion influx (not net uptake, which is the resultant of influx and efflux) during short exposures of root tissue to the labeled ion of interest. In long-term (time scale of hours) measurements, tracer efflux from the root compromises the influx measurements. Roots or other tissues are exposed to several concentrations of the ion above and below the anticipated K_m value, and either by direct fit of the data to a rectangular hyperbola or by linearization of the data, K_m and V_{max} values can be obtained. Another method that has frequently been employed is to allow roots to deplete the ion over a period of hours. The rate of ion uptake is then computed at particular concentrations from the slope of the concentration versus time plot at various concentrations. This method suffers the same errors associated with efflux but also introduces another potential source of error. Unless the system is at steady state, gene expression and altered transport kinetics can be induced or repressed within hours of imposed changes. Thus, the data collected over a period of 6–12 h compound effects of both concentration change and changes in the plant system.

6.2.3 Effects of Nutrient Status on Kinetic Constants

As examples of effects of nutrient status on kinetic constants, the V_{max} for $^{32}P_i$ influx in P-deficient excised barley roots increased fourfold, without statistically significant effects on the K_m (Barber 1972). Likewise, Lefebvre (1980) and Lee (1982) confirmed that withholding P_i caused four- and twofold increases in V_{max} for $^{32}P_i$ influx, without significant effects on K_m values. Only the findings of Cartwright (1972), using sterile excised barley roots, conflict with this consensus. Cartwright reported that although $^{32}P_i$ influx was increased by P deprivation, the V_{max} for high-affinity P_i influx was unaffected by this treatment, and the K_m for P_i influx was reduced from 18.9 to 5.9 μM. Whether the effect was on V_{max} or K_m, the resultant effect was to increase potential uptake rates.

Effects of nutrient supply upon the V_{max} and K_m for the uptake of other ions is shown in Table 6.1. As was the case for effects of P supply, different studies, sometimes using different plants or different growth methods, have produced different results, as was the case for NH_4^+. Thus, the V_{max} for $^{13}NH_4^+$ influx into roots of rice plants and citrus roots increased and K_m values declined at low rates of NH_4^+ supply (Wang et al. 1993; Cerezo et al. 2001), whereas the data for NH_4^+ influx into sterile roots of *Arabidopsis thaliana* confirmed the effect of NH_4^+ supply on the V_{max} for NH_4^+ influx, but failed to show significant effects on K_m (Rawat et al. 1999). These differences may be the result of species differences or methodological differences – the *Arabidopsis* plants were grown under sterile conditions in Magenta boxes and required an exogenous supply of sucrose in the nutrient media.

Using barley plants, Lee (1982) examined the effects of withholding SO_4^{2-} or Cl^- on the kinetic constants for $^{35}SO_4^{2-}$ and $^{36}Cl^-$ influx into the roots. In the case of SO_4^{2-} deprivation, the V_{max} increased ~14-fold without significant effects on the K_m value. Cl^- deprivation increased the V_{max} 2.6-fold, and also substantially reduced the K_m. In summary, withholding a particular ion almost universally results in increased values for V_{max}. In specific cases, for example, K^+ and Cl^-, the K_m values are also reduced, but this response has not

Table 6.1. Effects of nutrient supply upon kinetic constants (V_{max} and K_m) for ion uptake by roots of various plant species. The symbol ↔ indicates that there was not a significant effect upon the K_m

Ion	Plant	Treatment	V_{max}	K_m	Reference
$^{86}Rb^+$ (K^+)	Barley	K deprivation	↑	↓	Glass (1976)
		K re-supply	↓	↑	
$^{13}NH_4^+$	Rice	Low available NH_4^+	↑	↓	Wang et al. (1993)
		High available NH_4^+	↓	↑	
	Arabidopsis	Low available NH_4^+	↑	↔	Rawat et al. (1999)
		High available NH_4^+	↓	↔	
$^{35}SO_4^{2-}$	Barley	SO_4^{2-} deprivation	↑	↔	Lee (1982)
$^{36}Cl^-$	Barley	Cl^- deprivation	↑	↓	Lee (1982)

always been observed for other ions. Even prior to evidence based upon expression patterns of cloned genes and the development of antibodies against specific transporters, researchers explained the changes of V_{max} as the result of changes in the synthesis or turnover of the carrier proteins responsible for high-affinity influx (Pitman and Cram 1977; Glass and Siddiqi 1984; Clarkson and Luttge 1991). Changes of K_m in response to nutrient deprivation – for example, in the case of K^+ – were interpreted as the result of allosteric effects of accumulated K^+ on the K^+ transporter (Glass 1976).

6.2.4 Natural Selection and Kinetic Constants

Clearly, in an ecological context, it might be imagined that natural selection might operate to optimize nutrient uptake by lowering K_m values and/or increasing V_{max} values. However, when attempting to evaluate the putative selective advantage(s) accruing to different organisms by virtue of higher-affinity or higher-velocity transporters, there are so many other differences among species or individuals that it is difficult to specifically ascribe differences in growth rates or fitness to the kinetic properties of the transport systems per se. However, in *Aspergillus nidulans* there are only two NO_3^- transporters (NRTA and NRTB) and these have V_{max} values of 564 and 141 nmol mg^{-1} DW, respectively, and K_m values of 96 and 11 μM, respectively (Unkles et al. 2001). The two genes encoding these transporters have been cloned and we have obtained mutants expressing one or other of the two transporters. Thus, we have an organism in which only these kinetic parameters differ. We are currently attempting to determine if the high(er)-affinity, low(er)-capacity transporter (NRTB) confers selective advantage at low external $[NO_3^-]$, and if the low(er)-affinity, high(er)-capacity of the NRTA transporter confers selective advantage at high external $[NO_3^-]$. If this is the case, a clear rationale for this apparent gene duplication will have been provided. Alternatively, and despite our faith in natural selection, we may be forced to accept that gene duplication followed by gene differentiation may have occurred without selective advantage (Gutschick and BassiriRad 2003).

An examination of the derivation of K_m and V_{max} in elementary biochemistry textbooks indicates that it may not be possible to increase V_{max} without increasing K_m. In the following equations for the reaction of an enzyme *E* with a substrate *S*, the enzyme substrate complex and the product *P* are represented according to classical enzyme kinetics. The rate constants for component reactions are given above the arrows depicting forward and back reactions:

$$E + S \underset{k_1}{\overset{k_1}{\rightleftharpoons}} ES \underset{k_2}{\overset{k_2}{\rightleftharpoons}} E + P \tag{6.1}$$

$$V_{max} = k_2 [ES] \tag{6.2}$$

$$K_m = (k_2 + k_{-1}) / k_1 \tag{6.3}$$

From Eqs. (6.2) and (6.3), it appears that a high V_{max} may be obligatorily linked to a high value of K_m (i.e., low affinity), and likewise, high affinity may be obligatorily linked with low velocity. Extrapolating these equations to membrane transport processes, the same conclusion concerning the velocity and affinity tradeoff may apply. This may explain why most organisms that have been studied typically possess both low- and high-affinity transporters for the same ion. These differ in affinity much more than the above case for *Aspergillus* where both transporters have μM K_m values. In higher plants, low-affinity transporters for K^+, NH_4^+, NO_3^- and amino acids appear not to saturate even at high millimolar concentrations (Kochian and Lucas 1982; Reinhold and Kaplan 1984; Siddiqi et al. 1990; Wang et al. 1993). It might be argued that a high-affinity (low-capacity) transport protein could be produced in greater abundance to increase the total capacity, but clearly this would be costly in terms of N that is often limiting in natural systems. It may be more economic to make two classes of proteins.

6.2.5 High-Affinity and Low-Affinity Transporters

As stated above, kinetic analyses of nutrient uptake, conducted over a wide range of concentrations, has revealed the contribution of both high- and low-affinity transporters in most organisms (see Epstein 1976). These transport systems were originally referred to as systems I and II, respectively, but more recently they have been named according to their relative affinities for their substrates, as high-affinity transport systems (HATS) and low-affinity systems (LATS), respectively. HATS and LATS are thought to operate at low (<200 μM) and high (>1 mM) external ion concentrations, respectively. Comparative studies of the effects of altered nutrient provision upon the activity of high-affinity and low-affinity transporters appear to indicate that in general low-affinity transporters are relatively unresponsive. Thus, in the cases of increased provision of K^+, NO_3^- or NH_4^+, while high-affinity transporters were rapidly down-regulated, the low-affinity transporters showed little change in rates (Glass and Dunlop 1978; Siddiqi et al. 1990; Wang et al. 1993). This is not to suggest that low-affinity transporters are not down-regulated in the long term; the study by Clement et al. (1978) provides evidence to this effect for NO_3^- uptake. Rather, it indicates that the high-affinity transporters, which are probably of greater significance to nutrient acquisition under natural conditions, are subject to a much more rapid adjustment in response to altered availability of particular nutrients. Given the high concentrations of soil nutrients in Western-styled agricultural systems, and the potential to down-regulate high-affinity transporters, it is probable that the low-affinity transporters may serve to absorb nutrients as roots invade fresh soils. As soil nutrients are progressively depleted around the relatively older regions of the root, up-regulation of high-affinity transporters enables the plant to scavenge remaining resources. In densely competitive, established ecosystems (e.g., in permanent pastures and forests), low-affinity transporters may be irrelevant.

The time courses of down-regulation of influx following re-supply of a particular nutrient that has been withheld vary according to the external concentration of that nutrient, while rates of up-regulation vary according to the rapidity of tissue dilution of the nutrient by continued growth under nutrient-impoverished conditions. Down-regulation of HATS influx for NO_3^- or NH_4^+ may reduce influx by as much as 80% within 4–6 h after re-supply of NO_3^- or NH_4^+, respectively (Wang et al. 1993; Rawat et al. 1999; Zhuo et al. 1999). In a study of K^+ influx in barley roots, V_{max} and K_m values were significantly changed within 3–6 h after re-supplying K^+ to the K^+-depleted roots (Table 6.2). There are also species differences to consider. While the increased HATS influx of NO_3^- induced by first provision of NO_3^- generally requires 3–12 h to achieve maximum induction in fast-growing species such as barley, in the more slow-growing *Picea glauca* induction required 3 days (Kronzucker et al. 1995). Down-regulation of LATS requires substantially longer time intervals (Glass and Dunlop 1978; Kochian and Lucas 1982; Siddiqi et al. 1990). In the case of NH_4^+, down-regulation of LATS following re-supply of NH_4^+ may not occur at all (Wang et al. 1993; Rawat et al. 1999; Cerezo et al. 2001). This may explain the potential toxicity associated with provision of NH_4^+ when this nutrient is supplied as the sole source of N at concentrations above 1 mM (Glass et al. 2001; Britto et al. 2001).

Table 6.2. Changes of root K^+ concentration, V_{max} values and K_m values for K^+ influx after re-supplying K^+ to K-deprived barley roots. (Glass 1976)

Duration of K^+ re-supply (h)	Root $[K^+]$ ($\mu mol\ g^{-1}$ FW)	V_{max} ($\mu mol\ g^{-1}$ FW h^{-1})	K_m (mM)
0	26	7.9	0.036
3	65	6.2	0.054
6	94	3.9	0.66
12	131	1.9	0.133

6.3 From Black Box Kinetic Models of Regulation to Better Defined Mechanisms of Transport Regulation

A logical development of these physiological experiments might have been the isolation of the putative transport proteins, and evaluation of the hypothesis that increased/decreased transport resulted from changes of expression levels of these proteins. Indeed, several groups made valiant efforts in this direction (McClure et al. 1987; Fernando et al. 1992), to no avail. Instead, developments in molecular biology enabled workers to bypass the protein and proceed directly to the gene level. The following sections will deal with these molecular developments insofar as they may provide insights into the mechanisms responsible for regulating ion fluxes and maintaining ionic/nutrient homeostasis.

6.3.1 Molecular Basis of Nutrient Homeostasis

The identification of genes encoding plant ion transporters has relied on three basic strategies:

1. By using existing microbial transport mutants (commonly *Saccharomyces cerevisiae* mutants) incapable of growing on low levels of a particular inorganic nutrient to identify plant genes that are capable of rendering the microbial mutants competent to grow at low levels of that nutrient. Typically, the mutant strain is transformed, using plant cDNA libraries, and the transformants are screened on media that would not permit growth of the mutant. This strategy was effectively used to isolate plant genes encoding K^+, P_i, SO_4^{2-} and NH_4^+ transporters (see, e.g., Schachtman and Schroeder 1994).
2. By searching higher plant databases such as expressed sequence tags (ESTs) for sequences corresponding to conserved sequences of existing microbial ion transporters, it has been possible to isolate the higher plant homologues. For example, this method was successfully employed to characterize putative high-affinity K^+ transporters (Kim et al. 1998). The cloned genes can then be tested for function using the microbial mutants as described in point 1. Alternatively, oocytes of the African toad (*Xenopus laevis*) may be injected with mRNA corresponding to the cloned plant gene, and the oocytes tested for transport or electrophysiological activity in response to provision of the appropriate inorganic ion (Tsay et al. 1993).
3. By selecting for mutants that are capable of surviving treatment with toxic analogs of transported ions – for example, ClO_3^- has been used as a NO_3^- analog, and among microbes, methylamine (an ammonium analog) has been used effectively to isolate NH_4^+ transport mutants of the Mep family (Dubois and Grenot 1979).

In the following sections, examples of all three methods of cloning genes encoding ion transporters for different nutrients will be discussed, as well as what is known concerning the mechanisms responsible for regulating the expression of these genes. One issue becomes abundantly clear as a result of the application of the recent developments in molecular biology. The kinetic characterization of membrane transporters for various nutrients led many physiologists to assume that there might be two, or perhaps at most three transporters for each nutrient. Therefore, following the first successful cloning of genes encoding nitrate, potassium and other ion transporters, many naively considered that the challenge of molecular characterization might be all but met. Rather, molecular methods have opened a veritable Pandora's box. There are as many as 11 genes encoding nitrate transporters in *A. thaliana*, and all are expressed! Today, the task of identifying the functions of so many distinct transporters represents a major challenge.

6.3.1.1 Potassium

Determinations of cytosolic and vacuolar [K^+] by use of compartmental analysis (Memon et al. 1985) or by K^+-specific microelectrodes (Walker et al. 1996), and the responses of these compartments to K^+ deprivation have revealed that cytosolic K^+ is held constant at the expense of vacuolar K^+. As discussed in Fernando et al. (1992), the first responses of plant roots to K^+ deprivation is the net transfer of K^+ from the vacuole to the cytosol, and then to the shoot via the xylem. In addition, there is increased capacity for high-affinity transport at the plasma membrane. As K^+ stress intensifies, there is a reduction of the transfer of K^+ from roots to shoots. Thus, homeostasis is achieved through changes in K^+ transport at the root plasma membrane, the tonoplast, and the stele. While these responses refer specifically to the availability of potassium, they apply to other nutrients, too; these general responses are summarized in Fig. 6.1.

Using strains of *S. cereviciae* defective in K^+ absorption, genes encoding putative low-affinity K^+ channels, *KAT1* and *AKT1*, were cloned and characterized from *Arabidopsis thaliana* (Anderson et al. 1992; Sentenac et al. 1992) by transforming the defective strain with *Arabidopsis* cDNA libraries. Transformants were then screened on low-K^+ media. The predicted protein sequences of these plant genes are homologous with insect and mammalian Shaker channels, which are thought to be made up of four subunits, each consisting of six transmembrane segments. Using *Brassica napus*, Lagarde et al. (1996) reported that changes in tissue K^+ status failed to affect *AKT1* mRNA abundance. This observation confirms earlier physiological experiments that demonstrated a lack of down-regulation of low-affinity K^+ transport by roots (see above). One approach to confirming the functional role of the *AKT1* transporter was reported by Hirsch et al. (1998) using T-DNA insertional mutants of *A.*

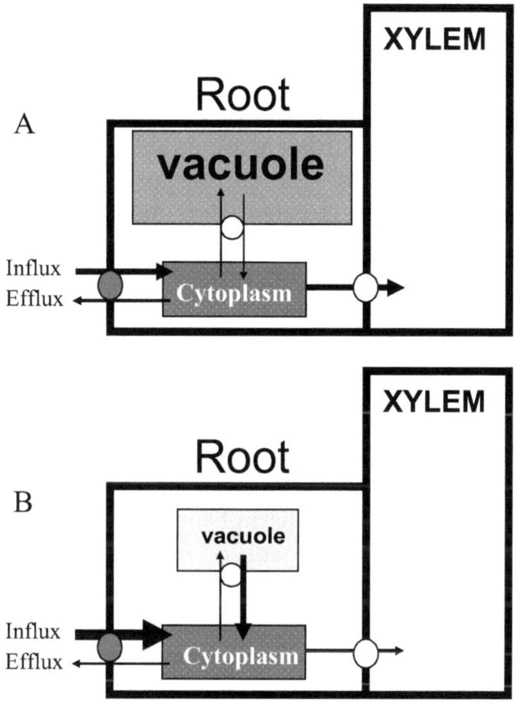

Fig. 6.1. A Replete plant root with vacuole filled with reserve nutrients and cytoplasm at prescribed concentration. The extent of individual fluxes is represented by the *thickness* of the *arrows*, while the *box size* for a particular compartment (e.g., vacuole) reflects the quantity of nutrients stored in that compartment. Note that ion influx, efflux and flux to the stele are significant, and that there is only a modest net exchange across the tonoplast. **B** Nutrient-deprived root shows a large increase in influx, while efflux and flux to the stele and to the vacuole are much reduced. There is substantial net transfer from the vacuole to the cytoplasm, with corresponding decline in vacuolar ion concentration, while this transfer sustains cytoplasmic ion concentration

thaliana, disrupted in the *AKT1* gene. Considering that the *AKT1* gene was thought to encode a low-affinity channel-like transporter, the demonstration that $^{86}Rb^+$ (a K^+ analog) influx at Rb^+ concentrations as low as 10 μM was also reduced in the mutant was unexpected. It was concluded that the transport system encoded by the *AKT1* gene might mediate what has previously been referred to as high-affinity uptake of K^+, in addition to low-affinity K^+ transport. However, it is also possible that the methodology used in these experiments caused suppression of high-affinity Rb^+ influx. Hirsch et al. (1998) grew their plants on MS media (50 mM K^+), and included 4 mM NH_4^+ (which is inhibitory to K^+ influx) in the influx medium. Furthermore, roots were excised from shoots and washed in K^+-free media for 17 h before measuring influx. The reported Rb^+ flux attributable to the *AKT1* transporter at 100 μM RbCl was calculated to be ~89 nmol g^{-1} FW h^{-1} in these experiments. This value represents ~1 % of the reported value of K^+ influx into roots of intact barley plants (Glass 1978). Therefore, if *AKT1* does represent a pathway for K^+ uptake at low external K^+ under the conditions of these experiments, it may represent only a very minor pathway under more natural conditions.

The technique that resulted in the isolation of *AKT1* and *KAT1* was used to isolate a putative high-affinity K^+ transporter gene (*HKT1*) from roots of

K^+-deprived wheat plants (Schachtman and Schroeder 1994). Unlike *AKT1*, this gene was up-regulated when K^+ was withheld from roots of barley and wheat, and down-regulated following re-supply (Wang et al. 1998). Although this gene appeared to be a likely candidate for the high-affinity K^+ transporter gene, further work established that this transporter functioned as a high-affinity K^+–Na^+ cotransporter at low $[Na^+]$, and as a low-affinity Na^+ transporter at elevated $[Na^+]$. Since most plants do not require Na^+ in order to transport K^+, the relevance of this transporter to K^+ transport in plants has been questioned (Walker et al. 1996). It may be that the increased abundance of *AKT1* mRNA following K^+ deprivation serves to increase the potential for the uptake of Na^+ as well as K^+. It is well documented that Na^+ may compensate (within the vacuole) for the limited availability of K^+ (Marschner 1986), and hence the uptake of both Rb^+ and Na^+ by barley roots is enhanced and reduced by K^+ deprivation and re-supply, respectively (Bange 1979). One method of evaluating the functional importance of a particular gene is to disrupt the gene by identifying mutants that have T-DNA insertions in the gene of interest, as was discussed above for the *AKT1* mutants. It will be interesting to investigate K^+ influx in such T-DNA insertional mutants, disrupted in the *HKT1* gene to determine the extent (if any) to which K^+ influx is modified.

Another class of putative high-affinity K^+ transporters that looks more promising than the *HKT1* family was identified by searching within the GenBank database of *A. thaliana* ESTs (Kim et al. 1998). A family of *KUP* genes was found to be highly homologous to the *KUP* and *HAK1* potassium transporters from *Escherichia coli* and *Schwanniomyces occidentalis*, respectively. *AtKUP1* and *AtKUP2* were capable of restoring K^+ transport in a mutant strain of *E. coli*, otherwise incapable of growing on low K^+, while overexpressing *AtKUP1* in suspension cells of *A. thaliana* gave rise to increased uptake of Rb^+, with an apparent K_m of ~22 μM. *AtKUP3* mRNA levels were strongly up-regulated in roots of plants grown on 2 mM K^+ compared to those grown on 40 mM K^+. This correlates with the pattern of repressible high-affinity K^+ influx (Fernando et al. 1992), suggesting that *AtKUP3* might encode this transport activity. In summary, despite the cloning of genes encoding K^+ channels and putative high-affinity transporters, little progress has been made in understanding the mechanisms responsible for the demonstrated physiological adaptations to K deprivation and/or excess.

6.3.1.2 Phosphate

Estimates of cytosolic P_i, generated by ^{31}P-NMR, demonstrated that during P deprivation, cytosolic P_i remained constant at the expense of vacuolar reserves (Lee et al. 1990). At the whole-plant level, Asher and Loneragan (1967) demonstrated that a wide range of plant species partitioned greater proportions of absorbed P to roots than to shoots under conditions of low available P provision. This is similar to the reduction of K^+ transport to the shoot referred to in the previous section. As was the case for K^+, the maintenance of a constant value for root cytosolic P_i must involve adjustments of fluxes at the plasma membrane, at the tonoplast, and to and from the stele.

The identification of an *A. thaliana EST*, showing significant sequence identity with the high-affinity phosphate transporters of *S. cereviciae*, *Neurospora crassa* and *Glomus versiforme*, led to the cloning of two putative high-affinity phosphate transporters (*AtPT1* and *AtPT2*) from *A. thaliana* (Muchhal et al. 1996; Smith et al. 1997). The proteins encoded by these genes, which were 78 % identical to each other, were expressed principally in roots, and expression levels increased in response to P starvation. Several more members of this family of transporters have subsequently been isolated from a variety of other higher plant species (Raghothama 1999). The availability of a yeast mutant lacking both the *PHO84* and *PHO89* (a Na^+/P_i cotransporter) genes has allowed the determination of the K_m for P_i of the tomato high-affinity P_i transporter, the value being 31 μM (Daram et al. 1998). It is worthwhile noting that although valuable qualitative data may be derived from heterologous expression systems such as for *Saccharomyces* or *Xenopus* oocytes, these are very different membrane systems than the native membranes in which these proteins are normally expressed. Therefore, some caution should be exercised in interpreting quantitative data derived from heterologous expression systems. Using cultured tobacco cells, for example, a K_m value of 3.1 μM was obtained for the *Arabidopsis PTH1* transporter (Mitsukawa et al. 1997). The high-affinity P_i transporters from a diverse group of higher plants conform to a general pattern of transcriptional regulation by feedback from plant P status (Mimura 1999). Expression of the tomato *LePT1* gene, encoding a high-affinity transporter, is localized in the root cap, the epidermis, and outer cortical cells (Muchhal and Raghothama 1999). Using antibodies raised against the *LePT1* protein, the same authors confirmed the anticipated increased level of protein expression associated with P deprivation, and demonstrated that the protein was localized in the plasma membrane. An interesting interaction between plant and mycorrhizal partners was demonstrated in a study of two high-affinity root P_i transporters of *Medicago truncatula* (Liu et al. 1998). While expression of the plant genes was responsive to P status, colonization of the roots by *G. versiforme* resulted in significant reduction of transcript abundance of both plant genes. Whether this result arose by direct signaling between the fungus and plant, or the simple effect of the normal down-regulation associated with elevated phosphate supply (in this case, provided by the fungus) is unclear.

A putative low-affinity P_i transporter, with a sequence that is conserved among species across all kingdoms, is predominantly expressed in green tissues, and transcript abundance was shown to be independent of P status (Daram et al. 1999). In roots, there was a small increase in mRNA in P-deprived plants. It is conjectured, therefore, that this transporter may be involved in internal P_i transfers.

6.3.1.3 Sulfate

Like other inorganic nutrients, the uptake of sulfate is subject to negative feedback control (Hart and Filner 1969; Smith 1975). Insofar as sulfur is typically absorbed in the form of sulfate (SO_4^{2-}), and is then reduced through sulfite (SO_3^{2-}) to sulfide (S^-) and subsequently incorporated into S-containing amino acids (cysteine, methionine) and the tripeptide glutathione (glutamyl cysteinyl glycine), there are numerous candidates for the role(s) of feedback inhibitors that might regulate SO_4^{2-} transport. Because it is a major storage and translocatable form of S, glutathione has been proposed as a likely candidate for this regulatory function (Herschbach and Rennenberg 1991; Lappartient and Touraine 1996).

The first higher plant SO_4^{2-} transporter was isolated by complementing a SO_4^{2-} transport-defective yeast mutant (YSD1) using cDNA libraries generated from roots of S-deprived *Stylosanthes hamata* (Smith et al. 1995) and *Hordeum vulgare* (Smith et al. 1997). Heterologous expression of these genes in the yeast mutant also allowed determination of K_m values for SO_4^{2-} uptake of 10, 11.2 and 100 μM for the *S. hamata* genes (*ShST1*, *ShST2* and *ShST3*, respectively), and of 6.8 μM for the *H. vulgare* gene (*HvST1*). Seven cDNAs encoding both high- and low-affinity SO_4^{2-} transporters have since been documented in *A. thaliana* (Saito 2000). Based upon their affinities for sulfate, the sulfate transporter genes have been classified into a high-affinity group expressed in roots (e.g., *ShST1* and *HvST1*), and lower-affinity transporters, e.g., *AST68* (from *Arabidopsis*) that are expressed in the vasculature of roots and shoots. While the former are probably responsible for absorption of soil solution SO_4^{2-}, the latter may function in loading and unloading xylem elements (Smith et al. 2000).

In earlier physiological studies using tobacco and other species (see, e.g., Herschbach and Rennenberg 1991), it was demonstrated that exogenous application of cysteine or glutathione (end products of S assimilation) reduced $^{35}SO_4^{2-}$ influx and xylem loading by 75 and 70%, respectively. Even though the inhibitory effect of cysteine (cys) was achieved at 1 μM, compared to 100 μM for glutathione (GSH), results for the simultaneous application of the inhibitor buthione sulfoximine (BSO), which blocks synthesis of GSH from cys, suggested that the cys effect was mediated via cys conversion to GSH. Based upon the rapidity of the GSH effect, it was suggested that GSH might act directly upon the SO_4^{2-} transporter (Herschbach and Rennenberg 1991). Transcript abundances of the high- and low-affinity transporter genes and sulfate uptake are increased by withholding SO_4^{2-}, and rapidly reduced following re-supply of SO_4^{2-} (Smith et al. 2000). In a study of SO_4^{2-} influx in *A.*

thaliana, Lappartient et al. (1999) reported that when S-starved roots were provided with SO_4^{2-} or cys, together with BSO, transcript abundance of *AST68* remained high, confirming that GSH, rather than SO_4^{2-}or cys, was responsible for regulating transcript abundance. Similar findings were reported by Vidmar et al. (1999), using barley. In addition to this negative effect of GSH, serving as a signal for plant S status, a positive effector, namely, *O*-acetylserine (OAS), has also been proposed (Smith et al. 1997). Provision of OAS (a precursor of cys) together with SO_4^{2-} to barley roots caused *HvST1* transcript abundance to increase, together with rates of SO_4^{2-} uptake and tissue levels of cys and GSH. Since *HvST1* is typically down-regulated by GSH, the OAS result suggests that OAS is capable of overriding the GSH signal.

Physiological studies of amino acid uptake by plant roots have demonstrated the existence of both high-affinity and low-affinity transporters (Reinhold and Kaplan 1984). Furthermore, in regions where soil concentrations of amino acids exceed other sources of N, it has been suggested that amino acids may represent the main source of N for plant growth (Keilland 1997). For example, roots of the alpine sedge *Kobresia* were found to absorb glycine faster than either NO_3^- or NH_4^+ (Raab et al. 1996). Measurements of soil solution nitrogen fractions revealed that amino acid concentrations were as high as 100 μM, and cysteine levels reached 15 μM late in the season. Thus, while inorganic sulfate is the primary source of sulfur for most plants, this element may also be absorbed in the form of sulfur amino acids. The existence of specific cysteine transporters in *S. cereviciae* (Olsen et al. 1999) and in *Synechocystsis* (Suginaka et al. 1998) provides further support for this argument.

6.3.1.4 Nitrogen

Based upon the large quantities of N required by plant species, in comparison to other essential elements (with the exceptions of C, H and O), it might be anticipated that under field conditions N would commonly limit plant growth, and that interspecies competition for this nutrient might be intense. This is generally the case, and it is certainly the case in the agricultural context that accounts for the fact that present-day global application of N fertilizer (~10^{11} kg/year) is roughly twice typical application rates for P or K (FAO statistic). As is the case for S, N exists in multiple forms in the environment, and root transporters reflect this multiplicity. Thus, transporters capable of absorbing nitrate, nitrite, ammonium, urea and amino acids have been physiologically defined. As has been demonstrated for other nutrients, both high- and low-affinity transporters have been characterized for NO_3^-, NH_4^+ and amino acids; in the case of NO_3^- influx, constitutive high-affinity, inducible high-affinity, and constitutive low-affinity transporters have been described in barley roots (Glass and Siddiqi 1995). Under natural conditions, NO_3^-, NH_4^+

and amino acids represent the major sources of N for plant growth (see Introduction and paragraph above). When plants simultaneously access more than one form of N, there would seem a priori to be a need to integrate the several potential signals of internal N status through some common intermediate, so as to properly regulate N uptake to match plant demand. Based upon this argument, Lee and Rudge (1986) proposed that amino acids such as glutamine or asparagine might serve this role. In the last decade, considerable progress has been achieved in cloning genes that encode N transporters. Indeed, so many genes have been identified that the next substantial challenge is to assign them all specific functions.

6.3.1.4.1 Nitrate

Kinetic analysis of NO_3^- influx indicates the presence of constitutive and inducible high-affinity transporters. In this regard, NO_3^- influx into roots is unique among all other nutrient transport systems, in that it is substantially increased or *induced* (by as much as 30-fold over uninduced (constitutive) values) by exposure to NO_3^- or NO_2^-. Typically, full induction requires 12–24 h, although the rapidity of induction depends upon external NO_3^- concentration (Siddiqi et al. 1989; Hole et al. 1990) and species identity. In slow-growing white spruce seedlings, full induction required 3 days (Kronzucker et al. 1995). Kinetic analysis of constitutive high-affinity NO_3^- influx in roots of barley and white spruce has revealed K_m values ranging from 7 to 20 μM (Lee and Drew 1986; Siddiqi et al. 1990; Aslam et al. 1992; Kronzucker et al. 1995). Following induction by NO_3^-, the K_m increased from 13 to 79 μM (Lee and Drew 1986; Siddiqi et al. 1990; Aslam et al. 1992; Kronzucker et al. 1995). This suggested that the inducible NO_3^- transport system was distinct from the constitutive transport system, on the basis that the K_m (affinity) of a particular transporter was essentially fixed and characteristic (but see below). However, an alternative explanation may be that the inducible transporters are expressed at very low abundance even prior to induction by NO_3^-. This would suggest that a single transport system mediating both constitutive and inducible high-affinity transport is up-regulated by exposure to NO_3^-. Given the plasticity of K_m values (see Table 6.2 for K^+ and Table 6.3 for NO_3^-), differences among measured K_m values may not represent strong evidence for discrete transport systems. Indeed, the case of NO_3^- transport in *Aspergillus nidulans* may illuminate this question. *A. nidulans* has two inducible high-affinity transporters that have been defined physiologically and genetically. When both genes are mutated, the fungus is completely unable to grow on or to absorb NO_3^-, demonstrating the absence of other NO_3^- transporters. Yet, wild-type mycelium shows a low level of transport activity when first exposed to NO_3^-, indicating that even before exposure to NO_3^-, these inducible high-affinity transporters are expressed at low abundance (Unkles et al. 2001).

Table 6.3. Kinetic constants for $^{13}NO_3^-$ influx prior to and after induction by two concentrations of NO_3^- after 1 or 4 days (from Siddiqi et al. 1990)

Duration of exposure to nitrate (days)	$[NO_3^-]$ (mM)	Root $[NO_3^-]$ (μmol g^{-1} FW)	K_m (μM)	V_{max} (μmol g^{-1} FW h^{-1})
0	0	Undetectable	20	0.34
1	0.1	40	79	9.4
4	0.1	76	45	3.6
4	10	97	30	1.7

At higher concentrations of NO_3^-, a class of low-affinity transporters becomes evident. Surprisingly, these show no evidence of substrate saturation, even at 50 mM external NO_3^- (Pace and McClure 1986; Siddiqi et al. 1990). Indeed, at first examination, plots of NO_3^- influx versus NO_3^- concentration from ~1 to 50 mM most closely resemble those anticipated for passive diffusion. Yet, thermodynamic evaluation of electrical potential differences has established that these low-affinity fluxes require active transport (Glass et al. 1992). Low-affinity NO_3^- influx is also inducible by NO_3^- in roots of *A. thaliana* (Okamoto et al., unpubl. data), whereas in barley roots it is constitutive (Siddiqi et al. 1990).

Among eukaryotic organisms, genes encoding putative high-affinity NO_3^- transporters – originally named *crnA* and *Nar* genes – were first isolated from *A. nidulans* and *Chlamydomonas reinhardtii*, respectively (Unkles et al. 1991; Quesada et al. 1994). By using degenerate primers based upon conserved sequences of these genes, higher plant genes were subsequently cloned from barley (Trueman et al. 1996; Vidmar et al. 2000a), *A. thaliana* (Zhuo et al. 1999; Filleur and Daniel-Vedele 1999), *Nicotiana plumbaginifolia* (Quesada et al. 1997) and soybean (Amarashinghe et al. 1998). In barley, as many as ten putative members of this *Nrt2* family of genes have been suggested (Trueman et al. 1996), while completion of the *A. thaliana* genome sequencing project has made it possible to identify seven members of the *Nrt2* family in this species. Some of these are preferentially expressed in roots, while others are primarily expressed in shoots. Localization studies using GUS fusion proteins have suggested that *NRT2.1* (the first characterized member of this family) is mainly expressed in the epidermis and cortex of *A. thaliana* roots (Vidmar et al., unpubl. data), while *NRT2.2* appears to be localized in epidermal cells bearing root hairs (Okamoto, unpubl. data). This area of research is moving so rapidly that in a relatively short period of time it may be possible to identify the functions of each member of the *NRT1* (see below) and *NRT2* family. A number of different approaches has been used in attempting to determine the individual functions of these genes. These include correlations between transcript abundance and physiological activity, localization of regions within the plant where the gene is expressed, and the selection of mutants with specific gene disruption.

Transcript abundance of the *NRT2.1* gene is highly correlated with high-affinity NO_3^- transport in barley, and in *A. thaliana* during the period following induction by NO_3^- (Vidmar et al. 2000a; Okamoto et al., unpubl. data), as shown in Figs. 6.2 and 6.3. Recently, a T-DNA insertional mutant of *A. thaliana* disrupted in the *Nrt2.1* and *Nrt2.2* genes have been shown to express only ~30 % of wild-type HATS, providing strong evidence of the importance of these two genes for the HATS in this species (Filleur et al. 2001). Also, expression studies using RT-PCR reveal that the *NRT2* family of genes contains two members (*NRT2.1* and *NRT2.2*) that are strongly induced by exposure to NO_3^- in roots of barley and *A. thaliana* (Okamoto et al., unpubl. data), while other members of the family are constitutively expressed. Constitutively expressed genes may encode transporters involved in the constitutive high-affinity NO_3^- transport observed when plants are first exposed to NO_3^-. They may also include genes that encode transporters that mediate internal NO_3^- fluxes, such as those across the tonoplast or from the root symplasm to the xylem. Because the *A. nidulans* system appears to possess only two NO_3^- transporters, it was relatively easy to generate mutants expressing only one of the *NRT2* genes at a time, and to characterize the role of each gene in NO_3^- transport. The large number of *NRT2* homologues in higher plants (possibly ten in barley and seven in *A. thaliana*) makes their characterization a much more formidable task. The clear difference between the multiple members of the *NRT2* family in, for example, *A. thaliana* and *H. vulgare*, and the two genes of *A. nidulans*

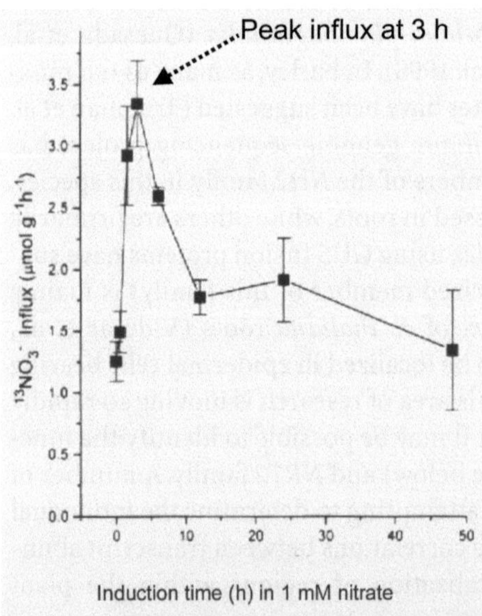

Fig. 6.2. $^{13}NO_3^-$ influx into roots of *Arabidopsis thaliana* previously grown on ammonium for 3 weeks, following induction with 1 mM NO_3^- for 48 h. Influx was measured at 100 µM NO_3^- to determine HATS activity (Zhuo et al. 1999). Note peak influx at 3 h

suggests that many of the multiple members are associated with internal transport of NO_3^- within the multicellular plant.

When NO_3^- is provided for a period of days, $^{13}NO_3^-$ influx and abundance of *NRT2* mRNA decline soon after peak induction, leading to lowered steady-state levels of both influx and *NRT2* mRNA. Experiments involving exogenous applications of amino acids with or without inhibitors of nitrate assimilation suggest that the amino acid glutamine (gln) is a major participant in this down-regulation of *NRT2* mRNA abundance and NO_3^- influx (Zhuo et al. 1999; Vidmar et al. 2000b). If this is indeed the case, then gln would represent an ideal "sensor" of N status, capable of integrating the signals arising from exogenous sources of NO_3^-, NO_2^-, NH_4^+, and gln, since they all must pass through the NO_3^- assimilation pathway leading to the formation of gln. Interestingly, many physiological studies have demonstrated that exogenously supplied glutamate (glu) and aspartate (asp) were much more potent down-regulators of NO_3^- influx than gln (Breteler and Arnozis 1985; Müller and Touraine 1992; Rodgers and Barneix 1993; Gessler et al. 1998; Vidmar et al. 2000b). The study by Vidmar et al. (2000b) revealed that this result is not at odds with a primary role for gln, because absorbed glu is rapidly converted to gln by the enzyme glutamine synthetase (GS). The next enzyme in this pathway (glutamate synthase) converts a molecule of gln, together with a molecule of α-ketoglutarate to two molecules of glu. Blocking this enzyme with the inhibitor azaserine increased cellular gln while decreasing glu. This treatment virtually eliminated expression of *NRT2.1* and NO_3^- influx in roots of *A. thaliana* and barley (Zhuo et al. 1999; Vidmar et al. 2000b), and provides con-

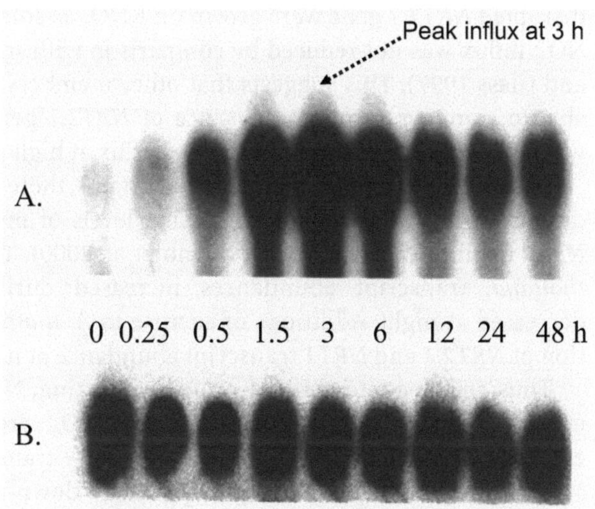

Fig. 6.3A, B. Transcript abundance of the *Arabidopsius thaliana NRT2.1* gene in roots of *Arabidopsis* plants, following provision of 1 mM nitrate to plants previously grown on ammonium for 3 weeks. **A** Transcript abundance; note peak abundance at 3 h. **B** Transcript abundance of the 18S gene (as control for the amount of RNA applied to the gel. f(Zhuo et al. 1999)

vincing evidence that it is gln, rather than glu, that regulates *NRT2.1* gene expression and NO_3^- influx.

By growing plants in the presence of chlorate (ClO_3^-), a toxic NO_3^- analogue, mutants defective in NO_3^- transport were isolated (Doddema and Telkamp 1979; Tsay et al. 1993), and a gene encoding a low-affinity NO_3^- transporter was cloned from *A. thaliana* (Tsay et al. 1993). This gene (originally named *CHL1*, now designated *NRT1.1*) is strongly induced by NO_3^- (Tsay et al. 1993), and expression in *Xenopus* oocytes gave physiological evidence that the gene encoded a low-affinity NO_3^- transporter (Tsay et al. 1993). Interestingly, the T-DNA disruption mutant corresponding to this gene has been demonstrated to show defective NO_3^- transport at low external [NO_3^-], suggesting that this gene may participate in high-affinity transport, as well as in low-affinity transport (R. Wang et al. 1998; Liu et al. 1999). This finding is perplexing in view of the fact that the *A. thaliana* genome contains at least seven members of the *NRT2* family, encoding putative high-affinity transporter genes. Even before completion of the *Arabidopsis* genome sequencing project, it was apparent that the *NRT1* family contained at least two members, one constitutive and one NO_3^--inducible (Lauter et al. 1996; Huang et al. 1999). It is now apparent that there are four such genes in *A. thaliana*, only one of which (*NRT1.1*) is NO_3^--inducible (Okamoto, unpubl. data).

The inducible *NRT1.1* gene, like its high-affinity counterparts (*NRT2.1* and *NRT2.2*) is down-regulated after peak levels of induction, as is $^{13}NO_3^-$ influx by the low-affinity transport system (Okamoto, unpubl. data). Details of the putative N pools responsible for this down-regulation of *NRT1.1* have not been worked out. The close correspondence between measured low-affinity NO_3^- fluxes and the pattern of *NRT1.1* expression during and following induction by NO_3^- suggests that low-affinity NO_3^- influx is largely controlled by this gene product, and that *NRT1.2*, *NRT1.3* and *NRT1.4* play minor roles in low-affinity root NO_3^- transport. Nevertheless, when plants possessing a T-DNA disrupted *NRT1.1* gene were grown on KNO_3 as sole source of N, low-affinity NO_3^- influx was not reduced by comparison with wild-type plants (Touraine and Glass 1997). This suggests that other members of the *NRT1* family were able to compensate for the absence of *NRT1.1* gene product. Interestingly, when plants were grown on NH_4NO_3, influx at high $[NO_3^-]_o$ was reduced.

In addition to regulation by tissue N status, there is evidence that carbohydrate supply may influence expression levels of members of the *NRT1* and *NRT2* families (Lejay et al. 1999; Ono et al. 2000). Thus, in tomato and in *A. thaliana*, transcript abundances increased during daylight hours and decreased at night. Additions of sucrose to *A. thaliana* prevented the reduction of *NRT2.1* and *NRT1* transcript abundance at night.

Thus, in the context of the ecological setting, N-deprived plants that are exposed to NO_3^- after a period without NO_3^- are able to rapidly induce expression of inducible high- and low-affinity transporter genes. As plant N status increases, accumulated gln appears to down-regulate gene expression, so that NO_3^- acquisition via the high-affinity systems is considerably reduced. At the same time, interactions with the carbohydrate supply will serve to fine-tune diurnal requirements for N transport. Nevertheless, if external [NO_3^-] is high (~1 mM or above), NO_3^- uptake will also occur via the low-affinity transporter system, which, in the case of *NRT1.1*, is also gradually subject to down-regulation. By these means, NO_3^- influx matches plant demand. The cycling of amino acids, notably gln (but also other amino acids), from shoot to root is probably responsible for integrating root uptake to match whole-plant demand (Cooper and Clarkson 1989; Schneider et al. 1996; Marschner et al. 1997; Gessler et al. 1998).

6.3.1.4.2 Ammonium

As stated in the Introduction, NH_4^+ may represent the main source of N for plant growth in many habitats. Even in agricultural soils where NO_3^- typically exceeds NH_4^+, the inhibitory effect of NH_4^+ on NO_3^- influx may reduce NO_3^- uptake. This inhibitory effect of NH_4^+ may operate at a number of levels. In the long term, newly absorbed NH_4^+ is rapidly converted within plant roots by the enzyme GS to gln, which is known to down-regulate *NRT2* gene expression (see above). Thus, through its conversion to gln, NH_4^+ may operate to block NO_3^- transport at the transcript level. However, Lee and Drew (1989) demonstrated that the provision of NH_4^+ to roots of barley plants actively absorbing NO_3^- inhibited NO_3^- influx within 3 min of application, a finding that could only be interpreted as a direct effect of NH_4^+ on the NO_3^- transporter. Kronzucker et al. (1999) have since demonstrated that NH_4^+ reduced NO_3^- influx and increased NO_3^- efflux in barley, the greater absolute effect being upon influx. This capacity to reduce NO_3^- uptake is widespread in nature (see Clarkson and Lüttge 1991). Thus, by both short-term and long-term effects, NH_4^+ suppresses NO_3^- uptake. These physiological studies are confirmed at the ecophysiological level; in a field study of N uptake by spruce and beech trees, conducted during 1994 and 1995, NO_3^- uptake was not detected in spruce, and detected on only one occasion in beech, despite the fact that soil [NO_3^-] was approximately six times (beech) and approximately ten times (spruce) higher than that of soil NH_4^+ concentration (Gessler et al. 1998). By contrast, NH_4^+ uptake by both species was observed in every field trial. The importance of NH_4^+ as a source of N has therefore probably been underestimated in the past. Nevertheless, high levels of NH_4^+ (when present as the sole source of N) are toxic to some species, while other species are relatively tolerant to NH_4^+ (Gerendas et al. 1997).

Influx measurements using $^{15}NH_4^+$ or $^{13}NH_4^+$ have established that two distinct categories of transporters (high- and low-affinity) contribute to NH_4^+ influx into plant roots (Ullrich et al. 1984; Wang et al. 1993; Kronzucker et al. 1996; Cerezo et al. 2001). While the high-affinity transporter displays saturable influx kinetics with respect to external NH_4^+, the low-affinity transport system was not saturated, even at 40 mM external NH_4^+ in rice (Wang et al. 1993) and 50 mM in white spruce and citrus roots (Kronzucker et al. 1996; Cerezo et al. 2001). Furthermore, whereas high-affinity NH_4^+ transport is subject to rapid regulation following perturbations of N supply to roots (Lee et al. 1992; Jackson and Volk 1992; Wang et al. 1993; Rawat et al. 1999), NH_4^+ influx via the low-affinity transport system remained high in rice, citrus and in *A. thaliana* (Wang et al. 1993; Rawat et al. 1999; Cerezo et al. 2001). Indeed, it has been suggested that the toxic effect of NH_4^+ in the mM concentration range may result from a failure to down-regulate influx, resulting in a steep rise in root respiration associated with active pumping out of NH_4^+. Thus, with high values of influx and efflux of NH_4^+, there is a futile cycling across the plasma membrane and high values of cytosolic NH_4^+ (Britto et al. 2001). Nevertheless, species that occupy habitats in which NH_4^+ predominates (e.g., paddy rice and many conifers) appear to be able to assimilate NH_4^+ as sole source of N without deleterious effects (Kronzucker et al. 1997).

Ninnemann et al. (1994) transformed a *Saccharomyces* double mutant (*mep1mep2*) that was incapable of normal growth on low [NH_4^+], with cDNA from *Arabidopsis*, and successfully cloned the first plant NH_4^+ transporter gene *AMT1* (now renamed *AtAMT1.1*). Since then, *AtAMT1.2* and *At AMT1.3*, members of the same family, have been cloned and studied (Gazzarrini et al. 1999). When N was re-supplied to N-deprived *A. thaliana* plants, *AMT1.1* transcript abundance declined rapidly (Rawat et al. 1999). However, when N was re-supplied in the presence of methionine sulfoximine (MSX), an inhibitor of the GS enzyme responsible for assimilating NH_4^+ into the amino acid gln, root [NH_4^+] increased 27-fold, and transcript abundance and $^{13}NH_4^+$ influx remained high. This result strongly suggests that *AMT1.1* expression is regulated by root gln, rather than by NH_4^+ (Rawat et al. 1999).

Two other members of the *AMT1* family of high-affinity NH_4^+ transporters have been cloned from *A. thaliana* (Gazzarrini et al. 1999) and from tomato (von Wiren et al. 2000). All three genes exhibit diurnal variation in abundance, consistent with a possible interaction with carbohydrate as well as N supply. Alternatively, the apparent carbohydrate effect may be explained as the indirect result of increased assimilation of N associated with increased availability of carbohydrate. This plentiful supply of carbohydrate might allow assimilation of amino acids to proceed without constraint, leading to a lowering of cytosolic amino acid concentrations. Such a sequence of events would, in turn, relieve the potential down-regulation of *AMT1* gene expression by glutamine. In tomato, *LeAMT1.3* was not detected in roots but was expressed in leaf tissue. While *LeAMT1.1* transcript remained at a low and constant level, *LeAMT1.2* showed highest expression level after the onset of the light period. By contrast, *LeAMT1.3* showed highest transcript levels in darkness (von Wiren et al. 2000). The leaf transporters may play a role in absorbing NH_4^+ from the xylem stream or in retrieving photorespiratory NH_4^+. In *A. thaliana*, *AMT1.1*, *AMT1.2* and *AMT1.3* were all expressed in roots, while *AMT1.1* and

AMT1.2 were also expressed in shoots. By expressing the *Arabidopsis AMT1* genes in yeast mutants defective in NH_4^+, it was possible to show that *AMT1.1* had a K_m of ~0.5 μM for ^{14}C-methylamine uptake, compared to values of ~40 μM for *AMT1.2* and *AMT1.3* (Gazzarrini et al. 1999).

As many as five members of the *A. thaliana AMT1* family have now been described (von Wiren, unpubl. data). High levels of expression of *AMT1.1*, *AMT1.2* and *AMT1.3* in roots of *A. thaliana* suggest that all three genes contribute to NH_4^+ uptake (Gazzarrini et al. 1999). Another approach to resolving the functional role of many different genes is the use of T-DNA insertional mutants, as was described above for the *AKT1* gene mutant. Insertional mutants disrupted in *AMT1.1* have been obtained from *A. thaliana* (Kaiser et al., unpubl. data). $^{13}NH_4^+$ influx analyses reveal that the mutant absorbs NH_4^+ at roughly 80 % of the rate of wild-type plants, indicating that other members of the *AMT1* family, or even other gene families are able to compensate for this defect.

Some success in localizing the *AMT1.1* gene product at the subcellular level has been achieved by use of antibodies generated against the *AMT1* protein. It has been possible to demonstrate that the *AMT1* protein is localized in the plasma membrane fraction of *A. thaliana* roots (von Wiren, unpubl. data). The signal was much weaker in plasma membrane preparations from leaves. By contrast, the antibodies raised against the *AMT1.3* protein reacted with plasma membrane fractions from leaves but not roots.

6.4 Summary

The last decade has witnessed spectacular progress in the cloning of genes involved in the uptake of the major inorganic nutrients. In virtually every case investigated, the elusive gene encoding transporter proteins for a particular nutrient and/or particular system (high- or low-affinity) proved to be but one member of a gene family expressed in both roots and shoots. Universally, it has been demonstrated that responses to perturbation of nutrient supply, involving up-/down-regulation of transport across the plasma membranes of root epidermal and cortical cells, are correlated with up-/down-regulation of the transcript abundance of some, but not necessarily all members of the gene family. Questions that await resolution include what unique or collective functions are performed by each member of a family of genes, and how are these functions integrated to satisfy whole-plant demands within the context of a constantly varying environment – this will necessitate concerted efforts, using physiological, biochemical and molecular approaches. Having bypassed the transporter proteins in moving directly to the gene level, there is an evident need to return to this level of biochemistry to answer questions concerning the regulation of transport function arising from post-translational modification of protein function. Evidence from physiological data certainly suggests that such effects may also participate in controlling transporter activity (Rawat et al. 1999; Fraisier et al. 2000; Vidmar et al. 2000b).

References

Amarasinghe BHRR, Debruxelles GL, Braddon M, Onyeocha I, Forde BG, Udvardi MK (1998) Regulation of *GMNRT2* expression and nitrate transport activity in soybean (*Glycine max*). Planta 206:44–52

Anderson JA, Huprikar SS, Kochian LV, Lucas WJ, Gaber RF (1992) Functional expression of a probable *Arabidopsis thaliana* potassium channel in *Saccharomyces cerevisiae*. Proc Natl Acad Sci USA 89:3736–3740

Asher CH, Loneragan JF (1967) Response of plants to phosphate concentration in solution culture. I. Growth and phosphorus content. Soil Sci 103:225–233

Aslam M, Travis RL, Huffaker RC (1992) Comparative kinetics and reciprocal inhibition of nitrate and nitrite uptake in roots of uninduced and induced barley (*Hordeum vulgare* L.) seedlings. Plant Physiol 99:1124–1133

Bange GGJ (1979) Comparison of compartmental analysis for Rb^+ and Na^+ in low and high-salt barley roots. Physiol Plant 46:179–186

Barber DA (1972) 'Dual isotherms' for the absorption of ions by plant cells. New Phytol 71:255–262

Barber SA (1995) Soil nutrient bioavailability: a mechanistic approach. Wiley, New York

Borkert CM, Barber SA (1983) Effect of supplying P to a portion of the soybean root system on root growth and P uptake kinetics. J Plant Nutr 6:895–910

Breteler H, Arnozis PA (1985) Effect of amino compounds on nitrate utilization by roots of dwarf bean. Phytochemistry 24:653–657

Brezeale JF (1906) The relation of sodium to potassium in soil and solution cultures. J Am Chem Soc 28:1013–1025

Britto DT, Siddiqi MY, Glass ADM, Kronzucker HJ (2001) Futile membrane ion cycling: a new cellular hypothesis to explain ammonium toxicity in plants. Proc Natl Acad Sci USA 98:4255–4258

Cartwright B (1972) The effect of phosphorus deficiency on the kinetics of phosphate absorption by sterile excised barley roots and some factors affecting the ion uptake efficiency of roots. Soil Sci Plant Anal 4:313–322

Cerezo M, Tillard P, Gojon A, Primo-Millo E, Garcia-Agustin P (2001) Characterization and regulation of ammonium transport systems in *Citrus* plants. Planta 214:97–105

Clarkson DT (1985) Factors affecting mineral nutrient acquisition by plants. Annu Rev Plant Physiol 36:77–115

Clarkson DT, Lüttge U (1991) Mineral nutrition: inducible and repressible nutrient transport systems. Progress in Botany 52. Springer, Berlin Heidelberg New York, pp 61–83

Clement CR, Hooper MM, Jones LHP (1978) The uptake of nitrate by *Lolium perenne* from flowing nutrient solutions. I. Effect of NO_3^- concentration. J Exp Bot 29:453–464

Cooper HD, Clarkson DT (1989) Cycling of amino-nitrogen and other nutrients between shoots and roots in cereals – a possible mechanism integrating shoot and root in the regulation of nutrient uptake. J Exp Bot 40:753–762

Cram WJ (1976) Negative feedback regulation of transport in cells. The maintenance of turgor, volume and nutrient supply. In: Lüttge U, Pitman MG (eds) Encyclopedia of plant physiology. New Series, IIA. Springer, Berlin Heidelberg New York, pp 284–316

Daram P, Brunner S, Persson BL, Amrhein N, Bucher M (1998) Functional analysis and cell-specific expression of a phosphate transporter from tomato. Planta 206:225–233
Daram P, Brunner S, Rausch C, Steiner C, Amrhein N, Bucher M (1999) *Pht2;1* encodes a low-affinity phosphate transporter from *Arabidopsis*. Plant Cell 11:2153–2166
Doddema H, Telkamp GP (1979) Uptake of nitrate by mutants of *Arabidopsis thaliana*, disturbed in uptake or reduction of nitrate. II. Kinetics. Physiol Plant 45:332–338
Drew MC, Saker LR (1984) Uptake and long-distance transport of phosphate, potassium and chloride in relation to internal ion concentrations in barley: evidence of non-allosteric regulation. Planta 160:500–507
Dubois E, Grenson M (1979) Methylamine/ammonia uptake systems in *Saccharomyces ceréviciae*: multiplicity and regulation. Mol Gen Genet 175:67–76
Epstein E (1976) Kinetics of ion transport and the carrier concept In: Lüttge U, Pitman MG (eds) Encyclopedia of plant physiology. New Series, IIB. Springer, Berlin Heidelberg New York, pp 70–94
Fernando M, Mehrokc J, Glass ADM (1992) De novo synthesis of plasma membrane and tonoplast polypeptides of barley roots during short-term K^+ deprivation. Plant Physiol 100:1269–1276
Filleur S, Daniel-Vedele F (1999) Expression analysis of a high-affinity nitrate transporter isolated from *Arabidopsis thaliana* by differential display. Planta 207:461–469
Filleur S, Dorbe MF, Cerezo M, Orsel M, Granier F, Gojon A, Daniel-Vedele F (2001) An *Arabidopsis* T-DNA mutant affected in Nrt2 genes is impaired in nitrate uptake. FEBS Lett 489:220–224
Fraisier V, Gojon A, Tillard P, Daniel-Vedele F (2000) Constitutive expression of a putative high-affinity nitrate transporter in *Nicotiana plumbaginifolia*: evidence for post-transcriptional regulation by a reduced nitrogen source. Plant J 23:489–496
Gazzarrini S, Lejay T, Gojon A, Ninnemann O, Frommer WB, von Wiren N (1999) Three functional transporters for constitutive, diurnally regulated, and starvation-induced uptake of ammonium into *Arabidopsis* roots. Plant Cell 11:937–947
Gerendas J, Zhu ZJ, Bendixen R, Ratcliffe RG, Sattelmacher B (1997) Physiological and biochemical processes related to ammonium toxicity in higher plants. Z Pflanzenernähr Bodenkd 160:239–251
Gessler A, Schneider S, Von Sengbusch D, Weber P, Hanemann U, Huber C, Rothe A, Kreutzer K, Rennenberg H (1998) Field and laboratory experiments on net uptake of nitrate and ammonium by the roots of spruce (*Picea abies*)and beech (*Fagus sylvatica*) trees. New Phytol 138:275–285
Glass ADM (1976) The regulation of K^+ influx into barley: an allosteric model. Plant Physiol 58:33–37
Glass ADM (1978) The regulation of K^+ influx into intact roots of barley (*Hordeum vulgare* (L.) cv. Conquest) by internal K^+. Can J Bot 56:1759–1764
Glass ADM (1983) Regulation of ion transport. Annu Rev Plant Physiol 34:311–326
Glass ADM, Dunlop J (1978) The influence of potassium content on the kinetics of potassium influx into excised ryegrass and barley roots. Planta 141:117–119
Glass ADM, Siddiqi MY (1984) The control of nutrient uptake rates in relation to the inorganic composition of plants. Adv Plant Nutr 1:103–147
Glass ADM, Siddiqi MY (1995) Nitrogen absorption by plant roots. In: Srivastava HS, Singh RP (eds) Nitrogen nutrition in higher plants. Associated Publ Co, New Delhi, pp 21–56
Glass ADM, Shaff JE, Kochian LV (1992) Studies of nitrate uptake in barley. IV. Electrophysiology. Plant Physiol 99:456–463
Glass ADM, Britto DT, Kaiser BN, Kronzucker HJ, Kumar A, Okamoto M, Rawat SR, Siddiqi MY, Vidmar JJ, Zhuo D (2001) Nitrogen transport in plants, with an emphasis on the regulation of fluxes to match plant demand. Pflanzen Boden 164:199–207

Gutschick VP, BassiriRad H (2003) Extreme events as shaping physiology, ecology, and evolution of plants: toward a unified definition and evaluation of their consequences. New Phytol 160:21–42

Hart JW, Filner P (1969) Regulation of sulfate uptake by amino acids in cultured tobacco cells. Plant Physiol 44:1253–1259

Herschbach C, Rennenberg H (1991) Influence of glutathione (GSH) on sulfate influx, xylem loading and exudation in excised tobacco roots. J Exp Bot 42:1021–1029

Hirsch RE, Lewis BD, Spalding EP, Sussman MR (1998) A role for the AKT1 potassium channel in plant nutrition. Science 280:918–921

Hole DJ, Emran A, Farres Y, Drew MC (1990) Induction of nitrate transport in maize roots and kinetics of influx measured with nitrogen-13. Plant Physiol 93:642–647

Huang NC, Liu KH, Lo HJ, Tsay YF (1999) Cloning and functional characterization of an *Arabidopsis* nitrate transporter gene that encodes a constitutive component of low-affinity uptake. Plant Cell 11:1381–1392

Jackson RB, Caldwell MM (1993) The scale of nutrient heterogeneity around individual plants and its quantification with geostatics. Ecology 74:612–614

Jackson WA, Volk RJ (1992) Nitrate and ammonium uptake by maize: adaptation during relief from nitrogen suppression. New Phytol 122:439–446

Kielland K (1994). Amino acid absorption by artic plants: implications for plant nutrition and nitrogen cycling. Ecology 75:2373–2383

Kielland K (1995) Landscape patterns of free amino acids in arctic tundra soils. Biogeochemistry 31:85–98

Kielland K (1997) Role of free amino acids in the nitrogen economy of arctic cryptogams. Ecoscience 4:75–79

Kim EJ, Kwak JM, Uozumi N, Schroeder JI (1998) *ATKUP1* – an *Arabidopsis* gene encoding high-affinity potassium transport activity. Plant Cell 10:51–62

Kochian LV, Lucas WJ (1982) Potassium transport in corn *Zea mays* 1. Resolution of kinetics into a saturable and linear component. Plant Physiol 70:1723–1731

Kronzucker HJ, Siddiqi MY, Glass ADM (1995) Kinetics of NO_3^- influx in spruce. Plant Physiol 109:319–326

Kronzucker HJ, Siddiqi MY, Glass ADM (1996) Kinetics of NH_4^+ influx in spruce. Plant Physiol 110:773–779

Kronzucker HJ, Siddiqi MY, Glass ADM (1997) Root discrimination against soil nitrate and the ecology of forest succession. Nature 385:59–61

Kronzucker HJ, Glass ADM, Siddiqi MY (1999) Inhibition of nitrate uptake by ammonium in barley. Analysis of component fluxes. Plant Physiol 120:283–291

Lagarde D, Basset M, Lepetit M, Conejero G, Gaymard F, Astruc S, Grignon C (1996) Tissue-specific expression of *Arabidopsis* AKT1 gene is consistent with a role in K^+ nutrition. Plant J 9:195–203

Lappartient AG, Touraine B (1996) Demand-driven control of root ATP sulfurylase activity and SO_4^{2-} uptake in intact canola – the role of phloem-translocated glutathione. Plant Physiol 111:147–157

Lappartient AG, Vidmar JJ, Leustek T, Glass ADM, Touraine B (1999) Inter-organ signaling in plants: regulation of ATP sulfurylase and sulfate transporter genes expression in roots mediated by phloem-translocated compound. Plant J 18:89–95

Lauter F-R, Ninnemann O, Bucher M, Riesmeier J, Frommer WB (1996) Preferential expression of an ammonium transporter and of two putative nitrate transporters in root hairs of tomato. Proc Natl Acad Sci USA 93:8139–8144

Lee RB (1982) Selectivity and kinetics of ion uptake by barley plants following nutrient deficiency. Ann Bot 50:429–449

Lee RB, Drew MC (1986) Nitrogen-13 studies of nitrate fluxes in barley roots. II. Effect of plant N-status on the kinetic parameters of nitrate influx. J Exp Bot 185:1768–1779

Lee RB, Drew MC (1989) Rapid, reversible inhibition of nitrate influx in barley by ammonium. J Exp Bot 40:741–752
Lee RB, Rudge KA (1986) Effects of nitrogen deficiency on the absorption of nitrate and ammonium by barley plants. Ann Bot 57:471–486
Lee RB, Ratcliffe RG, Southon TE (1990) Phosphorus-31 NMR measurements of the cytoplasmic and vacuolar inorganic phosphate content of mature maize roots: relationships with phosphorus status and phosphate fluxes. J Exp Bot 41:1063–1078
Lee RB, Purves JV, Ratcliffe RG, Saker LR (1992) Nitrogen assimilation and the control of ammonium and nitrate absorption by maize roots. J Exp Bot 43:1385–1396
Lefebvre DD (1980) The regulation of phosphate uptake by barley plants. MSc Thesis, University of British Columbia, Vancouver
Lejay L, Tillard P, Lepetit M, Olive FD, Filleur S, Daniel-Vedele F, Gojon A (1999) Molecular and functional regulation of two NO_3^- uptake systems by N- and C-status of *Arabidopsis* plants. Plant J 18:509–519
Liu H, Trieu AT, Blaylock LA, Harrison MJ (1998) Cloning and characterization of two phosphate transporters from *Medicago truncatula* roots – regulation in response to phosphate and to colonization by arbuscular mycorrhizal (AM) fungi. Mol Plant-Microbe Interact 11:14–22
Liu KH, Huang CY, Tsay YF (1999) *CHL1* is a dual-affinity nitrate transporter of *Arabidopsis* involved in multiple phases of nitrate uptake. Plant Cell 11:865–874
Marschner H (1986) Mineral nutrition of higher plants. Academic Press, London
Marschner H, Kirkby EA, Engels C (1997) Importance of cycling and recycling of mineral nutrients within plants for growth and development. Bot Acta 110:65–273
McClure PR, Ohmolt TE, Pace GM, Bouthyette P-Y (1987) Nitrate-induced changes in protein synthesis and translocation of RNA in maize roots. Plant Physiol 84:52–57
Memon AR, Saccomani M, Glass ADM (1985) Efficiency of potassium utilization by barley varieties: the role of subcellular compartmentation. J Exp Bot 36:1860–1876
Mimura T (1999) Regulation of phosphate transport and homeostasis in plant cells. Int Rev Cytol 191:149–200
Min X, Siddiqi MY, Guy RD, Glass ADM, Kronzucker J (1999) A comparative kinetic analysis of nitrate and ammonium influx in two early-successional tree species of temperate and boreal forest ecosystems. Plant Cell Environ 22:821–830
Mitsukawa N, Okumura S, Shirano Y, Sato S, Kato T, Harashima S, Shibata D (1997) Overexpression of an *Arabidopsis thaliana* high-affinity phosphate transporter gene in tobacco cultured cells enhances growth under phosphate-limited conditions. Proc Natl Acad Sci USA 94:7098–7102
Muchhal US, Raghothama KG (1999) Transcriptional regulation of plant phosphate transporters. Proc Natl Acad Sci USA 96:5868–5872
Muchhal US, Pardo JM, Raghothama KG (1996) Phosphate transporters from the higher plant *Arabidopsis thaliana*. Proc Natl Acad Sci USA 93:10519–10523
Müller B, Touraine B (1992) Inhibition of NO_3^- uptake by various phloem-translocated amino acids in soybean seedlings. J Exp Bot 43:617–623
Ninnemann O, Jauniaux J-C, Frommer WB (1994) Identification of a high affinity NH_4^+ transporter from plants. EMBO J 13:3464–3471
Olsen L, Regenberg B, Gjermansen C Kielland-Brandt MC, Hansen J (1999) Cysteine uptake by *Saccharomyces cerevisiae* is accomplished by multiple permeases. Curr Genet 35:609–617
Ono F, Frommer WB, von Wiren N (2000) Coordinated diurnal regulation of low- and high-affinity nitrate transporters in tomato. Plant Biol 2:17–23
Pace GM, McClure PR (1986) Comparison of nitrate uptake kinetic parameters across maize inbred lines. J Plant Nutr 9:1095–1111
Pitman MG, Cram WJ (1977) Regulation of ion content in whole plants. In: Jennings DH (ed) Integration of activity in the higher plant. SEB Symp 31, pp 391–424

Quesada A, Galvan A, Fernandez E (1994) Identification of nitrate transporter genes in *Chlamydomonas reinhardtii*. Plant J 5:407–419

Quesada A, Krapp A, Trueman LJ, Danielvedele F, Fernandez E, Forde BG, Caboche M (1997) PCR-identification of a *Nicotiana plumbaginifolia* cDNA homologous to the high-affinity nitrate transporters of the crnA family. Plant Mol Biol 34:265–274

Raab TK, Lipson DA, Monson RK (1996) Non mycorrhizal uptake of amino acids by roots of the alpine sedge *Kobresia myosuroides* – implications for the alpine nitrogen cycle. Oecologia 108:488–494

Raghothama KG (1999) Phosphate acquisition. Annu Rev Plant Physiol Plant Mol Biol 50:665–693

Rawat SR, Silim SN, Kronzucker HJ, Siddiqi MY, Glass ADM (1999) *AtAMT1* gene expression and NH_4^+ uptake in roots of *Arabidopsis thaliana*: evidence for regulation by root glutamine levels. Plant J 19:143–152

Reinhold L, Kaplan A (1984) Membrane transport of sugars and amino acids. Annu Rev Plant Physiol 35:45–83

Reisenauer HM (1966) Mineral nutrients in soil solution. In: Altman PL, Dittmer DS (eds) Environmental biology. Fed Am Soc Exp Biol, Bethesda, pp 507–508

Robinson D (2001) Root proliferation, nitrate inflow and their carbon costs during nitrogen capture by competing plants in patchy soil. Plant Soil 232:41–50

Rodgers CO, Barneix AJ (1993) The effect of amino acids and amides on the regulation of nitrate uptake by wheat seedlings. J Plant Nutr 16:337–348

Saito K (2000) Regulation of sulfate transport and synthesis of sulfur-containing amino acids. Curr Opin Plant Biol 3:188–195

Schachtman DP, Schroeder JI (1994) Structure and transport mechanism of a high-affinity potassium uptake transporter from higher plants. Nature 370:655–658

Schneider S, Gessler A, Weber P, von Sengbusch D, Hanemann U, Rennenberg H (1996) Soluble N compounds in trees exposed to high loads of N: a comparison of spruce (*Picea abies*) and beech (*Fagus sylvatica*) grown under field conditions. New Phytol 134:103–114

Sentenac H, Bonneaud N, Minet M, Lacroute F, Salmon JM, Gaymard F, Grignon C (1992) Cloning and expression of a plant potassium ion transport system. Science 256:663–665

Siddiqi MY, Glass ADM, Ruth TJ, Fernando M (1989) Studies of the regulation of nitrate influx by barley seedlings using $^{13}NO_3^-$. Plant Physiol 90:806–813

Siddiqi MY, Glass ADM, Ruth TJ, Rufty TW (1990) Studies of the uptake of nitrate in barley. I. Kinetics of $^{13}NO_3^-$ influx. Plant Physiol 93:1426–1432

Smith IK (1975) Sulfate transport in cultured tobacco cells. Plant Physiol 55:303–307

Smith FW, Ealing PM, Hawkesford MJ, Clarkson DT (1995) Plant members of a family of sulfate transporters reveal functional subtypes. Proc Natl Acad Sci USA 92:9373–9377

Smith FW, Ealing PM, Dong B, Delhaize E (1997) The cloning of two *Arabidopsis* genes belonging to a phosphate transporter family. Plant J 11:83–92

Smith FW, Hawkesford MJ, Ealing PM, Clarkson DT, Vandenberg PJ, Belcher AR, Smith FW, Rae AL, Hawkesford MJ (2000) Molecular mechanisms of phosphate and sulphate transport in plants. Biochim Biophys Acta 1465:236–245

Stark JM, Hart SC (1997) High rates of nitrification and nitrate turnover in undisturbed coniferous forests. Nature 385:61–64

Suginaka K, Yamamoto K, Ashiida Kono Y, Saw Y, Shibata H (1998) Cysteine uptake for accumulation of glutathione by the cyanobacterium *Synechocystis* strain PCC 6803. Biosci Biotechnol Biochem 62:424–428

Touraine B, Glass ADM (1997) Nitrate and chlorate fluxes in the chl1-5 mutant of *Arabidopsis thaliana*. Does the *CHL1-5* gene encode a low affinity nitrate transporter? Plant Physiol 114:137–144

Trueman LJ, Richardson A, Forde BG (1996) Molecular cloning of higher plant homologues of the high-affinity nitrate transporters of *Chlamydomonas reinhardtii* and *Aspergillus nidulans*. Gene 175:223–231

Tsay Y-F, SchroederJI, Feldmann KA, Crawford NM (1993) A herbicide sensitivity gene *CHL1* of *Arabidopsis* encodes a nitrate-inducible nitrate transporter. Cell 72:705–713

Ullrich WR, Larsson M, Larsson C-M, Lesch S, Novacky A (1984) Ammonium uptake in *Lemna gibba* G1, related membrane potential change and inhibition of anion uptake. Physiol Plant 61:369–376

Unkles SE, Hawker KL, Grieve C, Campbell EI, Montague P, Kinghorn JR (1991) *crnA* encodes a nitrate transporter *in Aspergillus nidulans*. Proc Natl Acad Sci USA 88:204–208

Unkles SE, Zhuo D, Siddiqi MY, Kinghorn JR, Glass ADM (2001) Apparent genetic redundancy facilitates ecological plasticity for nitrate transport. EMBO J 20:6246–6255

Vidmar JJ, Schjoerring JK, Touraine B, Glass ADM (1999) Regulation of the hvst1 gene encoding a high-affinity sulfate transporter from *Hordeum vulgare*. Plant Mol Biol 40:883–892

Vidmar JJ, Zhuo D, Siddiqi MY, Glass ADM (2000a) Isolation and characterization of *HvNRT2.3* and *HvNRT2.4*, cDNAs encoding high-affinity nitrate transporters from roots of *Hordeum vulgare*. Plant Physiol 122:783–792

Vidmar JJ, Zhuo D, Siddiqi MY, Schjoerring JK, Touraine B, Glass ADM (2000b) Regulation of *HvNRT2* expression and high-affinity nitrate influx in roots of *Hordeum vulgare* by ammonium and amino acids. Plant Physiol 123:307–318

von Wiren N, Lauter FR, Ninnemann O, Gillissen B, Walch-Liu P, Engels C, Jost W, Frommer WB (2000) Differential regulation of three functional ammonium transporter genes by nitrogen in root hairs and by light in leaves of tomato. Plant J 21:167–175

Walker DJ, Leigh RA, Miller AJ (1996) Potassium homeostasis in vacuolate plant cells. Proc Natl Acad Sci USA 93:10510–10514

Walker NA, Sanders D, Maathius FJM (1996) High-affinity potassium uptake in plants. Science 273:977–978

Wang M, Siddiqi MY, Ruth TJ, Glass ADM (1993) Ammonium uptake by rice roots. II. Kinetics of $^{13}NH_4^+$ influx across the plasmalemma. Plant Physiol 103:1259–1267

Wang R, Liu D, Crawford NM (1998) The *Arabidopsis CHL1* protein plays a major role in high-affinity nitrate uptake. Proc Natl Acad Sci USA 95:15134–15239

Wang TB, Gassmann W, Rubio F, Schroeder JI, Glass ADM (1998) Rapid up-regulation of *HKT1*, a high affinity K^+ transporter gene in roots of barley and wheat following withdrawal of K^+. Plant Physiol 118:651–659

Wolt JD (1994) Soil solution chemistry: applications to environmental science and agriculture. Wiley, New York

Zhuo D, Okamoto M, Vidmar JJ Glass ADM (1999) Regulation of a putative high-affinity nitrate transporter (*Nrt2;1At*) in roots of *Arabidopsis thaliana*. Plant J 17:563–568

7 Root Architecture and Nutrient Acquisition

J.P. Lynch

7.1 Introduction

Excavation of plants from soil readily reveals that the shape of root systems is quite complex. The elegant drawings of Kutschera (e.g., Fig. 7.1; Kutschera and Lichtenegger 1992), collected painstakingly over many years of fieldwork, provide beautiful illustrations of the complexity and diversity of the spatial configuration of root systems, or 'root architecture'. The possible importance of this complexity and diversity has been considered in the scientific literature for over a century (e.g., Anonymous 1887; Cannon 1911; Weaver 1926). Speculation has primarily focused on the role of root architecture in the acquisition of water (e.g., Pages et al. 1997) and nutrients (reviewed below), and in support of the shoot (Coutts 1983; Ennos and Fitter 1992; Bailey et al. 2002), although other important functions include storage, signaling processes (Aiken and Smucker 1996), disease and insect resistance, and belowground symbioses (e.g., Hetrick 1991).

Although the focus of this chapter is the role of root architecture in nutrient acquisition, root architecture can best be understood as an aggregate of traits that co-optimize plant adaptation to several current and potential constraints. By analogy, the functional importance of shoot architecture has been fruitfully analyzed as co-optimizing solutions to the needs for structural support, light capture, and propagule dispersal (Niklas 1997). For roots, architectural patterns may co-optimize the acquisition of immobile soil resources such as phosphorus, the acquisition of mobile soil resources such as water, and support of the shoot. It must also be considered that roots exist in a hostile environment in which any given root may be lost through abiotic or biotic stress, so redundancy may be a virtue in root architecture.

The term 'root architecture' is used to refer to the spatial configuration of a root system, excluding anatomical or morphological features of individual roots such as root hairs, but including assemblages of connected roots or entire root systems (Fitter 1991a; Lynch 1995). Root architecture is dynamic and environmentally plastic. It results from an aggregate of specific physio-

Ecological Studies, Vol. 181
H. BassiriRad (Ed.) Nutrient Acquisition by Plants
An Ecological Perspective

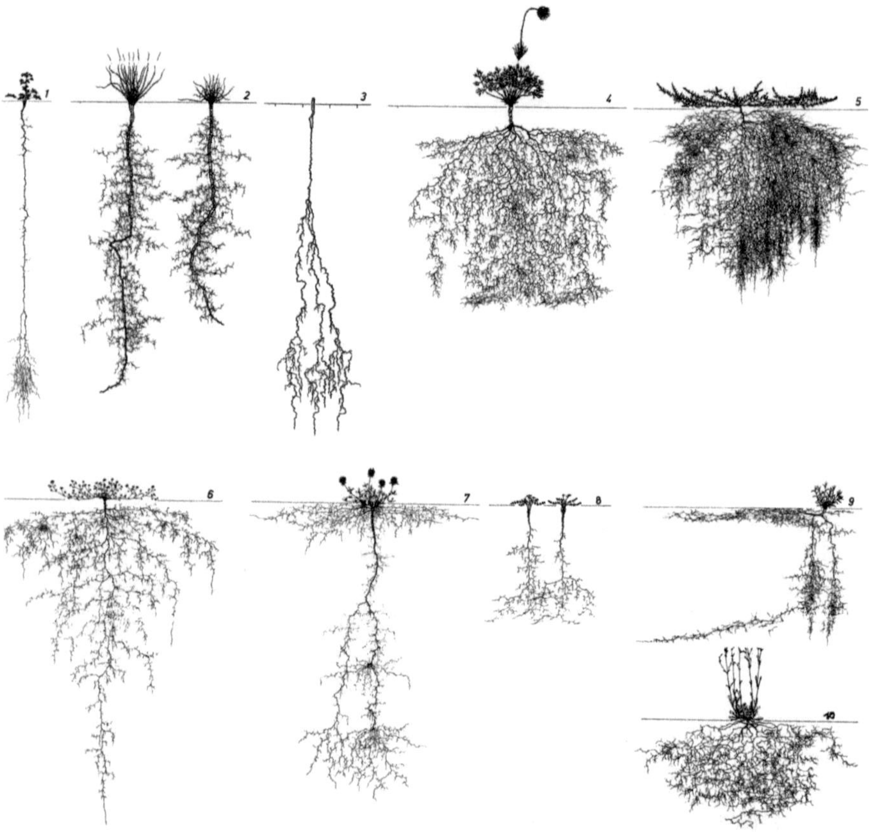

Fig. 7.1. Root architecture of several dicot species observed in the field. (Kutschera and Lichtenegger 1992)

logical traits, acting in coordination or independently, controlling the orientation, elongation, branching, and persistence of individual roots. Root architecture has a unique importance among root characteristics in that it places other traits in a specific spatial and, therefore, environmental context. For example, the effect of ion transporters on nutrient acquisition depends on the ion concentration in the soil solution (Barber 1995; Comerford, Chap. 1, this Vol.), which varies greatly in the soil, both inherently and as a result of root activity, and so is a function of root location and thus root architecture. Similarly, the production of root exudates may have very different effects on nutrient acquisition, depending on the chemical characteristics of the soil near the root, which again is a function of root location and thus root architecture.

In considering published evidence for the role of root architecture in nutrient acquisition, one is confronted with the fact that despite extended and laborious observation over many decades, as well as imaginative and insightful theoretical analyses, convincing empirical data on the topic are scarce. Root architecture is exceedingly complex, difficult to measure, and difficult to manipulate experimentally. In most cases, researchers have drawn inferences from correlations and coincidences without having eliminated possible confounding factors. Measurement of root architecture has few standard conventions, and is often attempted through more accessible, yet more indirect and potentially problematic metrics such as root topology (e.g., Fitter 1987; Fitter et al. 1991) or root distribution with depth. The plasticity of root architecture in response to many aspects of the soil environment, such as physical impedance, soil biota, water availability, oxygen status, and nutrient availability, makes the use of controlled environments, especially when the interest is nutrient acquisition, experimentally challenging. Ideally, it would be possible to link a specific architectural trait to acquisition of a specific nutrient in a specific soil environment, and therefore to plant fitness, in the absence of confounding influences from the environment as well as from other plant traits. The literature contains very few examples that approach this ideal. Thus, despite a long history of study, the role of root architecture in nutrient acquisition remains poorly understood. Nonetheless, root architecture touches upon many other aspects of plant biology that have been well researched and documented in literature that cannot be adequately reviewed here. Other chapters in this volume discuss topics that could be considered to be components of root architecture, namely, root life span (Eissenstat and Volder, Chap. 8) and root morphological plasticity (Robinson, Chap. 3). This being the case, it is not the intention of this chapter to provide an exhaustive review of a growing yet disparate, tangential, and imprecise body of literature that holds few points of consensus. Rather, this chapter will highlight links between root architecture, nutrient acquisition, and fitness for typical terrestrial plants (i.e., not dealing with extreme or unusual adaptations), as well as thoughts on fruitful avenues for exploration. Throughout this chapter, examples from common bean are used for the sake of continuity, and because this species is fairly well characterized as a model system for the analysis of the importance of root architecture in phosphorus acquisition.

7.1.1 Architectural Types

Root architecture has become more complex over the course of plant evolution. Bryophytes have rhizoids, which are relatively shallow, unbranched, and architecturally simple. *Equisetum* has a more complex root architecture with branching nodes that appear to be homologous with its shoot architecture (Fig. 7.2; Kutschera and Lichtenegger 1992). Ferns include larger and even arborescent species, which presumably must explore complex soils, but still have relatively simple root architecture (Fig. 7.3). The basic dicot root system

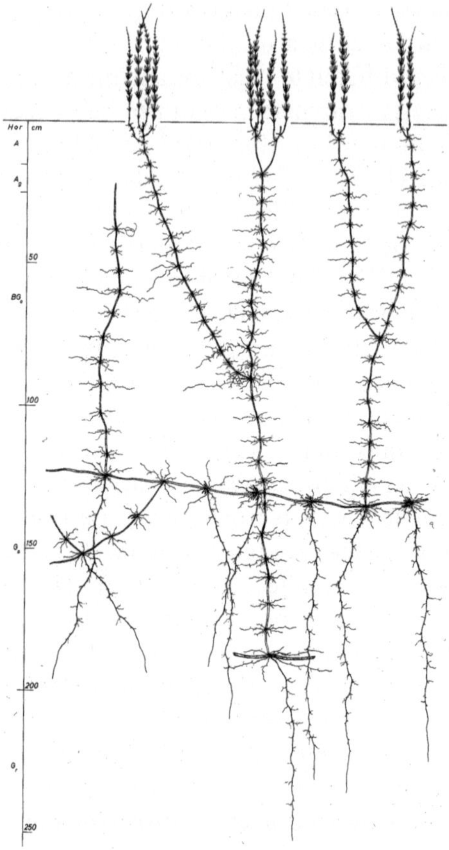

Fig. 7.2. Root architecture of the primitive plant *Equisetum palustre*. (Kutschera and Lichtenegger 1992)

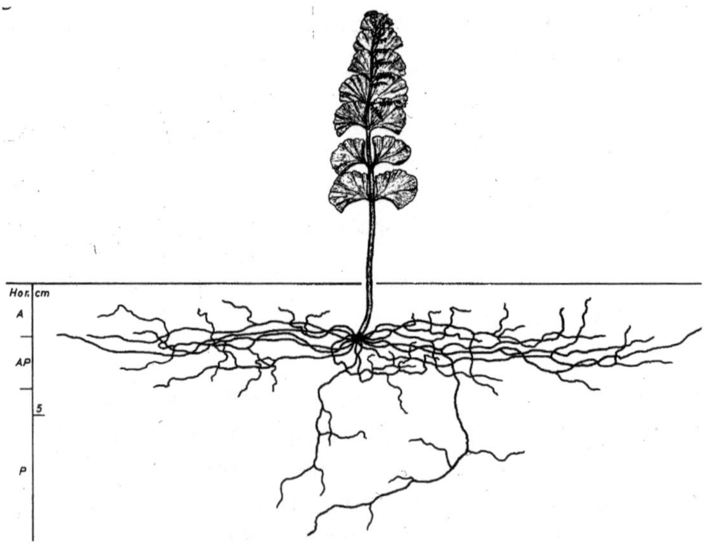

Fig. 7.3. Root architecture of the fern *Botrychium lunaria*. (Kutschera and Lichtenegger 1992)

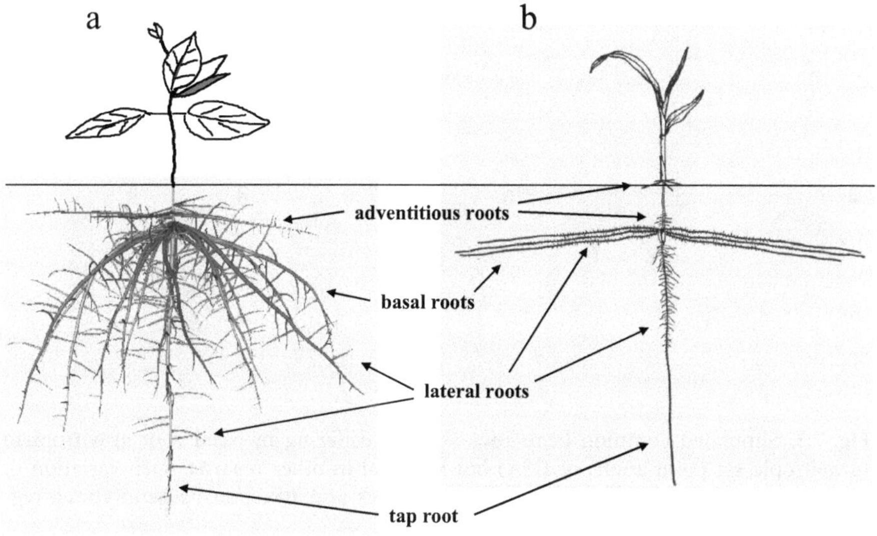

Fig. 7.4a, b. Herbaceous root systems, typified by common bean, a dicot (**a**), and maize, a monocot (**b**)

is shown in Fig. 7.4a. A large variety of architectures can result from the dicot root system, because of opportunities for branching and secondary growth among distinct root classes. The basic monocot root plan (Fig. 7.4b) is a modification of the dicot plan, which permitted the specialization of adventitious roots – those arising from the belowground mesocotyl, which are analogous to adventitious roots arising from subterranean shoots of dicots such as the hypocotyl in common bean, and those arising between shoot nodes, including those from shoot tillers, especially in grasses. Based on genetic, physiological, and anatomical evidence, Zobel (1986) has proposed that there are five distinct types of angiosperm roots – tap, lateral, adventitious, basal, and 'collateral' – although the nature of 'collateral' roots, which are small, short-lived, and determinate roots found in maize, remains controversial (Zobel 1996). Changes in the orientation, elongation, branching, and in the case of dicots, secondary growth of each of the four main root classes result in a remarkable diversity of possible architectures. Systems have been proposed that categorize the root systems of diverse species into distinct types (e.g., Cannon 1949), but considering the large genotypic and phenotypic plasticity of root architecture, the utility of such categories is uncertain.

Distinct classes of roots have distinct architectures, genetic control, and responses to environmental variables (Rundel and Nobel 1991; Lynch and van Beem 1993; Zobel 1996). Therefore, architecture is typically non-uniform within a root system, and overall root system architecture may be substantially changed by relatively slight changes in the growth of a single root class.

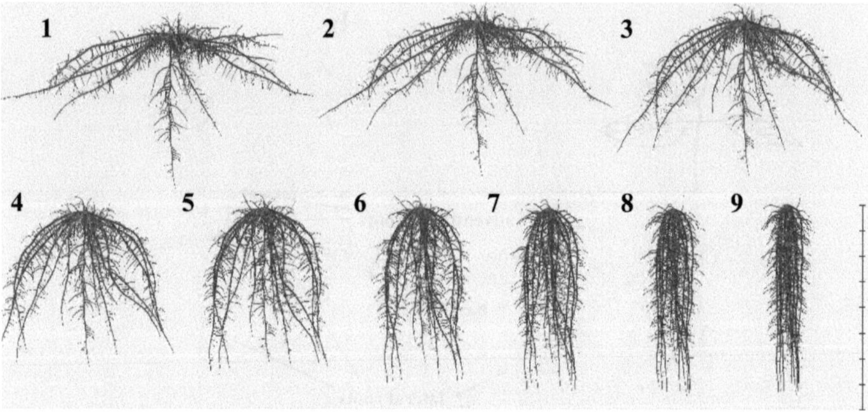

Fig. 7.5. Simulated common bean root systems differing in basal root gravitropism (gravitropic set point angle, or GSA) but identical in other regards. Such variation in basal root gravitropism is evident among genotypes, and in responsive genotypes is regulated by phosphorus availability. (Ge et al. 2000)

The modular basis of root system architecture may contribute to its large genetic variation, since traits regulating the emphasis on distinct root classes and their interplay may change architecture without changing an individual root class. For example, if each of six distinct root classes (tap root, basal roots, adventitious roots, and their respective laterals) has one of two states relative to an architectural trait (for example, graviresponsive or non-graviresponsive), we would have 2^6=64 possible resulting architectures. Conceivably, the same core set of genes could be regulating the trait, so that a simple 'on-or-off' genetic trait can result in 64 architectural phenotypes, because of its differential expression in six root classes. Indeed, numerous studies with crop plants show that root architecture varies substantially within species, even among closely related genotypes (O'Toole and Bland 1987; Lynch and van Beem 1993; Izumi et al. 1997; Fig. 7.5). The genetic complexity of root architecture is both a challenge for root researchers and an opportunity for crop improvement.

7.1.2 Classes of Nutrient Resources

From the point of view of root acquisition, nutrient resources can be divided into several functional groups, including (1) soluble nutrients that are drawn to the root through mass flow of water, including nitrate, sulfate, calcium, and magnesium; and (2) nutrients that move primarily by diffusion, including phosphate, potassium, ammonium, and micronutrient metals (Comerford, Chap. 1, this Vol.). Nutrients that move by mass flow of water also move by diffusion, and in some cases diffusive flux can exceed the mass flow of nutrients such as nitrate (Kage 1997). Two other functional categories could be identified as (3) nutrients acquired through microbial symbioses, such as nitrogen for leguminous plants (Sprent, Chap. 5, this Vol.), phosphorus and zinc for arbuscular mycorrhizal plants, and nitrogen for ectomycorrhizal and ericoid mycorrhizal plants (Smith and Read 1997), and (4) nutrients acquired through mobilization processes in the rhizosphere, such as iron in high pH soils, or phosphorus mobilization by proteoid roots (Barber 1995; Marschner 1995). Root architecture is likely to have important, yet distinct, effects on all these processes. For soluble nutrients, architectural traits that facilitate water uptake would be important. Since mass flow of soluble nutrients may occur over several cm in soil, architectures permitting extensive soil exploration, access to subsoil water, or in dryer habitats exploiting pulses of water in surface horizons (e.g., 'rain roots'; Rundel and Nobel 1991) would also facilitate the acquisition of soluble nutrients. For diffusion-mobile nutrients, architectural traits that co-localize roots with nutrient availability, and that permit soil foraging at a finer scale would be important. For nutrients mobilized in the rhizosphere, architectural traits that place roots in the appropriate soil microenvironment would be important, especially with reference to soil chemistry. For example, root phosphatase secretion in response to phosphorus limitation would be most effective in organic-rich horizons, while organic acid secretion, another common response to phosphorus limitation, would be most effective in horizons in which phosphorus availability was limited by iron and aluminum oxides. For nutrients obtained through microbial symbioses, root architecture may be important in root colonization and in determining the microenvironment of the symbiosis, although these relationships are poorly understood. The role of root architecture in nutrient acquisition is therefore most usefully evaluated in terms of specific nutrients, if possible in the context of specific soil environments. That being said, generalizations made in this review refer primarily to the acquisition of macronutrients, which move largely by mass flow and diffusion in most soils (Barber 1995).

7.2 Intrinsic and Extrinsic Roles of Root Architecture in Nutrient Acquisition

The roles of root architecture in nutrient acquisition can be divided into 'intrinsic' roles that pertain to the internal functions of plants, without particular reference to their environment, as opposed to 'extrinsic' roles that pertain specifically to interactions with the soil environment. In considering the value of architectural traits for both intrinsic and extrinsic functions, it is commonly assumed or postulated that the growth and maintenance of root systems represent a significant investment of plant resources. Using a microeco-

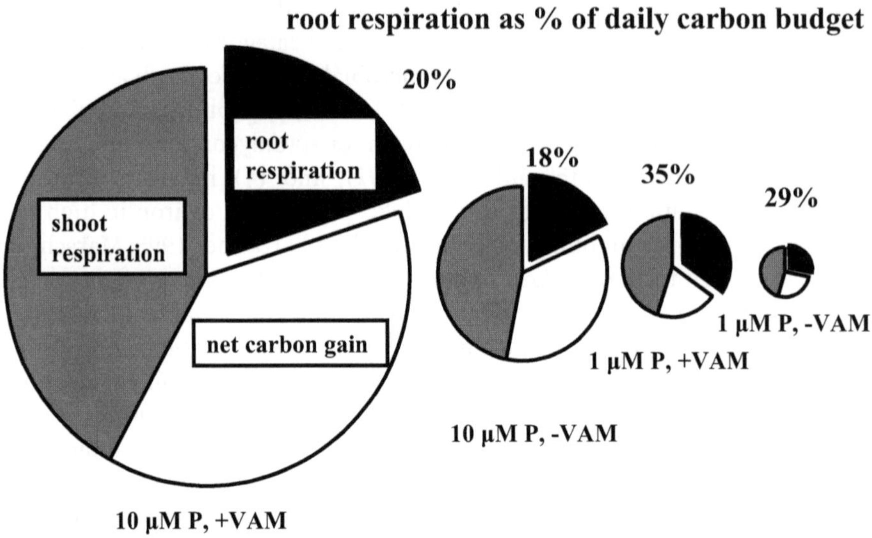

Fig. 7.6. Allocation of diurnal carbon assimilation in common bean plants at 35 days after planting, as affected by phosphorus availability and mycorrhizal status. The size of each *circle* represents total carbon fixed in a 24-h period, and subdivisions represent amount of carbon used in root (+rhizosphere) respiration, shoot respiration, and net carbon gain. Each value is the mean of six replicates. Phosphorus stress increased the proportion of the diurnal carbon budget devoted to root respiration. (Nielsen et al. 1998a)

nomic paradigm of plant resource allocation, resource investment in roots is assumed to be subject to evolutionary pressures for efficiency and optimality – in other words, that investment in roots should yield satisfactory 'returns' in terms of resource acquisition (Bloom et al. 1985). Although this model may not apply to all nutrients (Rubio et al. 2003b), for the present discussion I accept the premise that root systems use resources parsimoniously. Although the full costs and benefits of specific root traits are difficult to quantify, measurements of root carbon costs (carbon being a convenient currency for most analyses), including costs for construction, maintenance, exudates, and turnover, have shown that root carbon costs can be significant portions of whole-plant carbon budgets, i.e., 50–80 % in natural ecosystems (Caldwell and Richards 1986; Lambers 1987), and 40–65 % in crops (Whipps 1990; Gorissen 1996; Gregory et al. 1996). Under conditions of nutrient limitation, relative allocation of photosynthate to roots generally increases (Fig. 7.6; Nielsen et al. 1998a), as does relative allocation of nutrients to roots (Jeschke et al. 1997). By influencing the distribution of root biomass among tissues of distinct ages and classes, root architecture affects the partitioning of root carbon costs among construction, maintenance, uptake and exudation (Fig. 7.7; Nielsen et al. 1994). Comparison of bean genotypes differing in root architecture as well

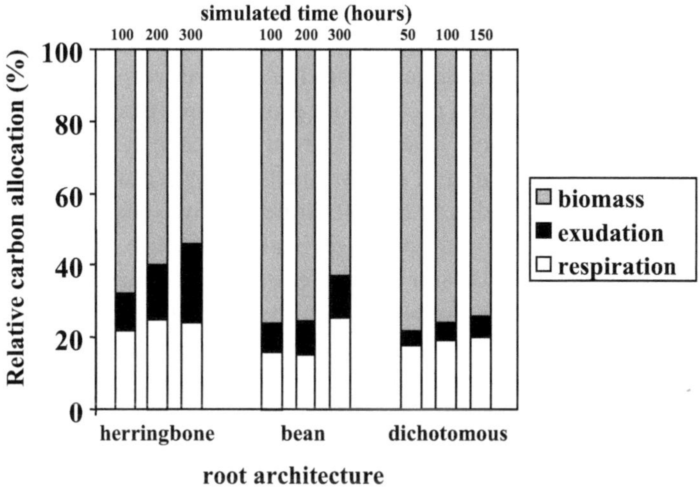

Fig. 7.7. Effect of root architecture on the distribution of root carbon costs among respiration, exudation, and biomass deposition, as simulated in *SimRoot*. The 'herringbone' architecture is just one taproot with first-order laterals, the 'bean' architecture is model 5 of Fig. 7.5, and the 'dichotomous' architecture was created by dichotomous branching, or symmetrical bifurcation of the root apices. (Nielsen et al. 1994)

Fig. 7.8. Relationship of the rate of root respiration and root relative growth rate (RGR) of four common bean genotypes, grown under low, medium, and high phosphorus conditions at 28 and 42 days after planting. Each value is the mean of four replicates. The two genotypes with poor adaptation to low phosphorus availability (DOR 364 and Porillo Sintetico, shown as *open symbols*) had significantly higher respiratory rates per unit of root growth than did the two genotypes with good adaptation to low phosphorus availability (G1937 and G19839, shown as *closed symbols*). (Nielsen et al. 2001)

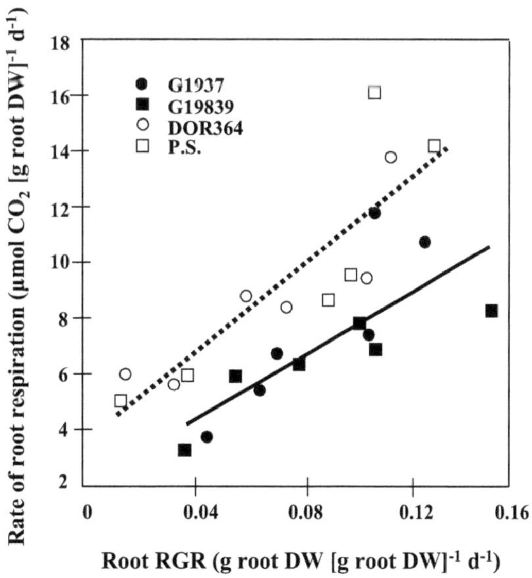

as tolerance to low phosphorus availability showed that, although both tolerant and intolerant genotypes allocated a similar fraction of their daily carbon budget to roots, the tolerant genotypes had a lower respiratory requirement for root growth, permitting greater root growth and thus greater phosphorus 'return' per unit of carbon 'invested' in roots (Nielsen et al. 2001; Fig. 7.8). In contrast to leaves, roots do not appear to efficiently remobilize nutrients during senescence, which adds to their lifetime resource cost (Snapp and Lynch 1996; Fisher et al. 2002), although cortical senescence may permit partial nutrient resorption (Robinson 1990). Because of the magnitude of resource allocation to roots, it is reasonable to assume that root architecture traits have developed that optimize the relationship of resource allocation to resource acquisition by the root system. This concept of efficiency, or the ratio of inputs to outputs of the root system, is implicit or explicit in most analyses of the role of root architecture in nutrient acquisition.

7.2.1 Intrinsic Roles of Root Architecture in Nutrient Acquisition

7.2.1.1 Efficiency of Vascular Networks

Fitter (1991b) has proposed that root architecture, or more precisely, root topology, affects nutrient acquisition efficiency by influencing the transport efficiency of materials in the root system, mainly assimilates from the shoot to growing root tips, and minerals transported to the shoot. This hypothesis is interesting and provocative, but remains unverified by empirical measurement. Current understanding of phloem transport ascribes the preponderance of metabolic costs to the loading of solutes in source tissues and their unloading in sink tissues, rather than long-distance transport, which is driven by osmotically generated pressure flow (OGPF; Ward et al. 1998). Since the metabolic costs of phloem transport per centimeter of sieve tube are minor compared with the costs of loading and unloading solutes at source and sink tissues, topological effects on transport efficiency may be slight. The standard model of xylem transport does not involve direct metabolic costs. It is not obvious how root architecture might influence water flow in roots. Possible complications are that (1) there is a lack of correlation of root diameter with the diameter distribution of functional xylem elements (Richards and Passioura 1981, 1989; Tyree and Ewers 1996; Ewers et al. 1997), (2) branch junctions may complicate water flow patterns (Shane et al. 2000), (3) radial resistance may match or exceed axial resistance to water flow in roots (Steudle and Peterson 1998), and (4) root water transport is strongly influenced by developmental maturation, which is only indirectly related to branching patterns (Doussan et al. 1998; Vercambre et al. 2002). Therefore, the mechanism for a hypothetical effect of root topology on transport efficiency is unclear. Empirical measurements of transport efficiencies in comparable root systems differing in topology are needed to evaluate this hypothesis.

7.2.1.2 Dispersal of Root Foraging

Another possible intrinsic function of root architecture in nutrient acquisition is in the dispersal of root foraging to minimize competition among roots of the same root system. This has been termed 'exploitation efficiency' by Fitter and colleagues (1991), with reference to the volume of soil exploited for phosphorus acquisition compared with the volume of tissue required to construct the root system, taking into account overlap between adjacent roots. Because phosphorus mobility is limited by diffusion, phosphorus is only slowly replenished in depleted zones near the root. Architectural patterns that place roots in close proximity to each other cause them to have overlapping depletion zones, which in terms of phosphorus acquisition by a single plant, is an inefficient use of resources. This is illustrated by geometric modeling of the effect of basal root gravitropism (Fig. 7.5) on inter-root competition for phosphorus in common bean (Ge et al. 2000). In this study, shallower basal root systems had less inter-root competition for phosphorus than deeper ones (Fig. 7.9), resulting in greater exploitation efficiency, and greater phosphorus acquisition from homogenous soils. These modeling results were supported by empirical studies with bean genotypes segregating for basal root shallowness. Shallower-rooted genotypes had superior phosphorus acquisition, even in homogenous soils (Fig. 7.10; Liao et al. 2001). An interesting feature of inter-root competition is that it is a function of the mobility of the nutrient in question. Modeling studies showed that as the effective diffusion coefficient

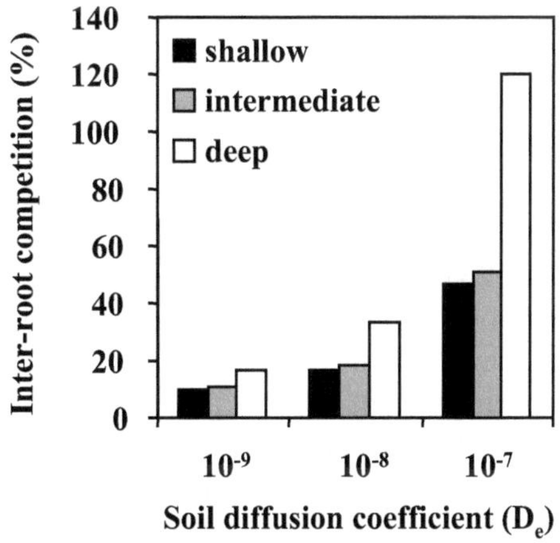

Fig. 7.9. Inter-root competition for phosphorus, expressed as % of depletion zone overlap, in simulated common bean root systems having shallow, intermediate, and deep basal roots (corresponding to models 1, 5, and 9 of Fig. 7.5), at three soil diffusivities. Shallower root systems are more dispersed than deeper root systems, and therefore suffer less inter-root competition for phosphorus. (Ge et al. 2000)

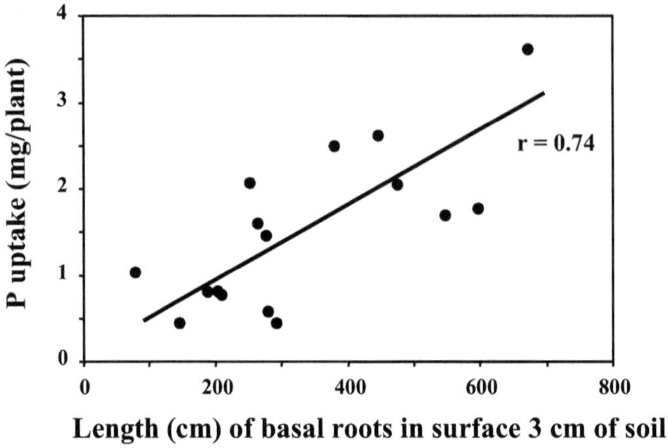

Fig. 7.10. Correlation of root shallowness and phosphorus acquisition in common bean genotypes growing in a low-phosphorus tropical soil (Hapludult). (Liao et al. 2001)

(D_e) increased from 10^{-9} to 10^{-7} cm^2 s^{-1}, inter-root competition increased nearly tenfold over 320 h of simulated root growth (Fig. 7.9; Ge et al. 2000). Likewise, inter-root competition for soluble nutrients such as nitrate may be much larger than competition for immobile nutrients such as phosphate. The effect of architectural traits on root dispersion, and thereby on exploitation efficiency, should therefore be considered in the context of specific nutrient mobilities (as distinct from nutrient availability), which vary among nutrients, within and among soils, and with time. Indeed, one function of architectural plasticity may be to adjust root foraging patterns to nutrient mobilities as they vary in time and space.

7.2.1.3 Structure of Root Demographics

A more complex but perhaps more important intrinsic role of root architecture in nutrient acquisition is in determining the proportion of root tissue in distinct root classes (e.g., basal root, adventitious root, tap root), distinct root orders (i.e., main root, first-order lateral, second-order lateral, etc.), and distinct stages of development (e.g., growing tips vs. mature regions, average distance from tip). This is because of the large anatomical, morphological, developmental, and physiological variation among root classes and orders (Eshel and Waisel 1996; Zobel 1996; Pregitzer et al. 2002). For example, in common bean we have observed substantial genetic variation for allocation to basal, tap, and adventitious roots, both constitutively (Lynch and van Beem 1993) and in response to low phosphorus availability (Fig. 7.11; Miller et al. 1998, 2003). In this species, root classes vary substantially for several traits related to phosphorus acquisition, including phosphorus uptake kinetics, aerenchyma, specific root length, metabolic construction cost, root hair density,

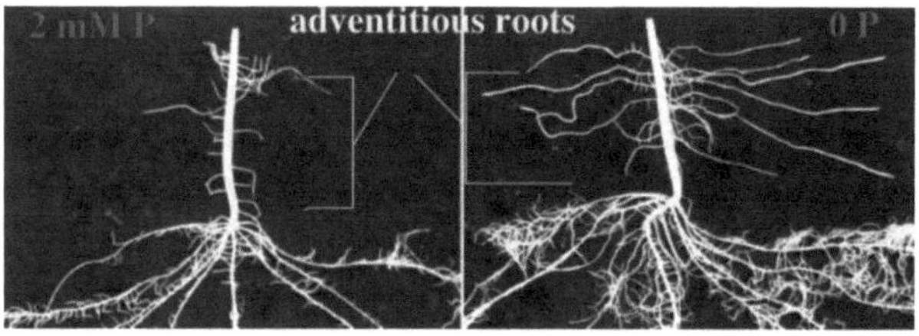

Fig. 7.11. Stimulation of adventitious rooting in common bean by low phosphorus availability. Plant at *left* received nutrient solution containing 2 mM phosphate, that at *right* received nutrient solution without phosphate. (Miller 1998)

Table 7.1. Some differences between adventitious roots and basal roots of common bean plants (genotype DOR 364), grown for 43 days in a low-phosphorus soil in the field in Costa Rica (each value is the mean of four replicates ± SE). Statistical analysis showed that adventitious and basal roots differed significantly for all the parameters shown. (Miller 1998)

Root type	Specific root length (m/g)	Construction cost (g glucose equiv./g)	Root hair length (mm root hairs per mm^{2+} root surface area)	Lateral root density (lateral roots per cm root)
Adventitious	142.2±8.6	0.404±0.013	31±5	3.87±0.27
Basal	75.7±3.4	0.451±0.011	46±6	4.52±0.32

and lateral root density (Table 7.1). Genotypes adapted to low phosphorus soils produce more adventitious roots in response to phosphorus stress than do non-tolerant genotypes. In this species, root architecture may regulate phosphorus acquisition partially by determining the relative proportion of roots with certain sets of properties for phosphorus acquisition, each with distinct cost/benefit relationships.

In long-lived herbaceous species and perennials, the effect of root architecture on root demographics could be very important in terms of root turnover and survivorship, which vary among roots of different classes and orders (Eissenstat 1997). For example, highly branched architectures may have a relatively large proportion of younger or higher-order roots, that may be more vulnerable to root herbivores or pathogens, and therefore may be more readily lost than more mature roots. For a comprehensive discussion of the importance of root turnover for nutrient acquisition, and of relevant differences among root types, the reader is referred to Eissenstat and Volder (Chap. 8, this Vol.).

7.2.1.4 The Importance of Scale

The value of an architectural trait for nutrient acquisition will depend on its scale, both in absolute terms and relative to the distribution of nutrient availability and mobility in a given soil. The importance of 'scale versus precision' in root foraging was pointed out by Grime (1991), and has been discussed at length by Berntson (1994) in terms of efficiency versus potential resource acquisition. Obviously, a simplistic efficiency parameter such as 'moles of phosphorus acquired per mole of carbon used by roots' is meaningless unless we also know something about the absolute size of the numerator and denominator of this term. Operational efficiency is the relationship of output versus input over the input range that the system actually experiences, and makes the most sense when considered with respect to a system that has organic integrity (i.e., a discrete unit with closely interrelated processes), and when the input has a direct relationship with the output of interest (Lynch 1998). Operational efficiency in this full sense is a valid context for the analysis of tradeoffs between extensive and intensive foraging for nutrients, just as the average fuel efficiency of an automobile under normal conditions of use is a valid measure of tradeoffs in vehicle design.

When diffusion-limited nutrients such as phosphorus are limiting, architectures that permit intensive or 'precision' foraging are efficient, whereas when more mobile resources such as water or nitrate are limiting, or where immobile resources such as phosphorus are sparsely distributed throughout the soil, architectures that permit extensive soil exploration are efficient. These processes can be seen in the responses of plants to nutrient stress. Phosphorus deficiency increases root hair length (Bates and Lynch 1996), density (Ma et al. 2001a), and geometry (Ma et al. 2001b), which can be interpreted as more precision foraging on a small scale. At a larger scale, phosphorus deficiency decreases lateral branching in plants such as bean, which permits a more extensive 'search' for fertile microsites (Borch et al. 1999). The distribution of nutrients in the soil in relation to the size of the plant will influence the potential 'returns' from soil exploration, and thereby the efficiency of extensive versus intensive foraging. For example, proteoid roots respond to phosphorus stress by forsaking extensive soil exploration in favor of intensive mining of rather randomly distributed patches of soil (Watt and Evans 1999), which is an efficient foraging strategy in soils that are uniformly infertile, and thus are not likely to reward extensive exploration. As another example, phosphorus stress increases topsoil foraging in bean (Lynch and Brown 2001), but does not have this effect in *Arabidopsis* (unpublished data), which makes sense considering that the diminutive size of *Arabidopsis* places virtually all of its roots in the topsoil already. Plant size will also have important effects on architectural patterns through allometric relationships (Niklas 1994; Berntson 1997; Cao and Ohkubo 1998; Cermak et al. 1998). Thus, scale-independent metrics such as fractal geometry (Tatsumi et al. 1989) and topology (Fitter 1991b) will have to be balanced with information on actual root system size and soil resource distribution, to understand the importance of root architecture for nutrient acquisition.

7.2.2 Extrinsic Roles of Root Architecture in Nutrient Acquisition

7.2.2.1 Exploiting a Non-uniform Environment

Root architecture may have important interactions with several extrinsic aspects of nutrient acquisition, including soil nutrient heterogeneity, inter-plant competition, and interaction with soil organisms. Of these, the best studied is the role of root architecture in the acquisition of nutrients that are not uniformly distributed in the soil (Fitter 1994). Robinson (Chap. 3, this Vol.) reviews root responses to variations in nutrient supply, and so my discussion here will be short.

Soil nutrient availability is generally quite heterogeneous, both spatially and temporally (Robinson, Chap. 3, this Vol.). Vertical stratification in nutrient bioavailability is usually substantial, especially in soils with well-developed profiles such as Alfisols, Ultisols, Oxisols, and Spodosols, which collectively support most of the terrestrial biomass. In more uniform soils, nutrient bioavailability varies with depth because of gradients in factors such as oxygen, temperature and water that influence nutrient mobilization. Root architecture is important in allocating root foraging to relatively discrete vertical strata with the greatest nutrient availability.

A good example of this is the role of topsoil foraging in phosphorus acquisition (Lynch and Brown 2001). As noted above, bean genotypes vary in basal root plagiogravitropism, resulting in root systems with varying degrees of topsoil foraging (Fig. 7.5). Geometric modeling showed that shallower root systems are inherently more efficient than deeper root systems, because the shallower systems disperse roots more widely, resulting in less inter-root competition (Ge et al. 2000). However, in stratified soils with greater phosphorus availability in the topsoil, this effect was overshadowed by the benefit of greater topsoil foraging by shallower roots (Fig. 7.12). These modeling results are supported by field and pot studies showing that root shallowness was significantly correlated with phosphorus acquisition from low-phosphorus tropical soil (Fig. 7.10; Bonser et al. 1996; Liao et al. 2001, 2004).

Perhaps the most intensively researched aspect of the role of root architecture in nutrient acquisition is root proliferation in localized patches of high nutrient availability (Robinson, Chap. 3, this Vol.). Elucidation of the physiological mechanisms regulating root architecture will be important in understanding how responses to environmental cues are integrated. In addition to the auxin and nitrate pathways discussed by Robinson (Chap. 3, this Vol.), eth-

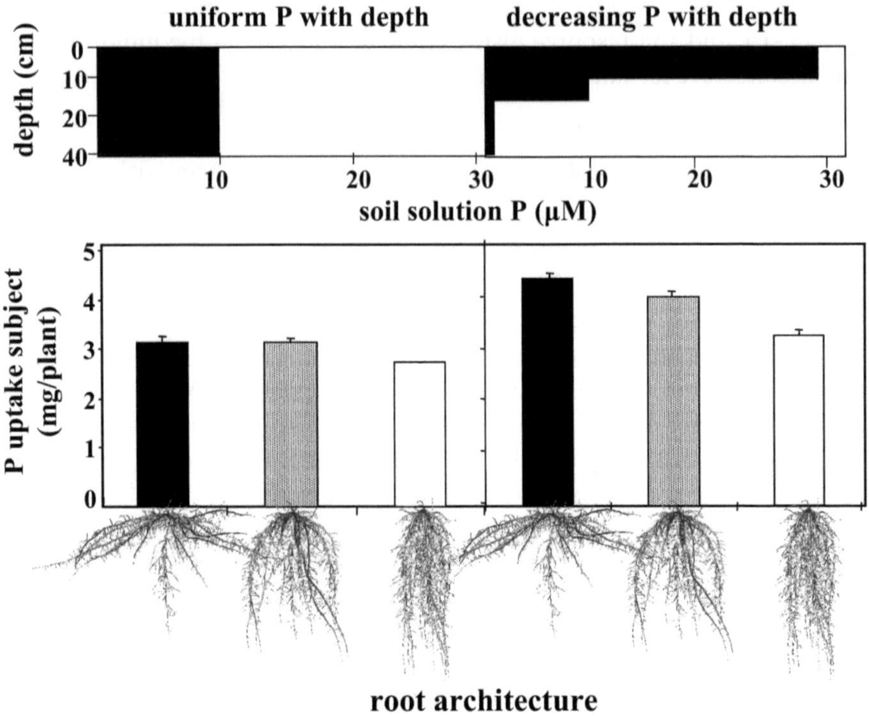

Fig. 7.12. Effect of root architecture on phosphorus acquisition from uniform and stratified soil in simulated common bean root systems. Each value is the mean of four replicates. Shallower root systems can acquire more phosphorus, especially when phosphorus is concentrated in the topsoil. (Ge et al. 2000)

ylene is involved in the mediation of root architectural responses to phosphorus availability (Lynch and Brown 1997). Low phosphorus availability stimulates ethylene production by roots, which helps maintain the elongation of the main root, thereby dispersing lateral roots over a larger volume of soil (Borch et al. 1999; Ma et al. 2003). This response may be beneficial in shifting root growth toward axial elongation rather than branching, thereby aiding extensive rather than intensive soil exploration, and potentially the discovery of more fertile soil domains.

The fact that species and genotypes vary in their ability to proliferate roots in localized nutrient patches suggests the presence of evolutionary tradeoffs in such proliferation. Root proliferation in localized nutrient patches would be most advantageous in situations of intense belowground competition, patchy nutrients, and the dominance of the patchy resource as a growth limitation. When these conditions do not occur, root proliferation in a nutrient patch may be counterproductive by diverting resources from more productive activities, such as the acquisition of other limiting resources. This may explain why many grassland species proliferate roots in response to localized nitrogen availability (e.g., Robinson et al. 1999), since in many grassland soils (Mollisols), nitrogen is a dominant edaphic constraint subject to intense belowground competition. In contrast, proteoid species from southwestern Australia proliferate lateral roots under conditions of *low* phosphorus availability, since these highly infertile soils have so few patches of phosphorus availability that it essentially must be mined in situ with protons and organic acids concentrated by local root proliferation (Watt and Evans 1999). The nature of the nutrient in question, especially its transience, and whether it moves to the root by mass flow or diffusion, are also important factors in determining whether the 'cost' of root proliferation results in an economically favorable 'return' in terms of nutrient acquisition. For example, root proliferation may be too slow to exploit a patch of nitrate, whereas a patch of decomposing organic matter, gradually releasing nitrogen compounds over a greater period of time, may be efficiently exploited by root proliferation (Hodge et al. 1999; Robinson et al. 1999). Thus, the importance of root plasticity for nutrient acquisition can best be interpreted within the context of a specific nutrient resource in a particular soil environment. Indeed, there is no evidence that roots proliferate in response to localized nutrient availability in general, but only to localized availability of nitrogen and phosphorus. In addition to better definition of the nutrient resource and the soil context, better definition of the plant genotype would also assist progress in understanding this topic. Common bean genotypes vary substantially in their architectural plasticity in response to phosphorus availability (Bonser et al. 1996). If other species have as much genotypic variation in architectural plasticity as does bean, it will be difficult to interpret the importance of architectural plasticity in an evolutionary or ecological context without assessing the plasticity of populations, rather than arbitrarily selected individuals, as is the norm in current literature.

In summary, in both native and agricultural systems, there is good, albeit largely circumstantial, evidence that nutrient acquisition is optimized by judicious allocation of root foraging in time and space to coincide with nutrient availability. Root architecture is a primary means to achieve this optimization, balanced with needs for the acquisition of water and support. Architectural optimization of nutrient acquisition by means of localized root proliferation in nutrient patches is perhaps the best studied aspect of root architecture. Better understanding of the physiological mechanisms of such plasticity, coupled with a more precise definition of both genotype and environment, will provide important conceptual advances in years ahead.

7.2.2.2 Interaction with Soil Organisms

Roots interact with a wide array of soil organisms that influence their root architecture, and thereby their ability to acquire nutrients (for an interesting recent example, see Yanni et al. 2001). Of these, perhaps the most important, and certainly the most intensely studied are mycorrhizal associations. Ectomycorrhizas profoundly alter the root architecture of the host plant (Smith and Read 1997), and there is some evidence that endomycorrhizas alter root architecture (Hetrick 1991; Schellenbaum et al. 1991), although it is difficult in such studies to control for the indirect nutritional effects of mycorrhizal symbioses on root architecture. In this volume, the role of mycorrhizas in nutrient acquisition is covered by Finlay (Chap. 9). Although typically root architecture refers to the spatial configuration of the root system, the case could be made that the architecture of the mycorrhiza, including fungal as well as plant components, is more relevant for the analysis of nutrient acquisition. Obviously, the measurement of mycorrhizal architecture in this sense would be very challenging. The observation that architectural traits in common bean are associated with phosphorus acquisition regardless of mycorrhizal status indicates that however important mycorrhizal architecture may be, it does not override the effects of root architecture as such. Possibly, hyphal foraging is concentrated in soil near the host root, so that root architecture in some way determines mycorrhizal architecture. The relationship of root architecture and mycorrhizal foraging is poorly understood and deserves further study.

Considerably less attention has been devoted to the effects of root herbivores and pathogens on root architecture. In addition to the fact that pathogens and herbivores may preferentially attack fine roots, thereby altering the functional balance of a root system, it is also to be expected that root architecture per se may influence resistance to attack by soil organisms. Considering the simple extremes of dichotomous vs. herringbone root topology, attacks randomly distributed throughout the root system are likely to have more destructive consequences for the herringbone than for the dichotomous root system. For example, consider an attack that severs the root system at a link of order 8. For the herringbone system (Fig. 7.13b), this will reduce root depth and length by about half, whereas in the dichotomous system (Fig. 7.13a), this will reduce length by about a quarter, without reducing root system depth. In other words, single attacks are more likely to excise large portions of the root system in the herringbone topology (or interfere with vascular connections to the shoot). This is obviously a gross simplification, albeit no more so than most analyses of root architecture based on topology. Thus, links of higher order may be thicker and more heavily defended than external links, and in a soil context, root architecture will also influence the position of roots in the soil, and therefore root exposure to different organisms, whose distributions in the soil vary tremendously. Nevertheless, it illus-

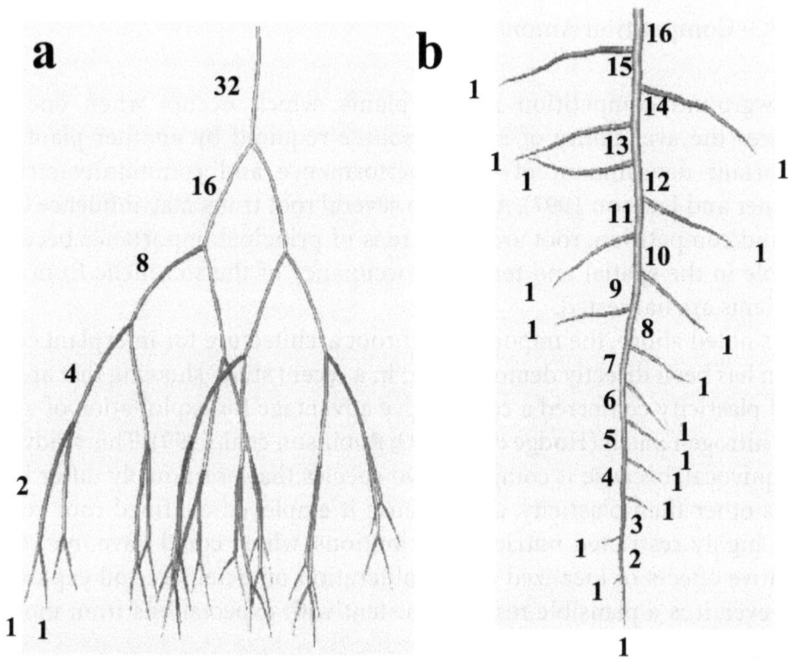

Fig. 7.13a, b. Example of root architectures with dichotomous (**a**) and herringbone (**b**) topology. *Link numbers* show the number of subtending links. Biotic attacks randomly distributed throughout the root system are likely to have more destructive consequences for the herringbone than for the dichotomous root system. For example, consider an attack that severs the root system at a link of order 8. For the herringbone system, this will reduce root depth and length by about half, whereas in the dichotomous system, this will reduce length by about a quarter, and will not reduce root system depth. In other words, single attacks are more likely to excise large portions of the root system (or interfere with vascular connections to the shoot) in the herringbone topology

trates that root architecture may have value in preserving root system functionality in a hostile environment. An interesting example of this is a study showing that VAM symbiosis changed the root architecture of strawberry, and thereby susceptibility to Phytophthora (Norman et al. 1996). Another interesting example is the case of rootworm damage to young maize root systems, which results in compensatory lateral branching and adventitious rooting, which in turn may enhance water uptake and resistance to drought later in the growing season (Riedell 1990; Riedell and Reese 1999). The role of root architecture in interactions with soil organisms remains an important and poorly understood aspect of root biology, and therefore nutrient acquisition.

7.2.2.3 Competition Among Plants

Belowground competition among plants, which occurs when one plant reduces the availability of a soil resource required by another plant, is an important determinant of plant performance and community structure (Casper and Jackson 1997). Although several root traits may influence belowground competition, root architecture is of principal importance because of its role in the spatial and temporal occupancy of the soil niche from which nutrients are harvested.

As noted above, the importance of root architecture for interplant competition has been directly demonstrated in a recent study showing that architectural plasticity conferred a competitive advantage for exploitation of a localized nitrogen patch (Hodge et al. 1999; Robinson et al. 1999). This study is not unequivocal, because it compares two species that presumably differ in root traits other than plasticity, and because it employed confined root volumes with highly restricted nutrient distributions, which could have masked any negative effects of localized root proliferation on extensive soil exploration. However, it is a plausible result consistent with expectations from modeling and theory.

With the exception of plasticity studies, there are few examples documenting effects of root architecture on competition for nutrients. A recent study employed an auxin mutant of *Arabidopsis* with reduced lateral branching to show that such branching is important for competition against wild-type plants for phosphorus, but not nitrate, which is more mobile (Fitter et al. 2002). This is a nice demonstration of the impact of a single gene on root competition, although it is difficult to interpret hormone mutants, since hormones are involved in so many physiological and developmental processes. Geometric modeling of bean roots indicates that basal root gravitropism could influence the competition for phosphorus between neighboring bean root systems (Rubio et al. 2001). This study examined competition for phosphorus, defined simply as overlap of phosphorus depletion zones (i.e., without a detailed analysis of the intensity of competition within the phosphorus depletion zone), among shallow, intermediate, and deep root systems (model numbers 1, 5, and 9 in Fig. 7.5). As expected, combinations of roots having the same architecture experienced greater competition than did combinations of contrasting architectures – for example, the deep–deep combination had twice as much inter-root competition as the shallow–deep combination (Fig. 7.14). Modeling results were supported by a field study that demonstrated that a shallow-rooted bean genotype had a competitive advantage over a deeper-rooted bean genotype under low phosphorus conditions (Fig. 7.,15; Rubio et al. 2003a). Comparisons of progeny from the deep- and shallow-rooted genotypes, segregating for root shallowness but otherwise genetically comparable, should permit a definitive assessment of the role of this architectural trait in interplant competition for phosphorus.

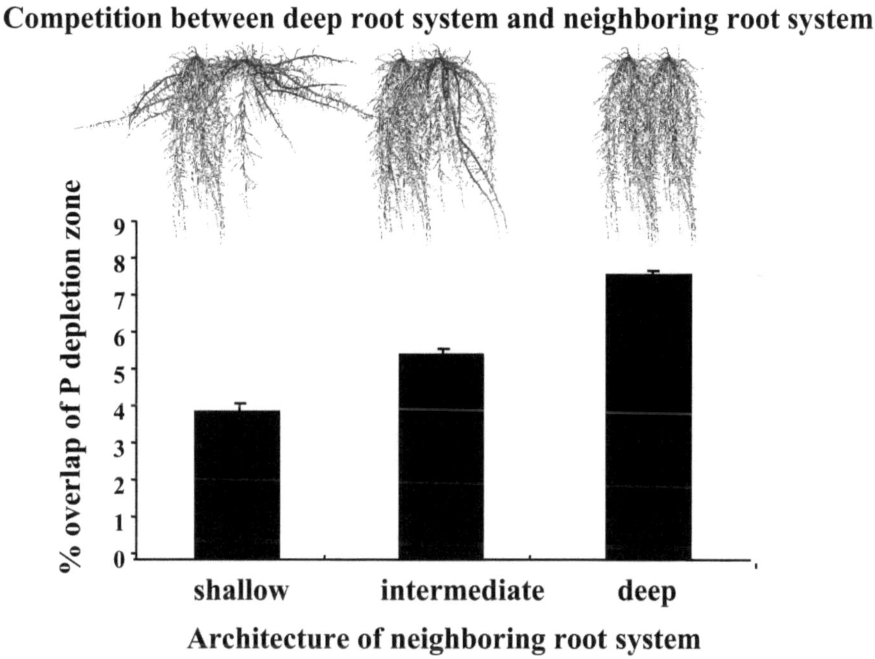

Fig. 7.14. Effect of root architecture on interplant competition for phosphorus. Competition was quantified in *SimRoot* and *DepZone* as overlap of the phosphorus depletion zones around roots of simulated bean root systems, assuming a D_e for phosphate of 1×10^{-8} cm^2 s^{-1}. One competitor has deep basal roots, while the other competitor has either deep, intermediate, or shallow basal roots. Root systems have identical topology and length, and soil phosphorus availability was assumed to be spatially uniform. Each value is the mean of three replicates. (Rubio et al. 2001)

The effects of root architecture on nutrient competition may be very different for distinct nutrients. Nutrients such as nitrate, drawn to the root primarily by mass flow of water, can be depleted in soil several centimeters away from a root (Barber 1995). In contrast, nutrients that move to the root primarily by diffusion, such as phosphate, are depleted only several mm away from a root. Therefore, interplant competition for nitrate may act over greater distances, but be less sensitive to the fine features of root architecture than may interplant competition for phosphorus. In simulations of competition for phosphorus between neighboring bean root systems, increasing the effective diffusion coefficient (D_e) from 10^{-9} to 10^{-7} enhanced interplant competition by a factor of eight (Rubio et al. 2001). Considering that D_e is a function of several dynamic soil variables, including water content (Tinker and Nye 2000), it is clear that the effects of root architecture on interplant competition must be considered in the context of specific nutrients in specific soils. This is clearly shown in the study by Fitter et al. (2002),

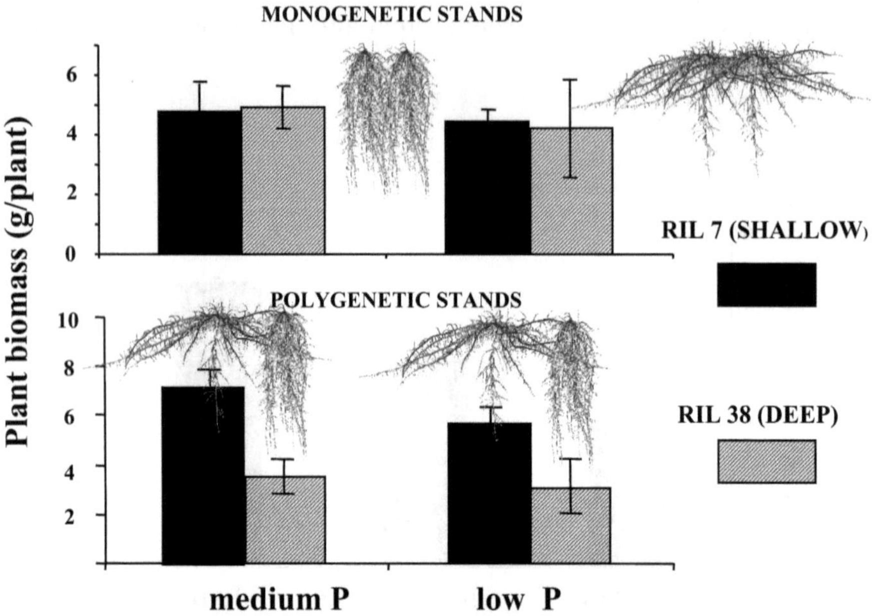

Fig. 7.15. Effect of root architecture on interplant competition between common bean genotypes in the field. Plants were grown in a low-phosphorus tropical soil in South China, either as pure stands or as mixtures of two recombinant inbred lines (RILs) from the same parents; RIL 7 has shallow basal roots while its sister line RIL 38 has deep basal roots. Plants were harvested 4 weeks after planting. Each value is the mean of four replicates. (Rubio et al. 2003a)

in which an architectural phenotype affects competition for phosphate but not for nitrate.

An interesting aspect of root architecture that directly influences inter-root competition is root avoidance (Huber-Sannwald et al. 1997; Mahall 1998; Schenk et al. 1999), which is not necessarily mediated by resource depletion (Huber-Sannwald et al. 1998). Root avoidance will tend to minimize root competition if it occurs over scales that are comparable to the scale of resource depletion. A better understanding of the physiological mechanism of root avoidance would be helpful in predicting its potential impact on nutrient competition. Thus, the niche separation observed within root systems (discussed above; see Fig. 7.1) also appears to occur among neighboring root systems.

It has been suggested that while architectural plasticity is advantageous in interspecific competition, it would not be important for breeding of crops for fertilized monocultures, in which plasticity would not affect nitrogen capture (Robinson et al. 1999). This is an important issue for the breeding of nutrient-efficient crops. In high-input agroecosystems, a substantial portion of applied nitrogen may be lost through leaching before roots fully explore the soil profile. Root plasticity may accelerate the capture of applied fertilizer before it is lost to the environment, where it represents a significant liability. Agroecosystems in many areas of the world are in fact polycultures, either of several crop species, diverse genotypes of one species, or crops in competition with non-crops (weeds or remnants of natural vegetation). The importance of polycultures in crop domestication may account for the observed genotypic variation in the architectural plasticity of crop species. For example, we have observed a surprising degree of genotypic variation in the architectural plasticity of common bean (e.g., Bonser et al. 1996), which may be related to the agroecosystems in which this species evolved. Wild bean plants are weedy vines that climb on other vegetation, and are therefore subject to intense interspecific competition (Kaplan and Kaplan 1988). It is believed that the initial cultivation of wild beans employed a system similar to the 'frijol tapado' system found in Central America, in which bean seeds are sown in slash residue on infertile forested slopes, and allowed to smother the re-growth of the native vegetation (Kettler 1996). In this system also, aggressive exploitation of nutrient resources under intense interspecific competition is probably advantageous. Traditional bean cultivation systems, which are among the most ancient crop cultivation systems on earth, are polycultures in which bean is grown with maize or other upright crops, or even as mixtures of diverse bean genotypes in one and the same field. These systems, still the most prevalent for bean cultivation on a global basis, have existed for some thousands of years, and are presumably where bean evolved into its present cultivated form. Genotypic variation in architectural plasticity in bean may therefore represent selection in polycultural (including polygenotypic) systems.

7.3 Practical Applications

An understanding of root architecture will be useful in addressing two of the great challenges facing humanity – how to sustainably feed a burgeoning population, and how to adapt to a changing world climate.

7.3.1 The Agroecology of Root Architecture – Improving Crop Nutrient Efficiency

In very simplified terms, there are two types of cropping systems with respect to nutrient availability, i.e., high-input and low-input (Lynch 1998). In both cases, the development of crop genotypes, and of cropping systems with superior nutrient efficiency, that is, with a superior ability to acquire nutrients and convert them into economic yield, would be very useful. In high-input systems characteristic of wealthy nations and some sectors of the developing world, the primary value of nutrient-efficient crops is in reducing environmental contamination from excessive fertilization (National Research Council 1989). This is especially important for the reduction of nitrate contamination of groundwater (Hallberg 1989; Juergens-Gschwind 1989), and phosphorus contamination of surface water resources (Ryden et al. 1973; Tunney et al. 1997). In low-input systems characteristic of developing countries (LDCs), the primary value of nutrient-efficient crops is in improving crop productivity, which in many cases is only 10–20 % of yields achieved with high inputs (Conway 1997; Webster and Wilson 1998).

Of the two types of systems, the need for nutrient-efficient crops in low-input systems is most compelling. Eighty-five percent of the earth's population lives in LDCs, and about half of the global workforce is engaged in LDC agriculture. The outlook for food security in these nations is worrisome (Evans 1998). Large numbers of disadvantaged people, especially children, are already malnourished (World Bank 2001). For the past 10 years, world grain production per capita has trended downward (Worldwatch Institute 2001). The Worldwatch Institute has identified the lack of response to fertilization as one of three principal threats to future food security (Brown and Kane 1994). Nutrient-efficient crops therefore have great potential value in addressing one of the central challenges facing humanity at this point in its history– how to sustainably feed itself.

Mechanisms of improved nutrient efficiency can be divided into traits enhancing the ability of a crop to extract nutrients from the soil – nutrient *acquisition* efficiency, and traits enhancing the ability of a crop to grow and yield with a given amount of nutrients – nutrient *utilization* efficiency (Marschner 1995). Although a full discussion of such traits is beyond the scope of this review, it is important to note that root architecture is promising as a trait with high genetic variability that has substantial impact on nutrient acquisition efficiency. Reports of intraspecific differences in crop macronutrient acquisition efficiency date back to 1887 (Anonymous 1887; Smith 1934). More recent literature contains numerous reports of genotypic variation in root growth and architecture that appears to be correlated with nutrient acquisition efficiency (e.g., O'Toole and Bland 1987; Zobel 1996).

An important historical example of breeding crops for improved nutrient efficiency is shown by line 1 in Fig. 7.16, in which the response of genotypes to high levels of nutrient input was improved. This was the basis for the 'Green Revolution', in which short-stature varieties of rice and wheat were developed that could respond to high rates of nitrogen fertilization without lodging (Borlaug 1972). An often-criticized aspect of the Green Revolution varieties is that Green Revolution technology disproportionately benefited producers that could afford fertilizers, thereby accentuating resource inequities among agricultural producers (Shiva 1991). An alternative approach is to enhance crop response at low nutrient availability (second

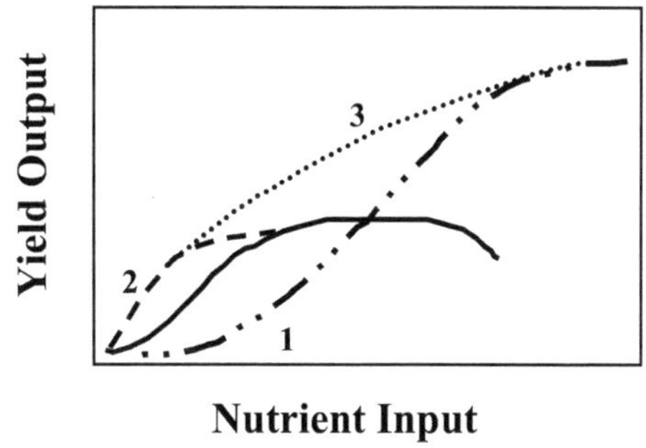

Fig. 7.16. Strategies for improved nutrient efficiency in agroecosystems. The *solid line* represents a standard yield response curve. *Curve 1* represents the Green Revolution, or improved yield response to high nutrient inputs, at the cost of reduced yield with low inputs. *Curve 2* represents improved adaptation to low-fertility soils. *Curve 3* represents improved response at all levels of inputs. Crop breeding for improved root architectural traits could contribute to the attainment of curve 3 by improving the physiological efficiency of nutrient acquisition from soil. (Lynch 1998)

curve in Fig. 7.16), which would be useful in the low-input systems that characterize much of Third World agriculture. Genotypes having superior yield response at all levels of nutrient availability (curve 3 in Fig. 7.16) would be highly valued in breeding programs. Curve 3 could result from traits permitting the acquisition of nutrients with reduced expenditure of plant resources, such as might be conferred by changes in root architecture (Fitter 1991b; Fitter et al. 1991; Lynch 1995).

A good example of the utility of specific root architectural traits in crop improvement is the case of phosphorus efficiency in common bean (Lynch and Brown 1998). As noted above, bean genotypes vary considerably in several traits related to topsoil foraging, including basal root gravitropism, and that confer adaptation to low phosphorus soils (Lynch and Brown 2001). Genetic (QTL) analysis identified several loci in the bean genome that account for variation in basal root shallowness as well as yield ability in tropical soils with low phosphorus availability (Liao et al. 2004). The genetic co-segregation of basal root shallowness and phosphorus efficiency is convincing evidence that architectural traits are important for phosphorus acquisition. Genetic markers such as these QTLs could be quite useful in breeding, because of the difficulty and expense of measuring phosphorus efficiency and architectural traits on large numbers of genotypes in the field. In this case, a simpler screening criterion is available, since the growth angle of basal roots of young seedlings in growth pouches is well correlated with topsoil foraging in the field (Liao et al. 2001). Indeed, root angles of 5-day-old seedlings in growth pouches were well correlated with the yield of contrasting genotypes over several field seasons under phosphorus stress (Fig. 7.17; Bonser et al. 1996). Genotypes with root architectural traits related to topsoil foraging are currently being used in bean breeding programs throughout Africa and Latin America to improve productivity in low-fertility, and espe-

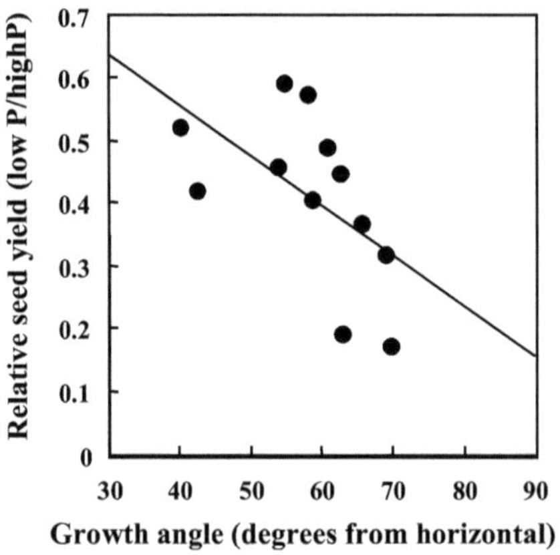

Fig. 7.17. Correlation between basal root growth angle and yield performance in low-phosphorus soils among common bean genotypes. Root growth angle was measure 5 days after germination in plastic growth pouches without added phosphorus. Yield in field studies in the tropics with low and high phosphorus availability was measured over several years by researchers at CIAT. (Bonser et al. 1996)

cially low-phosphorus soils (CIAT 1999). Based on the successful use of root architecture traits for phosphorus acquisition efficiency in bean, selection for similar traits has met with encouraging results in wheat (F. Zhang, personal communication), soybean (X. Yan and J.P. Lynch, unpubl. data), rice (Li et al. 2001), and maize (Zhu 2003). The combination of improved root-zone management tools and improved crop genotypes with superior architectural traits holds great promise for the development of agroecosystems that acquire nutrients more efficiently. The development of such agroecosystems, more productive with low inputs and more responsive and less polluting with higher inputs, would constitute a second Green Revolution, or a 'doubly green revolution' (Conway 1997).

7.3.2 Understanding and Managing Global Change

Bielenberg and BassiriRad (Chap. 11, this Vol.) specifically address global change and plant nutrient uptake, and the excellent and very comprehensive review by Pritchard and Rogers (2000) highlights the importance of root architecture in mediating plant response to elevated CO_2 and other aspects of global climate change. Therefore, the following comments are limited to a single point, i.e., the role of root architecture in mediating plant response to climate change will depend entirely on the nature of the edaphic constraints in specific ecosystems. Most terrestrial ecosystems have multiple edaphic constraints, including low availability of nitrogen, phosphorus, and often calcium, and excessive levels of aluminum and manganese, or salinity. Although the interaction of elevated CO_2 and nitrogen has received some attention, interactions with these other nutritional constraints have received little or no research attention. For example, changes in root architecture arising from elevated CO_2 may have very different impacts on the acquisition of phosphorus and calcium, two nutrients that are often limiting in forest soils, since phosphorus is diffusion-mobile and calcium moves by mass flow of water. Architectural changes that result in finer branching or root proliferation in the topsoil may increase phosphorus acquisition, whereas architectural changes that result in more extensive soil exploration, extending into deeper horizons with greater water availability, might increase calcium acquisition. Without an analysis of interactions with specific nutrients and edaphic limitations that prevail in the majority of native soils (and most agricultural soils in developing nations), it will be difficult to make general predictions about how elevated CO_2 will change root architecture, or how such changes are likely to affect nutrient acquisition in future atmospheres.

7.4 Future Prospects

Despite growing interest in root architecture, and growing evidence for its importance in nutrient acquisition, there exist very few cases documenting that a specific, genetically controlled architectural trait contributes to nutrient acquisition, and thereby to competitive or reproductive fitness within a specific environmental context. Such cases are needed, if this field of research is to progress from speculation and hypotheses to more precise conceptual models.

Molecular biology may be less helpful, at least initially, in this enterprise than it has been for understanding other physiological traits, because many architectural traits appear to be quantitatively controlled. Loss-of-function mutants may be helpful in identifying regulatory genes involved in root development, and architectural responses to specific environmental cues (e.g., Zhang and Forde 1997), but these will be difficult to apply to traits that are regulated by multiple genes, as appears to be the case with architectural traits in maize and bean (Zhu 2003; Liao et al. 2004; Yan et al. 2004). Compared to single-gene approaches, techniques in genomics that permit the analysis of multiple genes and gene families may be more fruitful for understanding the genetics of root architecture.

The field would benefit from better definition of architectural traits, both conceptually and experimentally. Root architecture is an aggregate of many distinct genetic and physiological traits that control the elongation, orientation, branching, and life span of individual root axes. These traits interact to produce patterns of root deployment and activity in time and space. Few studies have been able to address fundamental architectural traits, as opposed to aggregate spatiotemporal patterns. To return to the example of common bean, topsoil foraging is an architectural characteristic that is substantially influenced by both adventitious rooting and basal root gravitropism, and these are controlled by distinct sets of genes (unpublished data). Attempts to analyze the adaptive value or regulation of topsoil foraging may be confounded if it is not known that at least two distinct architectural traits contribute to this pattern of root deployment. Topsoil foraging achieved through adventitious rooting may be quite different – genetically, physiologically, and ecologically – than such foraging achieved through shallow basal roots.

The isolation of architectural traits is particularly problematic in studies of wild plants, for which segregating genotypes are rarely experimentally available. In addition to possible confounding effects of other architectural traits, confounding effects of differences in microbial symbioses, rhizosphere chemistry, ion uptake, etc., make it very difficult to evaluate the impact of architectural traits, in isolation, on the performance of contrasting wild species. Theoretical validation through simulation models (as in Robinson et al. 1999), or comparison of multiple species having contrasting traits, would be more robust than simple comparison of two species that differ in unspecified ways. Indeed, given the large genotypic variation within species for architectural and other traits related to nutrient acquisition, it is difficult to make robust generalizations from results for one population or genotype of one species and another population or genotype of another species – a broader genetic sampling is needed. For this reason, agroecological studies of crop plants, where specific traits can be isolated through genetic crossing, may be more fruitful than comparisons of wild plants in providing conclusive evidence for the utility of specific architectural traits for nutrient acquisition.

In addition to a clear definition of architectural traits, it is also important to clearly define the soil conditions in which these traits are observed. Part of the 'murkiness' in the relevant literature is due to uncertainty about the generality of architectural responses to poorly characterized soils, with unknown and possibly atypical combinations of biotic and abiotic stresses. By comparison, in the phytopathology literature, responses of specific plant genotypes to specific strains of specific pathogenic organisms are considered. A study of plant responses to a pathogen described only as 'a fungus' or 'a bacterium' would generally be considered too imprecise to interpret. Unfortunately, it is common in studies of root architecture that the soil conditions are only vaguely characterized, if at all, despite the fact that nutrient availability, water status, temperature, physical impedance, and many other soil factors influence root architecture. For this reason, field studies including many poorly defined or controlled factors should be confirmed in more controlled, less 'realistic' studies where specific factors can be varied.

Because of the difficulty in generating genetic contrasts in specific architectural traits, the difficulty in monitoring or controlling soil conditions, as well as the difficulty in measuring root architecture in the first place, geometric modeling is likely to play an important role in future research. A number of geometric models have been developed that simulate the growth, architecture, and some aspects of the physiology of root systems (Lynch and Nielsen 1996; Pages 1999). Simulation modeling has heuristic value in helping the modeler define relevant processes and interactions, in assessing the impact of single variables on system performance through sensitivity analysis, in evaluating the potential validity of empirical sampling and measurement techniques, and in suggesting issues and hypotheses for experimentation (Wullschleger et al. 1994). Experimentation can help refine the model, which can then suggest further experimentation, in an iterative process that eventually may lead to a model of sufficient validity for use in predictive scenarios, or as a module of more integrative models. As an aid to physiological research, simulation modeling has a very useful role in integrating information about specific processes or root axis responses into a complex, three-dimensional context. Although an adequate review of modeling literature is not possible here, two recent, noteworthy developments are the models of Hopmans and colleagues, integrating root architecture and soil water transport (Somma et al. 1998), and work by Pages and colleagues to develop a better understanding of the dynamics and morphogenetic bases of root architecture (Pages 1999).

A promising new technology permitting the noninvasive visualization of root architecture in soil is tomography, especially magnetic resonance imaging (MRI) and X-ray computer-aided tomography (Anderson and Hopmans 1994; Heeraman et al. 1997). These are particularly useful for the study of water flow in plant–soil systems, since water content and the status of H atoms in the sample can be distinguished. The availability of the technique for most root researchers is restricted, however, by limited access to suitable equipment, and the substantial expertise and cost required to obtain satisfactory results. It is also difficult with present technology to image fine roots in soil volumes that do not restrict the growth of mature plants. As these technical difficulties are inevitably overcome, tomography will play an increasingly important role in the study of root architecture. It is important to note, however, that the visualization of root architecture in soil still presents a geometric object that is difficult to quantify, summarize, or interpret.

A relatively new approach to the quantification of root architecture is fractal geometry. Fractal geometry is a system of geometry that is more suited to the description of complex natural objects than is standard Euclidean geometry (Mandelbrot 1983). A fractal is an object having a non-integer dimension, and root systems approximate fractal objects over a finite range of scales (Tatsumi et al. 1989). It is reasonable to hope that fractal geometry may provide quantitative summaries and functional insights into root architecture that have eluded researchers employing Euclidean geometry (Berntson et al. 1998; Ozier-Lafontaine et al. 1999). For example, the fractal geometry of a root system in three spatial dimensions can be estimated from measurements of the fractal geometry in one and two spatial dimensions, such as might be measured from soil cores and exposed trenches (Nielsen et al. 1997). If specific functional attributes can be linked to fractal geometry, then fractal properties may be useful in plant breeding (Masi and Maranville 1998; Nielsen et al.

1998b), and as indices of belowground functioning (Walk et al. 2004). Multifractal analysis may be useful in the description and analysis of architectural heterogeneity within root systems, resulting from, for example, morphological plasticity.

Progress in understanding the role of root architecture in nutrient acquisition is going to require an unusual degree of integration across scientific subdisciplines. The application of modeling, quantitative genetics, and advanced mathematics to problems that encompass plant physiology, soil science, microbiology, and ecology calls for creative interdisciplinary and multidisciplinary efforts at the frontiers of established techniques and dogmas. The reluctance of the scientific community to fund and publish such unorthodox research is not the least of the challenges to progress in this field. However, the very fact that root architecture is such an important factor in soil resource acquisition will inevitably continue to attract the attention of researchers from disparate fields to this complex, mysterious, and fascinating dimension of plant life.

7.5 Summary

Root architecture, the spatial configuration of roots in soil, is a highly complex trait resulting from several distinct classes of roots growing, branching, developing, and senescing in response to a highly complex and heterogeneous soil environment. Root architecture is senior to many other traits affecting nutrient acquisition, since it defines the spatial context, and therefore the biophysical environment in which other root traits are manifested. Over evolutionary time, root architecture has become more complex, and substantial genetic variation for root architecture is observed within and among species. The roles of root architecture in nutrient acquisition can be divided into 'intrinsic' roles that pertain to the internal functions of plants without particular reference to their environment, and 'extrinsic' roles that pertain specifically to interactions with the soil environment.

Intrinsic roles include the efficiency of vascular networks, the dispersal of root foraging, and the structure of root demographics. Although there is little empirical evidence for a role of root architecture in the efficiency of vascular networks, there is good evidence that root architecture has important roles in deploying root foraging activity to effectively encounter and exploit nutrient resources. Root architecture also has an important role in regulating the amount of competition for nutrients among roots of the same plant, and in determining the proportion of root tissue in distinct root classes, orders, and stages of development.

Extrinsic roles of root architecture in nutrient acquisition include important roles in exploiting a non-uniform environment, and in mediating interactions with other organisms, including competition with other plants for nutrient resources. There is good, albeit largely circumstantial, evidence that nutrient acquisition is optimized by judicious allocation of root foraging in time and space to coincide with nutrient availability, and root architecture is a primary means to achieve this optimization. Root architecture has complex and poorly understood interactions with soil organisms, including mycorrhizas, but such interactions are likely to be important for nutrient acquisition. Root architecture and topology should theoretically influence plant resistance to root herbivory and root pathogens, although little is known about this. Modeling and empirical studies show that root architecture is an important component of interplant competition for scarcely mobile nutrient resources such as phosphate.

An understanding of root architecture will be useful in addressing two of the great challenges facing humanity – how to sustainably feed a burgeoning population, and how to adapt to a changing world climate. There is growing evidence that root architectural traits may be employed in crop breeding programs to generate nutrient-efficient crops with superior productivity on the infertile soils that characterize many developing nations. Since most ecosystems are characterized by multiple soil constraints, nutrient acquisition, and therefore root architecture, will play an important role in mediating ecosystem response to climate change.

Future progress in understanding the role of root architecture in nutrient acquisition will be facilitated by the advent of several new analytical approaches. These include increasingly sophisticated geometric modeling, tools from mathematics and economics to evaluate shapes and their functional relevance, and insights from quantitative genetics.

References

Aiken RM, Smucker AJM (1996) Root system regulation of whole plant growth. Annu Rev Phytopathol 34:325–346

Anderson SH, Hopmans JW (1994) Tomography of soil-water-root processes. Vol SSSA Spec Publ no 36. American Society of Agronomy, Soil Science Society of America, Madison, WI

Anonymous (1887) Report of the Pennsylvania State College Agricultural Experimental Station. Official Document no 13

Bailey PHJ, Currey JD, Fitter AH (2002) The role of root system architecture and root hairs in promoting anchorage against uprooting forces in *Allium cepa* and root mutants of *Arabidopsis thaliana*. J Exp Bot 53:333–340

Barber SA (1995) Soil nutrient bioavailability: a mechanistic approach, 2nd edn. Wiley, New York

Bates TR, Lynch JP (1996) Stimulation of root hair elongation in *Arabidopsis thaliana* by low phosphorus availability. Plant Cell Environ 19:529–538

Berntson GM (1994) Modelling root architecture: are there tradeoffs between efficiency and potential of resource acquisition? New Phytol 127:483–493

Berntson GM (1997) Topological scaling and plant root system architecture: developmental and functional hierarchies. New Phytol 135:621–634

Berntson G, Lynch J, Snapp S (1998) Fractal geometry and plant root systems: current perspectives and future applications. In: Baveye P, Parlange J, Stewart B (eds) Fractals in soil science. Lewis, New York

Bloom AJ, Chapin FSI, Mooney HA (1985) Resource limitation in plants – an economic analogy. Annu Rev Ecol Syst 16:36:33–392

Bonser AM, Lynch J, Snapp S (1996) Effect of phosphorus deficiency on growth angle of basal roots in *Phaseolus vulgaris*. New Phytol 132:281–288

Borch K, Bouma TJ, Lynch JP, Brown KM (1999) Ethylene: a regulator of root architectural responses to soil phosphorus availability. Plant Cell Environ 22:425–431

Borlaug NE (1972) The green revolution, peace, and humanity. In: Speech delivered upon receipt of the 1970 Nobel Peace Prize, vol CIMMYT reprint and translation series no 3. Centro Internacional de Mejoramiento de Maiz y Trigo, El Batan, Mexico

Brown LR, Kane H (1994) Full house: reassessing the Earth's population carrying capacity. Norton, New York

Caldwell MM, Richards JH (1986) Competing root systems: morphology and models of absorption. In: Givnish TJ (ed) On the economy of plant form and function. Cambridge University Press, Cambridge, pp 251–273

Cannon WA (1911) The root habits of desert plants. Vol Publ no 131, Carnegie Institute of Washington, Washington, DC

Cannon W (1949) A tentative classification of root systems. Ecology 30:542–548

Cao K, Ohkubo T (1998) Allometry, root/shoot ratio and root architecture in understory saplings of deciduous dicotyledonous trees in central Japan. Ecol Res 13:217-227

Casper BB, Jackson RB (1997) Plant competition underground. Annu Rev Ecol Syst 28:545–570

Cermak J, Riguzzi F, Ceulemans R (1998) Scaling up from the individual tree to the stand level in Scots pine. I. Needle distribution, overall crown and root geometry. Ann Sci For 55:63–88

CIAT (1999) Bean project: annual report. Centro Internacional de Agricultura Tropical, Cali, Colombia

Conway G (1997) The doubly green revolution: food for all in the 21st century. Comstock, Cornell University Press, Ithaca, NY

Coutts MP (1983) Root architecture and tree stability. Plant Soil 71:171–188

Doussan C, Vercambre G, Pages L (1998) Modelling of the hydraulic architecture of root systems: an integrated approach to water absorption-distribution of axial and radial conductances in maize. Ann Bot 81:225–232

Eissenstat D (1997) Trade-offs in root form and function. In: Jackson L (ed) Ecology in agriculture. Academic Press, San Diego, pp 173–199

Ennos AR, Fitter AH (1992) Comparative functional morphology of the anchorage systems of annual dicots. Funct Ecol 6:71–78

Eshel A, Waisel Y (1996) Multiform and multifunction of various constituents of one root system. In: Waisel Y, Eshel A, Kafkafi U (eds) Plant roots: the hidden half. Marcel Dekker, New York, pp 175–192

Evans L (1998) Feeding the ten billion: plants and population growth. Cambridge University Press, Cambridge

Ewers FW, Carlton MR, Fisher JB, Kolb KJ, Tyree MT (1997) Vessel diameters in roots versus stems of tropical lianas and other growth forms. IAWA J 18:261–279

Fisher MCT, Eissenstat DM, Lynch JP (2002) Lack of evidence for programmed root senescence in common bean (*Phaseolus vulgaris*) grown at different levels of phosphorus supply. New Phytol 153:63–71

Fitter A (1987) An architectural approach to the comparative ecology of plant root systems. New Phytol 106:61–77

Fitter A (1991a) Characteristics and functions of root systems. In: Waisel Y, Eshel A, Kafkafi U (eds) Plant roots: the hidden half. Marcel Dekker, New York, pp 3–25

Fitter AH (1991b) The ecological significance of root system architecture: an economic approach. In: Atkinson D (ed) Plant root growth: an ecological perspective. Blackwell, New York, pp 229–243

Fitter AH (1994) Architecture and biomass allocation as components of the plastic response of root systems to soil heterogeneity. In: Caldwell MM, Pearcy RW (eds) Exploitation of environmental heterogeneity by plants. Academic Press, San Diego, pp 305–323

Fitter A, Stickland T, Harvey M, Wilson G (1991) Architectural analysis of plant root systems. 1. Architectural correlates of exploitation efficiency. New Phytol 118:385–382

Fitter A, Williamson L, Linkohr B, Leyser O (2002) Root system architecture determines fitness in an *Arabidopsis* mutant in competition for immobile phosphate ions but not for nitrate ions. Proc R Soc Lond Ser B Biol Sci 269:2017–2022

Ge ZY, Rubio G, Lynch JP (2000) The importance of root gravitropism for inter-root competition and phosphorus acquisition efficiency: results from a geometric simulation model. Plant Soil 218:159–171

Gorissen A (1996) Elevated CO_2 evokes quantitative and qualitative changes in carbon dynamics in a plant/soil system: mechanisms and implications. Plant Soil 187:289–298

Gregory PJ, Palta JA, Batts GR (1996) Root systems and root:mass ratio – carbon allocation under current and projected atmospheric conditions in arable crops. Plant Soil 187:221–228

Grime J (1991) Nutrition, environment and plant ecology: an overview. In: Porter J, Lawlor D (eds) Plant growth: interactions with nutrition and environment. Cambridge University Press, Cambridge, pp 249–267

Hallberg G (1989) Nitrate in ground water in the United States. In: Follett RF (ed) Nitrogen management and ground water protection. Elsevier, New York, pp 35–74

Heeraman DA, Hopmans JW, Clausnitzer V (1997) Three dimensional imaging of plant roots in situ with X-ray computed tomography. Plant Soil 189:167–179

Hetrick BAD (1991) Mycorrhizas and root architecture. Experientia 47:355–362

Hodge A, Robinson D, Griffiths BS, Fitter AH (1999) Why plants bother: root proliferation results in increased nitrogen capture from an organic patch when two grasses compete. Plant Cell Environ 22:811–820

Huber-Sannwald E, Pyke DA, Caldwell MM (1997) Perception of neighbouring plants by rhizomes and roots: morphological manifestations of a clonal plant. Can J Bot 75:2146–2157

Huber-Sannwald E, Pyke DA, Caldwell MM, Durham S (1998) Effects of nutrient patches and root systems on the clonal plasticity of a rhizomatous grass. Ecology 79:2267–2280

Izumi Y, Kono Y, Yamauchi A, Iijima M (1997) Genotypic variation in the development of seminal root system of rice under different culture conditions *in vitro*. Jpn J Crop Sci 66:427–435

Jeschke W, Kirkby E, Peuke A, Pate J, Hartung W (1997) Effects of P deficiency on assimilation and transport of nitrate and phosphate in intact plants of castor bean (Ricinus comunis L.). J Exp Bot 48:75–91

Juergens-Gschwind S (1989) Ground water nitrates in other developed countries (Europe) – relationships to land use patterns. In: Follett RF (ed) Nitrogen management and ground water protection. Elsevier, New York, pp 75–138

Kage H (1997) Relative contribution of mass flow and diffusion to nitrate transport towards roots. Z Pflanzenernähr Bodenkd 160:171–178

Kaplan L, Kaplan L (1988) *Phaseolus* in archeology. In: Gepts P (ed) Genetic resources of *Phaseolus* beans. Kluwer, Dordrecht, pp 125–142

Kettler JS (1996) Weeds in the traditional slash/mulch practice of frijol tapado: indigenous characterization and ecological implications. Weed Res 36:385–393

Kutschera L, Lichtenegger E (1992) Wurzelatlas Mitteleuropäischer Grünlandpflanzen. Band 2. Pteridophyta und Dicotyledoneae. Teil 1. Morphologie, Anatomie, Ökologie, Verbreitung, Soziologie, Wirtschaft. Gustav Fischer, Stuttgart

Lambers H (1987) Growth, respiration, exudation and symbiotic associations: the fate of carbon translocated to the roots. In: Gregory PJ, Lake JV, Rose DA (eds) Root development and function. Cambridge University Press, Cambridge, pp 125–146

Li HB, Xia M, Wu P (2001) Effect of phosphorus deficiency stress on rice lateral root growth and nutrient absorption. Acta Bot Sin 43:1154–1160

Liao H, Rubio G, Yan XL, Cao AQ, Brown KM, Lynch JP (2001) Effect of phosphorus availability on basal root shallowness in common bean. Plant Soil 232:69–79

Liao H, Yan X, Rubio G, Beebe SE, Blair MW, Lynch JP (2004) Genetic mapping of basal root gravitropism and phosphorus acquisition efficiency in common bean. Funct Plant Biol 31:959–970

Lynch JP (1995) Root architecture and plant productivity. Plant Physiol 109:7–13

Lynch JP (1998) The role of nutrient efficient crops in modern agriculture. J Crop Prod 1:241–264

Lynch JP, Brown KM (1997) Ethylene and plant responses to nutritional stress. Physiol Plant 100:613–619

Lynch JP, Brown K (1998) Root architecture and phosphorus acquisition efficiency in common bean. In: Lynch J, Deikman J (eds) Phosphorus in plant biology: regulatory roles in ecosystem, organismic, cellular, and molecular processes. American Society of Plant Physiologists, Rockville, MD

Lynch JP, Brown KM (2001) Topsoil foraging – an architectural adaptation of plants to low phosphorus availability. Plant Soil 237:225–237

Lynch JP, Nielsen K (1996) Simulation of root system architecture. In: Waisel Y, Eshel A, Kafkafi U (eds) Plant roots: the hidden half. Marcel Dekker, New York, pp 247–258

Lynch JP, van Beem JJ (1993) Growth and architecture of seedling roots of common bean genotypes. Crop Sci 33:1253–1257

Ma Z, Bielenberg DG, Brown KM, Lynch JP (2001a) Regulation of root hair density by phosphorus availability in *Arabidopsis thaliana*. Plant Cell Environ 24:459–467

Ma Z, Walk TC, Marcus A, Lynch JP (2001b) Morphological synergism in root hair length, density, initiation and geometry for phosphorus acquisition in *Arabidopsis thaliana*: a modeling approach. Plant Soil 236:221–235

Ma Z, Baskin TI, Brown KM, Lynch JP (2003) Regulation of root elongation under phosphorus stress involves changes in ethylene responsiveness. Plant Physiol 131:1381–1390

Mahall BE (1998) Inter-root communications and the structure of desert plant communities. In: Flores HE, Lynch JP, Eissenstat D (eds) Radical biology: advances and perspectives on the function of plant roots. American Society of Plant Physiologists, Rockville, MD, pp 265–280

Mandelbrot BB (1983) The fractal geometry of nature. WH Freeman, New York

Marschner H (1995) Mineral nutrition of higher plants, 2nd edn. Academic Press, London

Masi CEA, Maranville JW (1998) Evaluation of sorghum root branching using fractals. J Agric Sci 131:259–265

Miller CR (1998) Root architecture of common bean (*Phaseolus vulgaris* L.): adaptive nature and dynamic response to low phosphorus. MSc Thesis, Pennsylvania State University, University Park, Pennsylvania

Miller CR, Nielsen KL, Lynch JP, Beck D (1998) Adventitious root response in field grown common bean: a possible adaptive strategy to low phosphorus conditions. In: Flores

HE, Lynch JP, Eissenstat D (eds) Radical biology: advances and perspectives on the function of plant roots. American Society of Plant Physiologists, Rockville, MD, pp 394–396

Miller CR, Ochoa I, Nielsen KL, Beck D, Lynch JP (2003) Genetic variation for adventitious rooting in response to low phosphorus availability: potential utility for phosphorus acquisition from stratified soils. Funct Plant Biol 30:973–985

National Research Council (1989) Alternative agriculture. National Academy Press, Washington, DC

Nielsen KL, Lynch J, Jablokow AG, Curtis PS (1994) Carbon cost of root systems: an architectural approach. Plant Soil 165:161–169

Nielsen KL, Lynch JP, Weiss HN (1997) Fractal geometry of bean root systems: correlations between spatial and fractal dimension. Am J Bot 84:26–33

Nielsen KL, Bouma TJ, Lynch JP, Eissenstat DM (1998a) Effects of phosphorus availability and vesicular-arbuscular mycorrhizas on the carbon budget of common bean (*Phaseolus vulgaris*). New Phytol 139:647–656

Nielsen KL, Miller CR, Beck D, Lynch JP (1998b) Fractal geometry of root systems: field observations of contrasting genotypes of common bean (*Phaseolus vulgaris* L.) grown under different phosphorus regimes. Plant Soil 206:181–190

Nielsen KL, Eshel A, Lynch JP (2001) The effect of phosphorus availability on the carbon economy of contrasting common bean (*Phaseolus vulgaris* L.) genotypes. J Exp Bot 52:329–339

Niklas K (1994) Plant allometry: the scaling of form and process. University of Chicago Press, Chicago

Niklas K (1997) The evolutionary biology of plants. The University of Chicago Press, Chicago

Norman JR, Atkinson D, Hooker JE (1996) Arbuscular mycorrhizal fungal-induced alteration to root architecture in strawberry and induced resistance to the root pathogen *Phytophthora fragariae*. Plant Soil 185:191–198

O'Toole JC, Bland WL (1987) Genotypic variations in the root system. Adv Agron 41:91–140

Ozier-Lafontaine H, Lecompte F, Sillon JF (1999) Fractal analysis of the root architecture of *Gliricidia sepium* for the spatial prediction of root branching, size and mass: model development and evaluation in agroforestry. Plant Soil 209:167–180

Pages L (1999) Root system architecture: from its representation to the study of its elaboration. Agronomie 19:295–304

Pages L, Bruckler L, Pellerin S, Lafolie F (1997) Root architecture and water uptake. L'eau dans l'espace rural: production végétale et qualité de l'eau. Institut National de la Recherche Agronomique, Avignon

Pregitzer KS, DeForest JL, Burton AJ, Allen MF, Ruess RW, Hendrick RL (2002) Fine root architecture of nine North American trees. Ecol Monogr 72:293–309

Pritchard SG, Rogers HH (2000) Spatial and temporal deployment of crop roots in CO_2-enriched environments. New Phytol 147:55–71

Richards R, Passioura J (1981) Seminal root morphology and water use of wheat. II. Genetic variation. Crop Sci 21:253–255

Richards R, Passioura J (1989) A breeding program to reduce the diameter of the major xylem vessel in the seminal roots of wheat and its effect on grain yield in rain-fed environments. Aust J Agric Res 40:943–950

Riedell WE (1990) Rootworm and mechanical damage effects on root morphology and water relations in maize. Crop Sci 30:628–631

Riedell WE, Reese RN (1999) Maize morphology and shoot CO_2 assimilation after root damage by western corn rootworm larvae. Crop Sci 39:1332–1340

Robinson D (1990) Phosphorus availability and cortical senescence in cereal roots. J Theor Biol 145:257–265

Robinson D, Hodge A, Griffiths BS, Fitter AH (1999) Plant root proliferation in nitrogen-rich patches confers competitive advantage. Proc R Soc Lond Ser B Biol Sci 266:431–435

Rubio G, Walk T, Ge ZY, Yan XL, Liao H, Lynch JP (2001) Root gravitropism and below-ground competition among neighbouring plants: a modelling approach. Ann Bot 88:929–940

Rubio G, Liao H, Yan X, Lynch JP (2003a) Topsoil foraging and its role in plant competitiveness for phosphorus in common bean. Crop Sci 43:598–607

Rubio G, Zhu J, Lynch JP (2003b) A critical test of the two prevailing theories of plant response to nutrient availability. Am J Bot 90(1):143–152

Rundel PW, Nobel PS (1991) Structure and function in desert root systems. In: Atkinson D (ed) Plant root growth: an ecological perspective. Blackwell, Oxford, Spec Publ no 10, Br Ecol Soc, pp 349–378

Ryden JC, Syers JK, Harris RF (1973) Phosphorus in runoff and streams. Adv Agron 25:1–45

Schellenbaum L, Berta G, Ravolanirina F, Tisserant B, Gianinazzi S, Gianinazzi-Pearson V, Fitter AH (1991) Influence of endomycorrhizal infection on root morphology in a micropropagated woody plant species (*Vitis vinifera* L.). Ann Bot 68:135–141

Schenk HJ, Callaway RM, Mahall BE (1999) Spatial root segregation: are plants territorial? Adv Ecol Res 28:145–180

Shane MW, McCully ME, Canny MJ (2000) Architecture of branch-root junctions in maize: structure of the connecting xylem and the porosity of pit membranes. Ann Bot 85:613–624

Shiva V (1991) The green revolution in the Punjab. Ecologist 21:57–60

Smith SN (1934) Response of inbred lines and crosses in maize to variations of nitrogen and phosphorus supplied as nutrients. J Am Soc Agron 26:785–804

Smith SE, Read DJ (1997) Mycorrhizal symbiosis, 2nd edn. Academic Press, San Diego

Snapp SS, Lynch JP (1996) Phosphorus distribution and remobilization in bean plants as influenced by phosphorus nutrition. Crop Sci 36:929–935

Somma F, Hopmans JW, Clausnitzer V (1998) Transient three-dimensional modeling of soil water and solute transport with simultaneous root growth, root water and nutrient uptake. Plant Soil 202:281–293

Steudle E, Peterson CA (1998) How does water get through roots? J Exp Bot 49:775–788

Tatsumi J, Yamauchi A, Kono Y (1989) Fractal analysis of plant root systems. Ann Bot 64:499–503

Tinker P, Nye P (2000) Solute movement in the rhizosphere. Oxford University Press, New York

Tunney H, Carton OT, Brookes PC, Johnston AE (eds) (1997) Phosphorus loss from soil to water. CAB, Wallingford

Tyree MT, Ewers F (1996) Hydraulic architecture of woody tropical plants. In: Tropical forest plant ecophysiology. Chapman & Hall, New York, pp 217–243

Vercambre G, Doussan C, Pages L, Habib R, Pierret A (2002) Influence of xylem development on axial hydraulic conductance within *Prunus* root systems. Trees Struct Funct 16:479–487

Walk TC, van Erp E, Lynch JP (2004) Modelling applicability of fractal analysis to efficiency of soil exploration by roots. Ann Bot 94:119–128

Ward J, Kuhn C, Tegeder M, Frommer W (1998) Sucrose transport in higher plants. Int Rev Cytol 178:41–71

Watt M, Evans JR (1999) Proteoid roots. Physiology and development. Plant Physiol 121:317–323

Weaver J (1926) Root development of field crops. McGraw-Hill, New York

Webster CC, Wilson PN (1998) Agriculture in the tropics, 3rd edn. Blackwell, , Oxford

Whipps J (1990) Carbon economy. In: Lynch J (ed) The rhizosphere. Wiley, Chichester, pp 59–97

World Bank (2001) World development report: attacking poverty. Oxford University Press, Oxford

Worldwatch Institute (2001) State of the world 2001. WW Norton, New York

Wullschleger S, Lynch J, Berntson GM (1994) Modeling the belowground response of plants and soil biota to edaphic and climate change - what can we expect to gain? Plant Soil 165:149–160

Yan X, Liao H, Beebe SE, Blair MW, Lynch JP (2004) QTL mapping of root hair and acid exudation traits and their relationship to phosphorus uptake in common bean. Plant Soil (in press)

Yanni YG, Rizk RY, Abd El-Fattah FK, Squartini A, Corich V, Giacomini A, de Bruijn F, Rademaker J, Maya-Flores J, Ostrom P, Vega-Hernandez M, Hollingsworth RI, Martinez-Molina E, Mateos P, Velazquez E, Wopereis J, Triplett E, Umali-Garcia M, Anarna JA, Rolfe BG, Ladha JK, Hill J, Mujoo R, Ng PK, Dazzo FB (2001) The beneficial plant growth-promoting association of *Rhizobium leguminosarum* bv. *trifolii* with rice roots. Aust J Plant Physiol 28:845–870

Zhang H, Forde BG (1997) An *Arabidopsis* MADS box gene that controls nutrient-induced changes in root architecture. Science 279:407–409

Zhu J (2003) Composite interval mapping and physiological function of root traits conferring phosphorus efficiency in maize (*Zea mays* L.). PhD Diss, Pennsylvania State University, University Park, PA

Zobel RW (1986) Rhizogenetics (root genetics) of vegetable crops. Hortic Sci 21:956–959

Zobel R (1996) Genetic control of root systems. In: Waisel Y, Eshel A, Kafkafi U (eds) Plant roots: the hidden half, 2nd edn. Marcel Dekker, New York, pp 21–30

8 The Efficiency of Nutrient Acquisition over the Life of a Root

D.M. EISSENSTAT and A. VOLDER

8.1 Introduction

Our understanding of root physiology and morphology has been strongly influenced by short-term studies of very young roots of seedlings in solution culture (Clarkson 1985). Few species have been studied in detail, and most of these are annual crop species. Consequently, our perception of root physiology is narrowly based on only a few economically important species grown under managed conditions. While short-term physiological and anatomical studies of seedling roots are important, these studies provide little insight into the amounts of nutrients acquired by a root over its lifetime.

Plants vary widely in their root life span. In some species including apple, kiwi and grape, a sizeable fraction (e.g., 30 %) of the roots may live only a few weeks (Reid et al. 1993; Comas et al. 2000; Wells and Eissenstat 2001). In other species, like sugar maple or citrus (Hendrick and Pregitzer 1992; Eissenstat et al. 2000; Bouma et al. 2001; Tierney and Fahey 2001), the fine roots may live about a year or more. Species with long-lived absorptive roots likely exhibit mechanisms of nutrient acquisition that differ from those of species with short-lived roots. How have plants evolved to use limited energy reserves for the acquisition of mineral nutrients?

Nutrient availability in soil is highly heterogeneous in space and time. Consequently, efficient deployment of roots for nutrient acquisition can be strongly influenced by the ability of the root system to constantly relocate its most absorptive elements in the most favorable soil. Under many circumstances, root life span is probably closely linked to root efficiency (Caldwell 1979; Yanai et al. 1995; Eissenstat and Yanai 1997). Short life spans would be expected for thin roots of high uptake capacity, high maintenance respiration, and occurring in soils with fairly short-lived patches or where depletion zones readily form. Mechanistic modeling using an optimization or cost-benefit approach can be a powerful tool in explaining patterns of variation of roots

Ecological Studies, Vol. 181
H. BassiriRad (Ed.) Nutrient Acquisition by Plants
An Ecological Perspective

within a root system, among species, and in response to environmental changes.

In this chapter, we will review the current literature associated with nutrient acquisition efficiency over the life of the root. Root length and total absorptive surface area strongly influence nutrient acquisition, but do not completely explain differences among plants. As the root ages, changes in absorptive capacity, nutrient concentrations in the rhizosphere, as well as the costs to maintain the root length, root hairs, and mycorrhizal hyphae will influence the effectiveness of this absorptive surface area in nutrient acquisition. A complete study of root costs includes the energy required for root construction, growth, ion uptake and maintenance, as well as costs associated with the formation and maintenance of the mycorrhizal symbiosis. Benefits include acquisition of water and mineral nutrients by the root and its associated symbionts. Anatomical changes in roots over their life span will also be discussed. Lastly, we will illustrate the use of root efficiency modeling, and how this can be used to help explain different patterns of root foraging for nutrients. While our knowledge of root physiology as a function of root age is very limited, this information is critical to a better understanding of different plant strategies of nutrient acquisition in different environments.

8.2 Root Length and Absorptive Surface Area

8.2.1 Importance

The supply of water and ions to root surfaces can generally be increased by greater exploration of the soil by the roots and their associated mycorrhizal hyphae (Tinker and Nye 2000). Simulation models of solute transport to the root, and uptake kinetics at the root surface clearly indicate the importance of root length in explaining observed nutrient acquisition (Silberbush and Barber 1983). There also is considerable experimental evidence of the importance of root length for nutrient acquisition, especially in herbaceous plants. Nutrient mobility and the amount of root competition affect the extent that root length contributes to nutrient acquisition (Comerford, Chap. 1, this Vol.). For example, Andrews and Newman (1970) have demonstrated that, if some of the roots of all wheat seedlings in a pot are pruned, then P but not N uptake is diminished. If only some of the wheat plants are partially root-pruned, then the remaining plants take up more N but not P. These results indicate the importance of total root length for P acquisition (a relatively immobile nutrient), and the importance of N in plant competition, because of the broader depletion zones of N than of P. Similarly, root proliferation in patches was shown to be advantageous for nitrate uptake only when the plants were competing with other plants (Robinson et al. 1999). These results lead to the prediction that immobile nutrients like P are more influenced by the total length of root than nutrients like nitrate under steady-state conditions of solute supply. Under non-steady state conditions where pulses of nutrients occur and plants compete, plants with more root length will acquire more nutrients (e.g., nitrate), all else being equal.

When nutrients are scarce or when root length density is low, root length alone may not be a good predictor of nutrient acquisition. Additional absorptive surface area may be contributed by root hairs and mycorrhizal hyphae. The importance of root hairs for P acquisition has been demonstrated by simulation modeling, and by comparing nonmycorrhizal species of different hair length (Itoh and Barber 1983a, b). In a recent study, Bates and Lynch (2000a) demonstrated the importance of root hairs more directly by comparing root hair mutants with wild-type *Arabidopsis thaliana* in soils of low P availability. There are also numerous examples, under controlled greenhouse conditions, of mycorrhizal colonization enhancing P acquisition (Smith and Read 1997). Factors affecting plant response to mycorrhizal colonization include plant demand for phosphorus, plant development of root surface area without mycorrhizas (root length, root hair length and abundance, root length density), and soil P supply. In the field, studies demonstrating improved P acquisition by mycorrhizas are often less compelling. This may occur only during relatively brief periods of a plant's life history, such as during flower or seedling establishment (Fitter 1986b; Wallace 1987). A notable exception is the bluebell (*Hyacinthoides non-scripta*), a vernal geophyte with a sparse, unbranched root system with relatively thick (0.5–1 mm) roots. Merryweather and Fitter (1996) found that reduced VA mycorrhizal colonization over a 2-year period caused a 35 % reduction of P concentration in roots, and a 42 % reduction in bulbs after 2 years. In fertile soil, mycorrhizas may have a negligible effect on nutrient uptake, leading to reductions in plant growth due to carbon consumption (see Sect. 8.3.3).

8.2.2 Inconsistencies Between Root Length, Mycorrhizal Colonization and Observed Nutrient Uptake

Some studies have indicated that root length, while probably important, cannot in itself describe variations in resource acquisition among species. For example, in a study of competition among two grass species and a shrub, Caldwell et al. (1991a, b) demonstrated that while the shrub *Artemisia tridentata* typically has less than half the root length density of the tussock grasses *Agropyron desertorum* and *Pseudoroegneria spicata*, the shrub can obtain as much P from enriched patches as *A. desertorum*, and considerably more P than *P. spicata*. Thus, factors other than absorptive surface area seem to be influencing competition for P.

8.2.3 Root Architecture and Variation Among Fibrous Roots in a Single Root System

Not all root length is equally effective in nutrient acquisition. A root system can be considered as being composed of interconnected modules, each having its own physiology and life history. The position of each root in relation to other roots in the plant, and in relation to soil solute supply affects the effectiveness of each element in nutrient acquisition. Traditionally, the root length of the "absorptive" portion of the root system is defined by choosing an arbitrary size limit (e.g., all roots <2 mm in diameter), with little consideration for position within the root system or developmental stage. Consequently, this definition treats roots of very different order, diameter, physiology and function as being equal. Because of branching, the position of a root relative to others in the root system can have a large influence on its function. Root order can be described using a topological system (Lynch, Chap. 7, this Vol.) similar to that used in streams; here, we use the convention where first-order roots are those that are external links with tips, and do not have laterals (Fitter 1986a; Berntson 1997). Using this scheme, second-order roots would be those that have first-order laterals only. In woody plants, most of the fine laterals that are produced each year tend to be determinate in length, growing no more than a few centimeters in perhaps a few days, and then remaining at this length the rest of their life. When they die, they often leave behind a scar on the higher-order root (Pregitzer et al. 2002). These roots typically do not undergo secondary thickening (Brundrett and Kendrick 1988; Eissenstat and Achor 1999) and so, over their lifetime, function almost entirely for water and nutrient acquisition. These are also the roots that typically form a symbiosis with mycorrhizal fungi. Herbaceous dicots and grasses may exhibit more indeterminate root growth, continually expanding through the soil until seed set, with much of the root system dying back at the end of the growing season. Often, the level of branching in herbaceous dicots and grasses is more limited, with only about three levels of branchings. The main portion of the absorptive root system, however, is still typically the first- and second-order roots.

In contrast to first- and second-order roots, higher-order roots that support lower-order laterals in perennial plants may be semi-permanent members of the root system, despite their relatively fine (i.e., less than 1 mm) appearance (Eissenstat et al. 2000; Wells and Eissenstat 2001; Pregitzer et al. 2002). The long life span of higher-order roots should not be a surprise, for if a higher-order root dies, then all the lower-order roots dependent on this higher-order root for transport of photosynthates also die. Thus, plants should allocate more energy for the defense of higher-order roots, as indicated by their accumulation of condensed tannins and periderm formation (Eissenstat and Achor 1999; Enstone et al. 2001). While these secondary compounds may be important for the defense of the root, they often impede water and nutrient acquisition (Peterson et al. 1999), and probably mycorrhizal colonization. Thus, while the fine roots of woody plants are typically defined in terms of roots less than 1–2 mm in diameter, many of the higher-order fine roots probably play only a limited role in nutrient acquisition.

8.2.4 Variation in Root Life Span in Different Environments and Among Species

In general, data are limited on root longevity of different species and in different environments (reviewed by Bloomfield et al. 1996; Eissenstat and Yanai 1997, 2002; Gill and Jackson 2000). There remains much uncertainty on general patterns, partly because of shortcomings in methodology. Mechanisms of root shedding are unknown. No abscission layer has been clearly identified, as commonly found in leaf petioles. It is very hard to pinpoint an exact time at which a root is dead, as demonstrated by tetrazolium staining experiments (Comas et al. 2000). A small lateral root typically dies over its entire length in a fairly short time period (i.e., often on the order of days), and commonly leaves a scar on the surface of the mother root (Pregitzer et al. 2002). Root death may often be passive (Eissenstat and Yanai 1997). One hypothesis is that once the root has outlived its usefulness, the plant stops investing in defense compounds for that root, basically leaving the root vulnerable to herbivores and pathogens. In common bean, for example, root death was found to be minimal for plants grown in sand culture with limited soil fauna, even during pod fill and pod dry-down (Fisher et al. 2002). In the field, however, nearly 50 % of the bean roots died by pod dry-down in rich agricultural soil. Weak pathogens likely are able to parasitise roots once carbohydrates and defense compounds are low.

Several environmental factors have been associated with diminished life span, including temperature (Watson et al. 2000; Gill and Jackson 2000), parasitism and herbivory (Maron 1998; Eissenstat et al. 2000; Wells et al. 2002), and drought (Hayes and Seastedt 1987). Soil fertility has been shown to both increase and decrease root longevity, possibly suggesting that other factors interact with soil fertility to determine root longevity (reviewed by Eissenstat and Yanai 2002; also see Burton et al. 2000). Similarly, the myriad effects of elevated CO_2 on plant nutrition, plant water use, and rhizosphere soil chemistry complicate the interpretation of root life span responses in CO_2-enrichment studies. Compared to plants grown under ambient conditions, plants exposed to elevated CO_2 may exhibit increased (Johnson et al. 2000; Arnone et al. 2000), decreased (Fitter et al. 1997; Thomas et al. 1999), or no change (Gavito et al. 2001; Pritchard et al. 2001) in root longevity.

A cost:benefit approach can be used to integrate the effects of various environmental factors on root longevity, assuming that root life span can be predicted by the length of time required to reach maximum lifetime efficiency (Yanai et al. 1995; Eissenstat and Yanai 1997, 2002). For example, roots of a fast- and slow-growing species of maple differ considerably in their root physiology, which influences the predicted time required to reach maximum root efficiency in different soil types (Fig. 8.1; Comas et al. 2002). Lifetime efficiency always increases when a root is young, as the initial construction cost is amortized over a longer period (Fig. 8.1). As the root ages, uptake typically declines, causing efficiency to decrease. The data in Fig. 8.1 show that in poorly buffered sandy soil (Lilly loamy sand), maximum root efficiency is reached in about 80–120 days, depending on the species. The main factor lowering the efficiency of older roots is decreased uptake caused by depletion of P in the rhizosphere. By contrast, in the more fertile loamy soil (Hagerstown silt loam) with its high buffering capacity, nutrient depletion in the rhizosphere is insufficient to cause a downturn in root efficiency with root age.

Consequently, we predict shorter root life spans in poorly buffered soils, especially in the fast-growing maple with its thinner roots, higher uptake capacity, and higher maintenance respiration. Ideally, root efficiency modeling requires information on root benefits in relation to root costs over the lifetime of the root, which was not available in the study by Comas et al. (2002). These data are only beginning to emerge, and there is still considerable uncertainty in these costs and benefits in natural soils of high temporal and spatial heterogeneity (see Sects. 8.3.2 and 8.4.4).

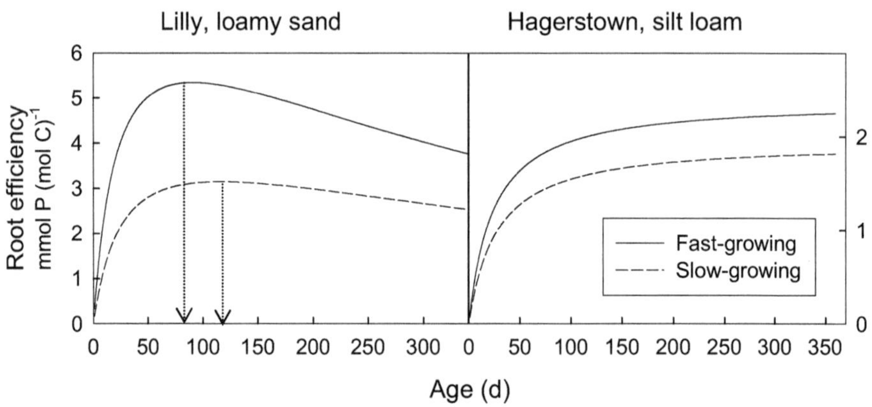

Fig. 8.1. Lifetime efficiency of first- and second-order roots of a fast- (*Acer negundo*) and slow-growing maple (*Acer sacharrum*; adapted from Comas et al. 2002) in two soils that differ widely in buffering capacity. *Arrows* indicate time to maximum efficiency where optimal life span is predicted. Root efficiency was calculated as cumulative P uptake per cumulative C cost, from root characteristics determined from seedlings in the greenhouse (Comas et al. 2002), and values for soil parameters from Kelly et al. (1992) and Bouma et al. (2001)

8.3 Costs of Root Production and Maintenance over a Root's Lifetime

8.3.1 Biomass Costs of Producing Root Absorptive Surface Area

8.3.1.1 Root:Shoot Biomass Partitioning

Typically, if soil resources such as nitrogen or phosphorus are limiting plant growth, plants increase the amount of root biomass allocation, and thus help maintain a "functional equilibrium" between shoot acquisition of C and root acquisition of mineral nutrients (Thornley 1972; Brouwer 1967, 1981; Gutschick and Pushnick, Chap. 4, this Vol.). At a whole-plant level, increasing the root:shoot ratio may enhance plant nutrient content, but it also increases biomass allocation to non-photosynthetic tissues. Therefore, it is not uncommon to find that while increased root-to-shoot ratios may help plant nutrient acquisition, the whole-plant relative growth rate is compromised (Gutschick and Pushnick, Chap. 4, this Vol.).

8.3.1.2 Specific Root Length

Root structure influences the allocation of root biomass for the production of root surface area. For example, species vary widely in their specific root length (SRL, root length:dry mass ratio; Eissenstat 1992), which strongly affects the amount of absorptive surface area for a given investment in root mass. There is also evidence that species of high SRL tend to more rapidly grow roots into disturbed soil or fertile soil patches (Eissenstat 1991; Fitter 1994; Bauhus and Messier 1999). Specific root length is controlled both by the length per unit volume (L/V; cm^{-2}) of the roots, and by the tissue density (root volume:dry mass ratio, V/M; g cm^{-3}) of the roots. Assuming cylindrical geometry, the L/V ratio is defined by the inverse of the square of the root radius. The radius of the terminal roots can vary an order of magnitude among species, from less than 0.05 mm in certain annual dicots, ericaceous species and grasses to greater than 0.5 mm among species in the Magnoliales, Cornaceae and Aliaceae (Fitter 1996). Tissue density can also vary widely among species, ranging from 0.09 to 0.14 in grasses (Ryser 1996; Bouma et al. 2003), and 0.05 to 0.20 in woody dicots (Wright and Westoby 1999; Eissenstat et al. 2000).

Specific root length varies widely with environmental conditions. However, the direction of change in SRL is not always predictable based on resource supply. Turf grasses, for example, can exhibit higher SRL in soil where water becomes less available (Huang and Fry 1998). Roots of *Festuca rubra*, by contrast, were found to have lower SRL when N was less available

(Paterson and Sim 2000). Because roots of different orders also vary widely in SRL, environmental effects on total root system SRL may be a result of differences in diameter or tissue density of the terminal laterals, or simply a result of greater production of new roots in a rapidly growing root system (Fitter 1985).

8.3.1.3 Root Hairs and Mycorrhizal Hyphae

Production of root hairs or extramatrical mycorrhizal hyphae can be a very efficient way by which a plant can increase absorptive surface for the same biomass allocation. While the diameter of the finest lateral roots of different plants may be typically on the order of 100–1,000 μm, root hairs are only 5–17 μm in diameter (Dittmer 1949), and the finest elements of extraradical mycorrhizal hyphal filaments are typically 2–7 μm (Smith and Read 1997). Assuming similar tissue densities among roots, root hairs and fungal hyphae, these differences in diameter suggest that for an equivalent biomass allocation, surface area production in the form of hairs and hyphae can exceed that of roots by factors on the order of 10–100.

As is the case for SRL, there is wide genotypic and phenotypic variation in the morphology of root hairs and fungal hyphae. Species vary widely in the density and length of root hairs. Root hairs range from less than 50 μm in length in onion and carrot to greater than 1,500 μm in spinach. Root hair density ranges from two hairs per millimeter of root in pine to about 100 hairs per millimeter of root in grasses and many annual dicots (Barber 1984; Tinker and Nye 2000). The frequency and size of root hairs also depend on environmental conditions. The direction of the response suggests increased foraging efficiency with greater limitation in soil nutrients. For example, root hairs are typically longer and denser at restricted supplies of P (Bates and Lynch 1996), nitrate (Bhat et al. 1979) and soil moisture (Mackay and Barber 1985).

Plants vary widely in the rate and extent to which their roots are colonized by mycorrhizal fungi (Smith and Read 1997; Finlay, Chap. 9, this Vol.). Environmental conditions that tend to depress colonization (and presumably extraradical hyphal growth) include high soil P, shade, and waterlogging. Vesicular-arbuscular (VA) mycorrhizal hyphal length can vary from 1 to 270 m m^{-1} root length in field soil (Miller et al. 1995; Smith and Read 1997). Different fungi and different host plants produce different amounts of mycelium. Ectomycorrhizal hyphal production has been less well studied. Reported ECM hyphal lengths in *Salix viminalis* (100–300 m m^{-1} root; Jones et al. 1990) and *Pinus taeda* (504 m m^{-1} root; Rousseau et al. 1994) are at the high end of the range reported in the VA mycorrhizas cited above. Unlike VA mycorrhizas, ectomycorrhizal roots develop a fungal sheath that may represent as much as 30–40 % of the biomass of these roots (Smith and Read 1997).

8.3.2 Costs of Root Respiration

8.3.2.1 Construction Cost and Growth Respiration

The cost of tissue production is commonly referred to as construction cost, and includes the C or energy content of the root plus the energy or respiration involved in tissue formation (i.e., growth respiration). Often, root construction cost is assumed to be a one-time cost that occurs when the root is born (Yanai et al. 1995) – of course, an oversimplification. In reality, as a root ages, it becomes more lignified and suberized but, for the ephemeral fine root system, it is normally considered acceptable to treat the costs for this continued construction as negligible compared to the initial construction costs.

It is often assumed that high construction costs indicate a long life span, but this assumption may be incorrect. Tissues with shorter life spans may have less carbon invested in defense compounds such as tannins and lignins, but often have more proteins to support rapid nutrient acquisition. For example, Poorter et al. (1991) compared root construction costs for fast- and slow-growing grass species, and concluded that fast-growing species tended to have less defense compounds, but higher protein concentrations, making construction costs similar among the plant species tested. Similar results have been found in comparisons of fast- and slow-growing trees (Comas et al. 2002).

Growth respiration is the respiration involved in reducing carbon compounds to bring about the formation of new plant matter. Specific respiratory costs for growth depend on the chemical composition of the roots. The main compounds of plant biomass are proteins, carbohydrates, lipids, lignin, organic acids, soluble phenols, and minerals. Compounds such as proteins, lipids, and lignins have higher respiratory costs of construction than do carbohydrates such as cellulose and starch. Penning de Vries et al. (1974) calculated values of oxygen consumption associated with the costs of producing one unit of each of these compounds. Assuming an ADP:O ratio of 2.37 for 24 herbaceous species, Poorter and Bergkotte (1992) calculated the oxygen consumption needed for the production of 1 g root biomass for each species. The respiration rate varied between 5.5 and 8 mmol O_2 g^{-1} dry root mass. They also found that fast-growing species had higher respiratory costs for growth, due mainly to the higher protein contents of these roots. Thus, there is little evidence that slow-growing, well-defended root tissues are more expensive to produce than fast-growing tissues (on a gram dry weight basis). This is because of the high amounts of proteins in fast-growing tissues, which offset their lack of defense compounds.

Only a few studies have examined differences in root growth respiration or construction costs in relation to environmental conditions. In general, environmental effects on these costs are modest. Wullschleger et al. (1997)

reported that, compared to trees grown at ambient CO_2, *Lirodendron tulipifera* and *Quercus alba* trees exposed to elevated CO_2 exhibited less than 5% lower root construction cost. In nonmycorrhizal Volkamer lemon seedlings, plants deficient in P exhibited about 6% higher root construction costs than those sufficient in P (Peng et al. 1993). Mycorrhizal colonization increased construction costs of Volkamer lemon roots about 8% for plants of equivalent P nutrition.

8.3.2.2 Maintenance Respiration

Maintenance respiration is that fraction of root respiration that is needed to maintain existing, mature cells in a viable state. Relatively little is known about the biochemistry and physiology of maintenance processes. Protein and carbohydrate turnover and maintenance of membrane potentials are considered major "maintenance" processes requiring energy. A more practical approach is to define maintenance respiration as the residual respiration not attributable to growth or ion-uptake processes, which includes respiration associated with the metabolism of exudates as well as respiration used to induce and maintain mycorrhizal infection. Bouma et al. (2001) estimated that belowground C expended for maintenance may be at least as large as that expended for root construction and growth respiration. In a study comparing protein turnover in fast- and slow-growing grass species, Scheurwater (1999) concluded that 22–30% of maintenance respiration was used in protein turnover, regardless of species. This constituted 11–15% of daily root ATP production.

Root hair production seems to add little additional cost to root maintenance, while mycorrhizal colonization can have a substantial additional maintenance cost. In a study of *Arabidopsis* root hair mutants, there was no clear pattern of greater respiration per unit root length for genotypes with abundant root hairs relative to those with no or very limited root hair development (Bates and Lynch 2000b). In citrus, mycorrhizal colonization caused maintenance respiration to increase by 39% compared to nonmycorrhizal plants of similar P nutrition (Peng et al. 1993).

8.3.2.3 Ion-Uptake Respiration

Ion-uptake respiration is used to absorb and assimilate solutes. The absorption and assimilation of nutrients can account for 50–70% of the respiratory energy in roots (Poorter et al. 1991). As much as 90% of the respiration associated with ion uptake can be devoted to the uptake and assimilation of nitrogen ions (Veen 1981). Depending upon the form of solute, costs for ion uptake can vary. For example, Bloom et al. (1992) found different costs associated with the uptake and assimilation of NH_4^+ and NO_3^- in barley. Costs associated with net N uptake (the balance of N influx and N efflux) can vary considerably between species, and appear to be linked to RGR. Costs associated with net N uptake were up to four times higher in slow-growing grass species than in fast-growing species (Poorter et al. 1991; Scheurwater et al. 1998). Scheurwater et al. (1999) found that N influx was similar between fast- and slow-growing species, but slow-growing species had higher rates of N efflux during periods of darkness. Thus, lower respiratory costs for N uptake do not necessarily mean that the plant has used less energy for nitrogen influx, but rather that it has retained a higher percentage of the N acquired (lower efflux).

8.3.3 Cost of Mycorrhizal Colonization and Maintenance

Most plants can benefit from forming a symbiosis with mycorrhizal fungi, at least at some periods of the year, when the demand for nutrients exceeds the amount that can be supplied by the roots alone. However, the symbiosis of mycorrhizal fungi with roots can have an appreciable C cost. For plants of similar size and P nutrition, belowground C costs may be on the order of 10–20 % higher in both VA and ectomycorrhizal compared to nonmycorrhizal plants (reviewed by Smith and Read 1997). Increased metabolic activity of the root, and respiration of the associated fungi both cause greater belowground expenditure (Peng et al. 1993; Rygiewicz and Anderson 1994). Costs of mycorrhizas are typically highest during the colonization phase when C demands for the fungus are high (Eissenstat et al. 1993). Costs vary widely as a function of, amongst others, the species of fungi, the mycorrhizal dependency of the host plant, and soil fertility. Indeed, mycorrhizal fungi can be parasitic on plants when the cost of the symbiosis exceeds the benefits (e.g., Johnson et al. 1997; Graham and Eissenstat 1998), which may be most common in soils of high fertility and when the fungus rapidly colonizes plants.

8.3.4 Changes in C Costs as the Root Ages

Not much is known about changes in tissue construction costs as the root ages. Bouma et al. (2001) reported that there was no change in C and N contents of fine apple and citrus roots as they aged. However, there was a dramatic decline in root respiration (Bouma et al. 2001; Fig. 8.2; also see Comas et al. 2000). The decline in root respiration without a corresponding decrease in N concentration might indicate that some of the N was used for other purposes such as the production of immobile defense compounds associated with root pigmentation. The effect of root age on root chemical composition may have a significant bearing on nutrient acquisition, and deserves further study.

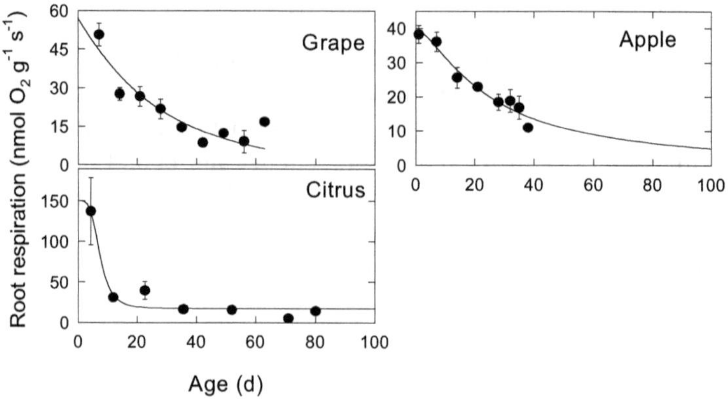

Fig. 8.2. Respiration of first- and second-order roots of Concord grape, apple (red chief delicious on M26 rootstock), and citrus (red grapefruit on sour orange rootstock) roots growing in the field. Roots of known age were collected from root boxes, and O_2 consumption (±SE) of the excised roots determined with an oxygen electrode. (Redrawn from Comas et al. 2000; Bouma et al. 2001)

Bouma et al. (2001) found that root respiration of fine citrus roots dropped to less than half of the initial levels within 10 days after the root was born (Fig. 8.2). Fine apple and grape root respiration also decreased rapidly; these were about half of the initial levels after approximately 20–30 days. Grape and apple respiration both exhibited an exponential decline with root age. However, citrus respiration remained quite stable after the initial decline (Fig. 8.2).

Mycorrhizal costs (and benefits) are likely to vary strongly with root age. The rapid colonization phase is a time of high C cost; transient whole-plant growth depression can occur at this time (Smith and Read 1997). In VA mycorrhizas, maintenance costs of mycorrhizal structures may decline with root age when arbuscules deteriorate and the root is occupied mainly by vesicles, the fungal storage organs. Older roots typically only have vesicles (Brundrett and Kendrick 1988). In ectomycorrhizal species, the fungus typically develops a more prominent hyphal sheath outside the root (Smith and Read 1997). Indeed, it is very likely that mycorrhizal fungi, especially ectomycorrhizal fungi, may have pronounced effects on the response curve of root costs as a function of root age and root longevity (e.g., Smith and Read 1997; Espeleta et al. 1999). Unfortunately, there is a notable lack of information on the costs (and benefits) of mycorrhizas as a function of root age, especially at the level of the individual root.

8.3.5 Other Costs

Other C costs involved with maintaining and expanding a root system include the exudation of C compounds from the roots, either through "active" exudation or sloughing of the cortex. Sloughing of the cortex has been known to increase microbial activity around the root, which may increase competition for nutrients between microbes and roots. Compounds that are exuded from the root can serve multiple functions – they can be used to decrease the pH around the root causing increased nutrient availability in calcareous soils (Hoffland et al. 1989; Grayston et al. 1997), to increase interplant root interference (alellopathy), to attract fungi/microbes (Grayston et al. 1997; Tarawaya et al. 1998; Yoder 2001), to improve nutrient uptake (chelation), and even to generate signals between plants or within root systems (Lambrecht et al. 2000). The quantity and quality of root exudates vary among life-forms (either herbaceous or woody), and with plant age. Seedlings have been reported to produce greater quantities and a more diverse range of carbohydrates than do mature trees, but mature trees exuded larger quantities of amino acids, amides and organic acids (Grayston et al. 1997). These differences suggest that root age may exert a strong effect on the type and amount of exudates from fine roots.

Marschner (1998) reviewed the role that root exudates can play in improving the efficiency of nutrient acquisition. Root exudates can facilitate the acquisition of P (via the release of acid phosphatases), Fe^{3+}, and other micronutrients (e.g., Zn^{2+}, Mn^{2+}, and Cu^{2+}). Other studies suggest that exudation increases when plants are grown under N-limited conditions. Paterson and Sim (2000) found that as much as 16% of the total daily assimilate was lost in the form of root exudates when *F. rubra* plants were grown under low N supply. This value decreased to 1.6% for plants grown under high N supply. These findings were supported by those of Hodge et al. (1996), who reported that 8.8 and 14% of plant carbon was released as exudate by *Festuca ovina* seedlings grown at high and low N over a 34-day growth period, respectively. In the same experiment, they found percentages for *Plantago lanceolata* to be 5.6 and 15% at high and low N, respectively. Other recent estimates also suggest that 7–15% of net carbon fixation is exuded during plant growth (Johansson 1992; Swinnen et al. 1995). Thus, root exudates can represent an appreciable carbon cost over the lifetime of a root, especially for nutrient-deficient plants in infertile soil.

8.4 Uptake of Mineral Nutrients over a Root's Lifetime

8.4.1 Initial Advantages to Root Growth

Root growth allows the plant to explore new areas in the soil. Either root or soil properties, or both, may limit uptake (Williams and Yanai 1996). The factors that make new roots better than old ones include diminished uptake capacity in older roots, and the depletion of nutrients in solution around active roots. For example, Bar-Tal et al. (1994) found that root pruning and subsequent production of new roots resulted in a N uptake that was higher than that of undisturbed plants with intact root systems. Yanai (1994) simulated the contribution of new root growth to total nutrient acquisition using a modified steady-state model of solute uptake. She found that root growth was most important for fast-growing plants and immobile nutrients, which is consistent with earlier predictions (Caldwell 1979). Moreover, root production and rapid rates of resource acquisition may be very important aspects of resource competition, especially in nutrient patches of short duration (Robinson et al. 1999).

8.4.2 Importance of Mycorrhizal Colonization

Mycorrhizal colonization allows plants to increase the amount of nutrients captured by the root system (reviewed by Smith and Read 1997). For example, calculations by Sanders and Tinker (1971) show that roots infected with mycorrhizal fungi can transport phosphate up to four times faster than uninfected roots. Vesicular-arbuscular mycorrhizas are most noted for increased acquisition of P, whereas ecto- and ericoid mycorrhizas can also increase N acquisition. Mechanisms of enhancement also vary, with VA mycorrhizas being primarily associated with enhanced acquisition solely of inorganic phosphate. By contrast, ecto- and ericoid mycorrhizal fungi also have the ability to increase the mineralization and acquisition of organic sources of P and N. Soil conditions, such as temperature and aeration, can greatly influence the potentially beneficial role of mycorrhizal fungi in overall nutrient acquisition. Little is known about the mechanisms by which nutrients absorbed by these fungi are transferred into the plants. In ectomycorrhizal systems, the root cortical cells may simply absorb inorganic phosphate leaked from hyphae in the Hartig net (Ashford et al. 1989), but in VA mycorrhizal systems the situation may be more complex.

The contribution of mycorrhizas to root nutrient acquisition has an important age component. When new roots grow into undisturbed soil with an established hyphal network, colonization may be very rapid. For example, Smith and Read (1997) reported that over 70 % of tomato (*Lycopersicon escu-*

lentum) root length exhibited internal fungal colonization inside the root cortex 2 days after transplantion into soil with established mycorrhizal plants. Arbuscules, the fungal organ primarily associated with nutrient exchange, typically form soon after colonization and reach a peak about 6 days after transplantion. However, arbuscules are usually short-lived, with a life span of only about 7 days, although slow-growing plants may exhibit somewhat longer-lived arbuscules (Smith and Read 1997). Older roots typically contain internal hyphae, which are probably less effective in nutrient transfer than arbuscules. External hyphal development begins as main branches that may branch into fan-like networks. Most estimates of time for development of external mycelium have been made at the whole-root-system level, and not at the level of an individual root in undisturbed soil. A timescale of about 1–2 months is typical for potted plants (Smith and Read 1997). Similar to colonization, the rate and extent of external hyphal development are strongly influenced by fungus and host species type, and environmental conditions such as light intensity and soil phosphorus concentration (Smith and Read 1997; Henry and Kosola 1999). Longevity of external hyphae is another important unknown, but is recognized to be strongly influenced by soil organisms such as collembolas, which may not be present in typical pot-culture experiments (McGonigle and Fitter 1988).

Similar to VA plants, ectomycorrhizal plants can have their roots colonized within 2–4 days of contact by the ectomycorrhizal hyphae, and have a fully developed Hartig net and mantle within 7 days (Smith and Read 1997). In the majority of ectomycorrhizal angiosperms (e.g., *Alnus*, *Betula*, *Eucalyptus*), the formation of the Hartig net is confined to the epidermal layers. In ectomycorrhizal gymnosperms, the Hartig net typically encloses several cortical cell layers as well as the epidermis (Smith and Read 1997). Downes et al. (1992) grew seedlings of *Picea sitchensis* in 20 × 20 cm transparent root observation chambers to follow morphological and anatomical changes of roots of known age, and inoculated with either the ectomycorrhizal fungus *Paxillus involutus* or *Tylospora fibrillosa*. Mycorrhizal roots over 50 days old tended to have diminished extramatrical mycelium. Roots 30 days old exhibited 30–50 % cortical cell death with a linear decline in living cortical cells until, by 110–140 days, only 5–15 % of the cortex was still alive. Hartig net senescence generally followed soon after cortical cell senescence. The coupled loss of the extramatrical mycelium and the interface of the cortical cells with the Hartig net clearly indicated loss of uptake capacity in older roots. Downes et al. (1992) concluded from these anatomical studies that ectomycorrhizal *Picea* seedling roots exhibited a major decline in function after 85 days.

8.4.3 Longevity of Root Tissues

8.4.3.1 Overview

There is a large literature relating anatomical features of the root to nutrient acquisition, including the development of the endo- and exodermis, of root hairs and mycorrhizal fungi, and of vascular tissue (Clarkson 1996). It should be noted that most of this research has focused on the seminal root of developing seedlings; much less attention has been given to anatomical development in the fine laterals. Even more problematic is the lack of work on older roots in the field. When one evaluates nutrient acquisition over the life of a root, one needs to consider also the anatomical changes that typically occur over that root's lifetime. Many features of a root, such as root hairs, mycorrhizal arbuscules and cortical cells, may persist only for a certain portion of a root's life.

Traditionally, ecologists have characterized absorptive roots, especially those of woody plants, by some arbitrary diameter limit (e.g., all roots <2 mm). While most of these roots may serve some role in the absorption of water and nutrients, many no longer have an epidermis, or even a cortex. For example, in loblolly pine (*Pinus taeda*), only 53 % of the root length of roots less than 2 mm in diameter was in a primary stage of development with no secondary vascular tissue, and most of these were mycorrhizal short roots (McCrady and Comerford 1998). Kramer and Bullock (1966) found that growing root tips typically accounted for less than 1 % of the total root surface area in stands of loblolly pine and yellow poplar (*Lirodendron tulipifera*). Okano and Omae (1996) found that white rootlets constituted 30 % of the root system dry mass of the tea tree (*Camellia sinensis* L.), accounting for 60 % of root respiration and 75 % of total nitrogen uptake. Roots that have formed a periderm typically lose their cortex, and may have greatly reduced capacity for water and nutrient absorption (e.g., Sands et al. 1982; van Rees and Comerford 1990). They also have lost their ability to form a symbiosis with mycorrhizal fungi. Even among fine roots that exhibit no periderm formation, in many species substantial mortality of epidermal and cortical cells may have occurred.

8.4.3.2 Longevity of Root Hairs and Epidermis

From a physiological perspective, root hairs can be very important for nutrient acquisition. Root hairs have been shown to exhibit intense H^+-ATPase activity, and are involved in the uptake of most major nutrients, including Ca^{2+}, K^+, NH_4^+, NO_3^-, Mn^{2+}, Zn^{2+}, Cl^-, and $H_2PO_4^-$ (Gilroy and Jones 2000). Using ^{32}P, it has been estimated that root hairs can satisfy as much as 60 % of

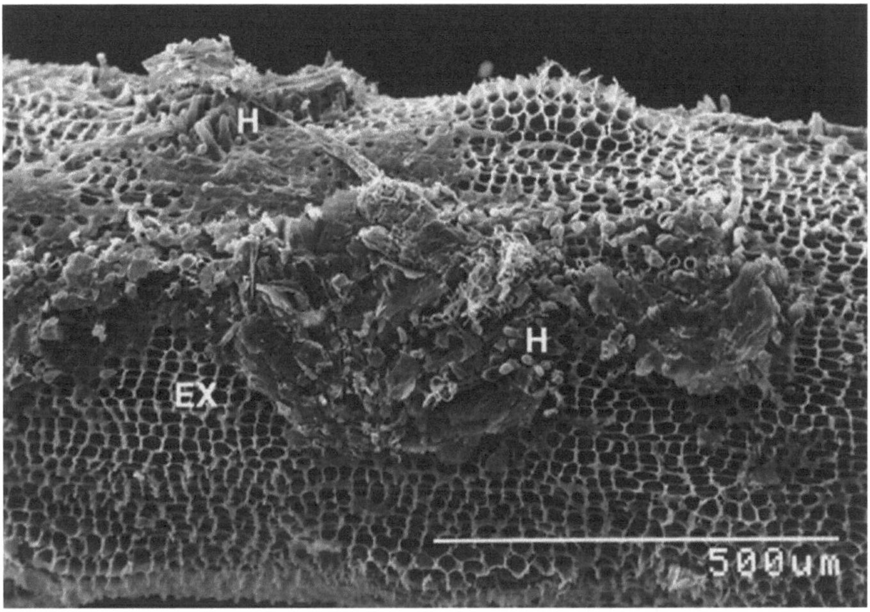

Fig. 8.3. Surface characteristics of typical sour orange roots collected from a Valencia orange rootstock trial (for details, see Eissenstat and Achor 1999). Most of the epidermis has sloughed away, exposing the radial walls of the exodermis (*EX*). The inner tangential walls of the exodermis are the effective barrier between the soil and the root. At this study site, citrus roots rarely have arbuscules, and only sparse portions of the root have root hairs (*H*), except in the very youngest roots. Presumably, most water and nutrient acquisition occurs from older roots, like the one illustrated here. Samples were fixed in glutaraldehyde and dehydrated in an ethanol series, critical-point dried with CO_2 and sputter coated with 80:20 gold:palladium for 90 s. Microscopy was performed with a Hitachi Model S530 at 100 ×

the plant P demand (Gahoonia and Nielsen 1998). From an ecological perspective, however, root hairs may be quite ephemeral and may not contribute greatly to the total nutrient acquisition over the lifetime of the root, especially for long-lived roots. In cereals, root hairs may live only from a few days to 2–3 weeks (Holden 1975; Henry and Deacon 1981; Fusseder 1987). Eventually the entire epidermis can die, especially for roots that form an exodermis. In citrus, healthy roots of the finest order often lack an epidermis (Fig. 8.3; Eissenstat and Achor 1999). Most northern hardwood species also typically do not have an epidermis on many of their fibrous roots (Brundrett and Kendrick 1988; Brundrett et al. 1990). Root hair death can be accelerated by exposure to dry soil (e.g., 10-day drought in *Cicer arietinum*; Spaeth and Cortes 1995). This is especially common in species such as corn (Stasovski and Peterson 1991), onion (Stasovski and Peterson 1993), and citrus (Eissenstat and Achor 1999), whose fine roots have a suberized exodermal layer just below the epidermis that protects the inner cortex. If a soil sheath forms around the root, however, root hairs may live considerably longer (Vermeer and McCully 1982).

8.4.3.3 Longevity of Cortex

In many fine lateral roots less than 2 mm in diameter, the cortex is an ephemeral tissue that dies sooner than the entire root. While cortical cell death can result from natural developmental processes associated with the expansion of the vascular cylinder as the root undergoes secondary development, cortical cell death is also common in roots that have not undergone secondary thickening. Most wetland plants, for example, exhibit cortical cell death and formation of aerenchyma in adventitious roots soon after initiation following flooding (e.g., within 2 weeks in *Rumex* spp.; Visser et al. 1996). Drought can also accelerate cortical cell death, especially in species like perennial ryegrass (*Lolium perenne*; Jupp and Newman 1987), grape (Mapfumo et al. 1994) and chickpea (Spaeth and Cortes 1995), which lack heavily suberized outer layers of tissues. In species with a heavily suberized and lignified exodermis (onion, Stasovski and Peterson 1993; citrus, Eissenstat and Achor 1999) or epidermis (e.g., kiwi; Lemon and Considine 1993), however, little cortical cell death may occur in roots exposed to dry soil.

As new, white roots begin to turn yellow and brown (associated with the accumulation of condensed tannins; McKenzie and Peterson 1995), much of the cortex dies in some species (grape, Richards and Considine 1981; apple, Head 1967; spruce, Downes et al. 1992; pine, Peterson et al. 1999, but see McCrady and Comerford 1998). In other species, especially those with a heavily suberized epidermis and exodermis, the cortex may remain viable for the entire life span of the root (e.g., citrus, Eissenstat and Achor 1999; sugar maple, Brundrett and Kendrick 1988). Peterson et al. (1999) suggest that root maturation from white to yellow/brown in the condensed tannin zone should diminish uptake capacity, with limited symplastic uptake via the passage cells in the endodermis. In addition, the death of cortical cells also is typically followed by a loss of mycorrhizal fungal organs used for nutrient exchange (e.g., Downes et al. 1992). Formation of a cork periderm should result in a more complete barrier to water and nutrient absorption. Nevertheless, water and nutrient absorption can still occur in roots with a cork periderm, presumably through occasional cracks in this otherwise impermeable layer (Kramer and Bullock 1966). Because most of the fine root system may be brown and devoid of living cortex, at least in trees, roots with a cork periderm undoubtedly still play an important role in water, and perhaps also nutrient absorption (Kramer and Bullock 1966).

8.4.3.4 Longevity of Xylem Vessels

Actively growing roots may have immature xylem a considerable distance (i.e., 15–40 cm) from the root tip (McCully 1999). Living xylem vessels with cross-walls are not open conduits – consequently, they are not effective in conducting water. In the fine lateral roots of corn, for example, up to 40 % of the vessels are closed and nonconducting (McCully 1999). It is not known at what age xylem vessels in lateral roots typically mature, especially in dicots. In grasses, the transition from closed to open large xylem vessels is associated with decreases in the relative water content of the cortical tissues, and death of the epidermis (McCully 1999).

8.4.4 Changes in Nutrient Uptake Capacity as the Root Ages

Young roots are generally considered to have larger uptake capacities (capacity to take up nutrients at the root surface when nutrients are not limiting; i.e., V_{max}) than older roots, and thus the proportion of young roots in the whole root system can strongly affect overall plant nutrient uptake. A number of studies have compared "woody" or "suberized" roots with white roots in various tree species (Van Rees and Comerford 1990, and references cited therein). While white roots typically have higher rates of water and mineral nutrient uptake than do woody roots, the magnitude of the difference varies considerably. For example, white roots have 74 % faster K^+ uptake than do woody roots in *Prunus*, but only 22 % faster P uptake (Atkinson and Wilson 1980). In contrast to the relatively small effects of root age on nutrient acquisition, Kramer and Bullock (1966) found that water uptake in *P. taeda* white roots was on average about five times faster than that recorded in woody roots.

Several studies have measured nutrient uptake activity along a root axis of young seedlings. In these studies, it was found that the young region directly behind the meristem exhibits the greatest nutrient uptake activity (Russell and Sanderson 1967; Clarkson et al. 1968; Clarkson 1991). Gao et al. (1998) developed a method to calculate the age of a whole root system, and estimated that nitrogen uptake capacity was decreased to less than 50 % within 10 days. Root systems with an integrated age older than 10–15 days maintained a steady, slowly decreasing nitrogen uptake rate as a function of age. This would suggest that plants with higher fine root turnover rates would maintain a higher nutrient uptake capacity, as high turnover rates would decrease the integrated age of a root system.

Only few studies have directly measured nutrient uptake on differently aged, first-order roots in older plants (Bouma et al. 2001; Volder et al. 2005). Bouma et al. (2001) measured ^{32}P uptake rates of single, excised first- and second-order roots for two tree species, apple and citrus (Fig. 8.4). They found

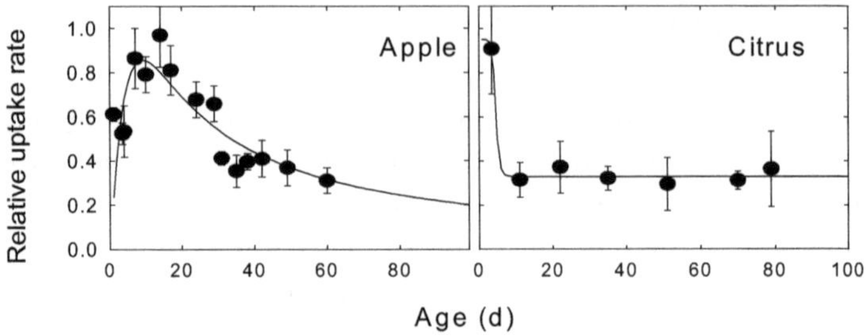

Fig. 8.4. Relative uptake rate of phosphorus as a function of root age in apple (red chief delicious on M26 rootstock) and citrus (red grapefruit on sour orange rootstock) roots growing in the field (from Bouma et al. 2001). Relative P uptake (±SE) was determined by measuring ^{32}P uptake from 1-cm excised root segments from mature trees at concentrations ranging from 1 to 1,000 μM P

marked differences between the two species, the longer-lived, coarser, citrus roots (median life span 300 days) showing a decreasing P uptake capacity about 4 days after birth, whereas apple roots (median life span 30 days) did not diminish P uptake capacity until they were about 25 days old. Maximum P uptake capacity was more than five times faster in apple than in citrus. After peak uptake capacity had been reached, citrus root values dropped to a stable level of about 35 % of maximum capacity that did not change further with root age, whereas apple roots exhibited a gradual but continuous decline with root age. A rapidly decreasing uptake capacity was also found for the uptake of NO_3^- in intact young roots of tomato (Volder et al., unpubl. data) and grape (Volder et al. 2005) seedlings grown in the laboratory, and 25-year-old mature grapevines in the field (Volder et al., unpubl. data). Within 5 days of root birth, N uptake was decreased by 50 % in all of these species. The pattern was best described by a logarithmic decrease with age for all the species, with very high NO_3^- uptake rates during the first 2 days and a rapid decline in uptake capacity thereafter.

In summary, nutrient uptake capacity typically diminishes with root age. The specific pattern of decline is affected by the type of mineral nutrient and the species of plant. Plants with long-lived roots likely have lower maximum uptake capacities than do plants with short-lived roots, but long-lived roots may exhibit less decrease in uptake capacity with age. Further study is needed to test these predictions.

8.5 The Efficiency of Nutrient Acquisition over a Root's Lifetime

8.5.1 Efficiency Concepts and Model Simulations

Root efficiency, E, can be defined as the rate of resource acquisition divided by the rate of carbon expenditure (Yanai et al. 1995; Eissenstat and Yanai 1997). Both daily efficiency (E_{daily}) and lifetime efficiency ($E_{lifetime}$) can be used to describe costs of nutrient foraging. The daily efficiency of the root, E_{daily}, represents the instantaneous efficiency at a single point in time, using appropriate rates of *uptake* (for example, mmol P g^{-1} root day^{-1}) and *cost* (mol C g^{-1} root day^{-1}). If daily costs and benefits are summed over time, then the integrated lifetime efficiency, $E_{lifetime}$, can be calculated. In our examples, cumulative *uptake* has units of mmol P or N/g root, while cumulative *cost* has units of mol C/g root. Both E_{daily} and $E_{lifetime}$, therefore, have units of mmol P or N/mol C. While $E_{lifetime}$ is theoretically most closely related to the overall efficiency of tissue deployment, E_{daily} can be a useful measure of instantaneous costs of nutrient uptake.

We have used the relationships between P uptake and root age (Fig. 8.4), and between respiration and root age (Fig. 8.2) to model E_{daily} and $E_{lifetime}$ of individual roots of citrus (Fig. 8.5; Bouma et al. 2001). Citrus groves are fertilized with P and other nutrients, and are commonly planted in sandy soils having low inherent fertility. If root P uptake does not deplete the soil, as might occur in a regularly fertilized grove, then efficiency remains high, and the optimal life span is infinite. Alternatively, if the P in the soil solution in the rhizosphere is depleted by the root over time, as predicted in the Candler fine sand, then E_{daily} of the root declines after about 35 days, and $E_{lifetime}$ peaks at about 120 days. The observed life span of citrus roots under low biotic pressure is about 300 days at this location (Eissenstat et al. 2000), suggesting that the roots remain effective in nutrient capture longer than predicted by the optimization model. Other factors not presently included in the efficiency model, such as P uptake by mycorrhizal fungi and P fertilization every 90 days, likely contribute to extend the P acquisition of older roots.

Some studies have also examined instantaneous or daily E levels. Scheurwater et al. (1998) measured the specific respiratory cost of nutrient uptake for a fast- (*Dactylis glomerata*) and a slow-growing species (*Festuca ovina*) in solution, where typically no depletion zones form. They found that the slow-growing grass had a respiratory cost per unit nitrogen taken up that was three times higher because of greater efflux. Clearly, there can be a considerable range in the efficiency of nutrient uptake among species grown in similar environments.

Volder et al. (2005) calculated E_{daily} in solution culture using nitrate uptake estimates from greenhouse-grown grape roots immersed in ^{15}N-labeled solu-

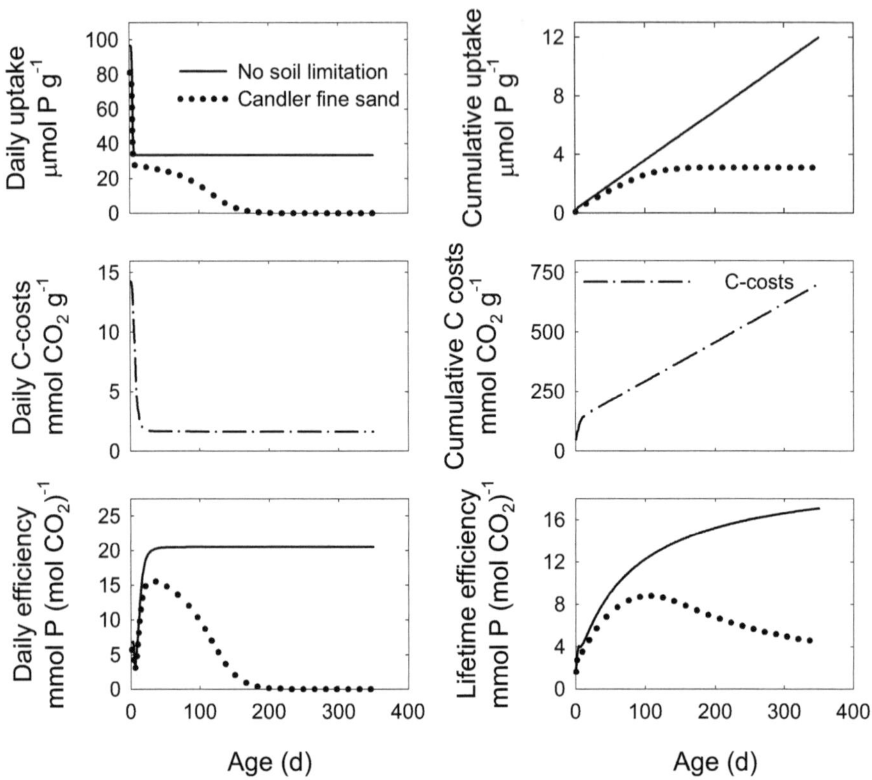

Fig. 8.5. Daily (E_{daily}) and lifetime efficiency ($E_{lifetime}$; mmol P mol^{-1} C) of citrus roots, based on C costs and P uptake. Carbon costs (*dashed line with one dot*) were calculated by multiplying the respiration rate (nmol O_2 g^{-1} s^{-1}) as a function of root age (Fig. 8.2) by a respiratory coefficient (RQ) of 1.1 mol CO_2 mol^{-1} O_2, and integrating over time with a time step of 1 day. Uptake capacity for P (i.e., I_{max}) was assumed to follow the age relationship shown in Fig. 8.4. Uptake affinity (K_m) was assumed not to change with root age. These data are slightly different from those published in Bouma et al. (2001) because of an incorrect effective diffusion coefficient in the original paper. The correct effective diffusion coefficient is 4.1×10^{-8} cm^2 s^{-1}. Carbon cost at age=0 is assumed to be equal to the carbon content of the root at age=1; the respiratory costs at day=1 are included. Soil P supply was assumed to be either not limiting (*solid line*) or limited by the soil characteristics of Candler fine sand where the trees were growing (*dotted line*). We assumed C costs were independent of P supply. Simulations suggest that unless soil depletion zones form around the root, limiting P uptake (see Candler fine sand simulation), $E_{lifetime}$ never reaches a maximum level and the root should not be shed. Thus, soil characteristics are at least as important as root characteristics in determining when $E_{lifetime}$ is maximized and when the roots should be shed

tion. Both N uptake and root respiration declined rapidly with root age (within 10 days, roots measured varied between 1 and 80 days in age). However, respiration declined slightly faster than N uptake, causing E_{daily} to increase with root age. In the field, these grape roots had median life spans varying in the range of 42–165 days, depending on year and time of birth (Anderson et al. 2003). The actual nitrate uptake rates from the soil are likely slower than those measured in solution culture. Respiration is largely a function of growth, ion uptake, and maintenance. For older first-order roots, growth respiration and ion-uptake respiration (based on low uptake capacity) are probably minor components of total root respiration. Thus, despite older roots having high efficiency, the large decrease in maintenance respiration suggests a gradual abandonment of the root by the plant, by way of not investing in defense compounds or in the energy required to maintain membrane potentials for an already ineffective root (in terms of nutrient uptake).

8.5.2 Problems Associated with Efficiency

8.5.2.1 Efficiency Versus Effectiveness

Higher efficiency does not always lead to higher plant fitness. In a competitive environment where resources are available only for short periods, rapid resource acquisition, rather than high efficiency, may be a key to plant success (e.g., Eissenstat and Caldwell 1988; Robinson et al. 1999). High expenditures for rapid resource acquisition may be an ecologically effective strategy if fitness of neighbors is diminished to a greater extent than is the fitness of the individual exhibiting rapid growth.

Plants may also overproduce tissues as a means of coping with herbivory, or as insurance against extreme events. The notion that plants might not be efficient in resource use was underscored in a review by Thomas and Sadras (2001). They argue that there may be many instances where plants may support large numbers of "unproductive" tissues that may provide secondary benefits for N storage, as a buffer against herbivory, and as a way of offloading excess C and other nutrients. For example, Thomas and Sadras (2001) speculate that plant species in fertile environments may exhibit high rates of leaf and root turnover, not in response to a reduced need for nutrient conservation (Berendse and Aerts 1987; Aerts 1990, 1999), but rather because of a greater need to offload excess resources associated with overproduction of carbohydrates. Consistent with the excess tissue hypothesis of Thomas and Sadras (2001) are the arguments that plants may use the alternative respiratory path as an "energy overflow" pathway (Lambers 1987), and the evidence from the elevated CO_2 literature that shows an average of 42 % enhanced soil respiration (root plus microbial respiration) in response to elevated CO_2 without an increase in shoot growth (Zak et al. 2000). Our view is that while there may be times when plants appear "wasteful", especially over short time spans, maintaining redundant absorptive tissues to offset the risks of herbivory or extreme weather events still follows general concepts of optimization in an uncertain environment. Consequently, broad economic analogies (sensu Bloom et al. 1985) of plant resource acquisition and allocation may be useful tools as a first approximation for interpreting plant responses to multiple resource limitations and strategies for tissue deployment. More specific and quantitative models of nutrient use efficiencies are discussed by Gutschick and Pushnick (Chap. 4, this Vol.).

8.5.2.2 Problems of Currency

Ideally, efficiency should be unitless, so that both benefit and cost are defined by one and the same currency. There are many definitions of nutrient acquisition efficiency (e.g., Koide and Elliott 1989; Koide et al. 1999), and some use a mineral nutrient (e.g., N or P) for expression of both benefit and cost. We are particularly interested in a definition that embraces both costs of construction and costs of maintenance over the lifetime of individual roots. Costs of ion uptake may also be included in this definition (Veen 1981; Bouma et al. 1996). While costs associated with tissue construction are amenable to the use of mineral nutrients as a currency, the relationship of mineral nutrients to the costs of tissue maintenance and ion uptake is more problematic. In choosing a currency to assess costs of root production and maintenance under a wide range of conditions, an estimator of the energy allocated for nutrient acquisition (e.g., carbon) tends to have the widest utility. Even in what might be considered open habitats, there is evidence that energy limits growth and reproduction. For example, it has been shown that partial shading leads to a loss of plasticity in P uptake capacity (Jackson and Caldwell 1992) and root proliferation (Bilbrough and Caldwell 1995) of the perennial tussock grass, *Agropyron desertorum*. Taken together, these studies highlight the value of C as a broadly defined currency of costs of nutrient acquisition. While we recognize that C supply does not always constrain growth and reproduction (e.g., Arnone and Körner 1995), as a general tool, no other single currency is as effective as C (or glucose equivalents thereof) for estimating the costs of root construction, maintenance, and ion uptake.

It also is reasonable to express efficiency in units of C benefit/C cost. This can be done by determining the ratio of N or P acquisition over a given time interval to the C gain over the same time interval (measured by whole-plant gas exchange or estimated by whole-plant C increase in growth) to obtain the exchange ratio during that time interval. It should be emphasized that these estimates only approximate the exchange ratio of a mineral nutrient with C, as storage and costs of transport are not generally considered. The reader is referred to Bloom et al. (1985) for a more complete discussion on exchange ratios of different currencies.

8.6 Heterogeneous Soil

8.6.1 Overview

Soil resources are unevenly distributed in space and time. New root growth is most advantageous when roots are deployed in the most favorable soil patches where soil nutrient availability is highest (Robinson, Chap. 3, this Vol.). Over the life of the root, soil nutrient availability can diminish for a number of reasons, including changes in temperature, soil moisture, and nutrient depletion by plants and microbes. A key to maximizing root efficiency is to reduce root costs as nutrients become less available, and to reallocate photosynthate to roots at more favorable soil locations. There are a number of studies that suggest that plants can adjust root activity to favor roots in patches of soil highest in nutrients, diminishing allocation of carbohydrates to those roots in less favorable soil (Huang and Eissenstat 2000). Pregitzer et al. (1993), for example, found that fine roots produced in hardwood forest lived longer in patches of water and water plus nitrogen than in untreated soil. The tendency to proliferate in patches varies among species, and is likely influenced by the costs of root construction. Consequently, nutrient acquisition over the life of the root is strongly affected by changes in soil nutrient availability, and root costs that may occur over the root's lifetime.

8.6.2 Effects of Water and Temperature

Water availability can dramatically affect soil nutrient availability, root physiology, and plant nutrient acquisition. If respiration does not decrease at a rate similar to those of water or nutrient uptake, then the roots become more costly to the plant. Moreover, dry surface soil is often associated with high soil temperature, potentially further increasing root costs. Plants can reduce root costs in dry soil by shedding roots quickly, or by reducing respiration. Desert succulents and many grass species cope with dry soil by rapidly shedding the fine lateral roots (Eissenstat and Yanai 1997). Immediately following drought, these species are capable of water uptake from main lateral roots and nodal roots (Lauenroth et al. 1987; Huang and Nobel 1993). New root growth soon follows, which in some species occurs within hours of rewetting (Lauenroth et al. 1987; Carmi et al. 1993); these roots may have a hydraulic conductivity three times higher than that of older roots (e.g., Huang and Nobel 1993).

Nutrient uptake also recovers quickly after drought in some species. For example, *Larrea tridentata* exhibited a fivefold increase in $\delta^{15}N$ as soon as 3 days after watering, compared to unwatered controls (BassiriRad et al. 1999). Nutrient uptake recovery in *Larrea* was attributed primarily to increased root growth, similar to that found in wheat (Brady et al. 1995) and barley (Shone and Flood 1983).

Some species rarely shed their roots in dry surface soil. Citrus roots, for example, when exposed to high temperatures and prolonged drought in sandy soil, strongly decrease respiration but exhibit little or no root shedding (Kosola and Eissenstat 1994; Bryla et al. 1997; Espeleta and Eissenstat 1998). In addition, citrus roots that were exposed to dry soil were unresponsive to an increase in soil temperature (i.e., respiration remained low, regardless of temperature), while roots in wet soil responded to changes in soil temperature by slowing rates at higher temperatures (roots exposed to temperatures>25 °C for more than 3 days had the same respiration rate as those at 25 °C; Bryla et al. 1997, 2001). Furthermore, citrus roots exposed to 43 days of drought were capable of rapidly resuming P and water uptake when watered and fertilized, indicating little or no loss of activity for P uptake by existing roots that could have been caused by exposure to dry soil (Eissenstat et al. 1999; also see Matzner and Richards 1996). Thus, we predict that species that construct coarse absorptive roots with a heavily lignified exodermis will tend to tolerate dry soils, reduce C losses by minimizing respiration at times of drought, and exhibit little loss in root activity after drought periods.

8.6.3 Effects of Nutrient-Rich Patches

In an extensive review of plant responses to non-uniform supplies of nutrients, Robinson (1994) found that two thirds of the measurements of root growth showed increased root proliferation in response to a localized nutrient supply. Overall root:shoot ratio generally increased (50 % of studies), or remained stable for heterogeneously supplied plants. Responses were dependent on time, plant species and nutrient, and plant nutrient status. Plants that were nutrient-deprived generally responded to localized nutrient enrichment by increasing uptake per unit root mass. Also, the extent to which a root system has access to patches of nutrients affects root uptake physiology – the smaller the fraction of the root system in the patch, the higher the uptake rate per unit root.

Not always does a higher proliferation rate in the patch equal a higher nutrient capture. For example, Hodge et al. (1998) found that capture of N from decomposing organic residue was not related to the degree of proliferation within the patch. Using two lupin species, *L. angustifolius* and *L. pilosus*, Dunbabin et al. (2001) found that only one species (*L. angustifolius*) had the capacity to increase nitrate uptake rate when locally supplied with high nitrate, compared to a uniformly high nitrate supply. Plants grown singly in pots may mask some of the advantages of root proliferation. When plants compete with neighbors, the advantages of rapid root proliferation for nitrogen capture may become more apparent (Robinson et al. 1999; Hodge et al. 1999).

Root foraging in nutrient-rich patches can be energetically expensive. For example, Cui and Caldwell (1997) examined the effects of partial shading on nitrate and phosphate acquisition from soil patches in arid-land plants, using a shading-treatment designed to simulate the kind of light competition that can occur in shrub-steppe vegetation. The impact of shading on nutrient foraging was quite dramatic. For example, unshaded *Artemisia* seedlings acquired 54% more P than did shaded plants in the uniform nutrient treatment, and 185% more in the patchy nutrient treatment. If shading affected only plant demand, then one would expect that shading would diminish P uptake similarly in the uniform and patchy treatments. Other studies in this same environment have demonstrated that partial shading can diminish both root proliferation in nutrient-rich patches (Bilbrough and Caldwell 1995), and enhancement of nutrient uptake kinetics (Jackson and Caldwell 1992).

The higher efficiency in nutrient-rich patches should lead to longer root life spans in these patches. Eissenstat et al. (2001), for example, showed that apple roots grown in permanent high-nitrogen patches lived longer and had a higher total nutrient gain than those in low-nutrient patches. Similar results were found in northern hardwood forests with the addition of water and nutrients (Pregitzer et al. 1993; Fahey and Hughes 1994). Note that contrasting patterns of root longevity in fertile vs. unfertile patches, where competition for carbohydrates occurs among different axes of the same plant, are not likely to be equivalent to comparisons of plants at sites of different fertility (see Eissenstat and Yanai 2002).

Patch duration can of course influence the potential benefits of root proliferation. Cui and Caldwell (1997) varied pulse duration from 0.5 to 72 h but applied the same total amount of nitrate to an arid-land tussock grass (*Agropyron desertorum*) and shrub (*Artemisia tridentata*). Plants that had been exposed to longer pulses acquired significantly more nitrate from a 30-min tracer pulse than those plants exposed to shorter pulses. However, longer pulse durations did not increase root nitrate uptake activity for these two species. The higher nitrate acquisition capacity in the longer-pulse treatments was primarily due to the greater root mass that had developed.

From an economical point of view, it makes sense to maintain roots longer in patches of higher nutrient enrichment, as this maximizes nutrient capture. However, the duration of the patch also influences optimal root life span (Fitter 1985). If the patches are short lived, as they often are in low-nutrient habitats, construction costs of the roots may not be recovered by the additional nutrient gain of patch proliferation. One might predict that enhanced nutrient uptake kinetics (i.e., increased V_{max}, decreased K_m, or both) might be a more viable option under these conditions (Fitter 1994; Eissenstat and Yanai 1997). In more permanent patches of high nutrients, one would expect plants to show a proliferation of roots, and to increase the life span of the roots.

8.7 Summary

Plants vary widely in root life span, which has important consequences for nutrient acquisition. The deployment of short-lived roots should correspond to a high surface area:mass ratio (small diameter, low tissue density), high nutrient uptake capacity, and a high respiration rate with a rapid decline in physiological capacity with age. The opposite traits should be associated with long-lived roots. Studies of patterns of nutrient acquisition and respiration over the life of a root are limited, but are generally consistent with this hypothesis. For long-lived roots, the importance of root hairs and possibly vesicular-arbuscular mycorrhizal fungi may be diminished if the epidermis and arbuscules persist only during a small fraction of the overall life of the root. Too much emphasis has been placed on the role of young roots of potted plants; the function of older roots in a fine root network deserves greater study.

Acknowledgments. Some of the work reviewed was supported by the United State Department of Agricutlure (NRI 9403081, 9735107) and the National Science Foundation (BSR 911824, IBN 9596050), for which the authors are grateful. We would like to thank Ruth Yanai and Tjeerd Bouma for their helpful reviews.

References

Aerts R (1990) Nutrient use efficiency in evergreen and deciduous species from heathlands. Oecologia 84:391–397

Aerts R (1999) Interspecific competition in natural plant communities: mechanisms, trade-offs and plant-soil feedbacks. J Exp Bot 50:29–37

Anderson LJ, Comas LH, Lakso AN, Eissenstat DM (2003) Multiple risk factors in root survivorship: a 4-year study in Concord grape. New Phytol 158:489–501

Andrews RE, Newman EI (1970) Root density and competition for nutrients. Oecol Plant 5:319–334

Arnone JA III, Körner Ch (1995) Soil and biomass carbon pools in model communities of tropical plants under elevated CO_2. Oecologia 104:61–71

Arnone JA III, Zaller JG, Spehn EM, Niklaus PA, Wells CE, Körner C (2000) Dynamics of root systems in native grasslands: effects of elevated atmospheric CO_2. New Phytol 147:73–85

Ashford AE, Allaway WG, Peterson CA, Cairney JWG (1989) Nutrient transfer at the fungus-root interface. Aust J Plant Physiol 16:85–97

Atkinson D, Wilson SA (1980) The growth and distribution of fruit tree roots: some consequences for nutrient uptake. In: Atkinson D, Jackson JE, Sharples RO, Waller WM (eds) Mineral nutrition of fruit tree roots: some consequences for nutrient uptake. Butterworths, London, pp 137–150

Barber SA (1984) Soil nutrient bioavailability: a mechanistic approach. Wiley, New York

Bar-Tal A, Feigin A, Rylski I, Pressman E (1994) Root pruning and N-NO_3 solution concentration effects on nutrient uptake and transpiration by tomato. Sci Hortic 58:77–90

BassiriRad H, Tremmel DC, Virginia RA, Reynolds JF, de Soyza AG, Brunell MH (1999) Short-term patterns in water and nitrogen acquisition by two desert shrubs following a simulated summer rain. Plant Ecol 145:27–36

Bates TR, Lynch JP (1996) Stimulation of root hair elongation in *Arapidopsis thaliana* by low phosphorus availability. Plant Cell Environ 19:529–538

Bates TR, Lynch JP (2000a) Plant growth and phosphorus accumulation of wild type and two root hair mutants of *Arabidopsis thaliana* (Brassicaceae). Am J Bot 87:958–963

Bates TR, Lynch JP (2000b) The efficiency of *Arabidopsis thaliana* (Brassicaceae) root hairs in phosphorus acquisition. Am J Bot 87:964–970

Bauhus J, Messier C (1999) Soil exploitation strategies of fine roots in different tree species of the southern boreal forest of eastern Canada. Can J For Res 29:260–273

Berendse F, Aerts R (1987) Nitrogen-use-efficiency: a biologically meaningful definition? Funct Ecol 1:293–296

Berntson GM (1997) Topological scaling and root system architecture: developmental and functional hierarchies. New Phytol 135:621–634

Bhat KKS, Nye PH, Brereton AJ (1979) The possibility of predicting solute uptake and plant growth response from independently measured soil and plant characteristics. VI. The growth and uptake of rape in solutions of constant nitrate concentration. Plant Soil 53:137–167

Bilbrough CJ, Caldwell MM (1995) The effects of shading and N status on root proliferation in nutrient patches by the perennial grass *Agropyron desertorum* in the field. Oecologia 103:10–16

Bloom AJ, Chapin FS III, Mooney HA (1985) Resource limitation in plants: an economic analogy. Annu Rev Ecol Syst 16:363–392

Bloom AJ, Sukrapanna SS, Warner RL (1992) Root respiration associated with ammonium and nitrate absorption and assimilation in barley. Plant Physiol 99:1294–1301

Bloomfield J, Vogt K, Wargo PM (1996) Tree root turnover and senescence. Root structure and sites of ion uptake. In: Waisel Y, Eshel A, Kafkafi U (eds) Plant roots: the hidden half, 2nd edn. Marcel Dekker, New York, pp 363–381

Bouma TJ, Broekhuysen AGM, Veen BW (1996) Analysis of root respiration of *Solanum tuberosum* as related to growth, ion uptake and maintenance of biomass. Plant Physiol Biochem 34:795–806

Bouma TJ, Yanai RD, Elkin A, Hartmond U, Flores-Alva DE, Eissenstat DM (2001) Estimating age-dependent costs and benefits of roots with contrasting lifespan: comparing apples and oranges. New Phytol 150:685–695

Bouma TJ, Hengst K, Koutstaal BP, van Soelen J (2003) Estimating root lifespan of two grasses at contrasting elevation in a salt marsh by applying vitality staining on roots from in-growth cores. Plant Ecol 165:235–245

Brady DJ, Wenzel CL, Fillery IRP, Gregory PJ (1995) Root growth and nitrate uptake by wheat (*Triticum aestvum* L.) following wetting of dry surface soil. J Exp Bot 46:557–564

Brouwer R (1967) Beziehungen zwischen Spross- und Wurzelwachstum. Angew Bot 41:244–250

Brouwer R (1981) Co-ordination of growth phenomena with a root system of intact maize plants. Plant Soil 63:65–72

Brundrett MC, Kendrick B (1988) The mycorrhizal status, root anatomy, and phenology of plants in a sugar maple forest. Can J Bot 66:1153–1173
Brundrett MC, Murase G, Kendrick B (1990) Comparative anatomy of roots and mycorrhizae of common Ontario trees. Can J Bot 68:551–578
Bryla DR, Bouma TJ, Eissenstat DM (1997) Root respiration in citrus acclimates to temperature and slows during drought. Plant Cell Environ 20:1411–1420
Bryla DR, Bouma TJ, Hartmond U, Eissenstat DM (2001) Influence of temperature and soil drying on respiration of individual roots in citrus: integrating greenhouse observations into a predictive model for the field. Plant Cell Environ 24:781–790
Burton AJ, Pregitzer KS, Hendrick RL (2000) Relationships between fine root dynamics and nitrogen availability in Michigan northern hardwood forests. Oecologia 125:389–399
Caldwell MM (1979) Root structure: the considerable cost of belowground function. In: Solbrig OT, Jain S, Johnson GB, Raven PH (eds) Topics in plant population biology. Columbia University Press, New York, pp 408–427
Caldwell MM, Manwaring JH, Jackson RB (1991a) Exploitation of phosphate from fertile soil microsites by three Great Basin perennials when in competition. Funct Ecol 5:757–764
Caldwell MM, Manwaring JH, Durham SL (1991b) The microscale distribution of neighbouring plant roots in fertile soil microsites. Funct Ecol 5:765–772
Carmi A, Plaut A, Sinai M (1993) Cotton root growth as affected by changes in soil water distribution and their impact on plant tolerance to drought. Irrig Sci 13:177–182
Clarkson DT (1985) Factors affecting mineral nutrient acquisition by plants. Annu Rev Plant Physiol 36:77–115
Clarkson DT (1991) Root structure and sites of ion uptake. In: Waisel Y, Eshel A, Kafkafi U (eds) Plant roots: the hidden half. Marcel Dekker, New York, pp 417–453
Clarkson DT (1996) Root structure and sites of ion uptake. In: Waisel Y, Eshel A, Kafkafi U (eds) Plant roots: the hidden half, 2nd edn. Marcel Dekker, New York, pp 483–510
Clarkson DT, Sanderson J, Russel RS (1968) Ion uptake and root age. Nature 220:805–806
Comas LH, Eissenstat DM, Lakso AN (2000) Assessing root death and root system dynamics in a study of grape canopy pruning. New Phytol 147:171–178
Comas LH, Bouma TJ, Eissenstat DM (2002) Linking root traits to potential growth rate in six temperate tree species. Oecologia 132:34–43
Cui M, Caldwell MM (1997) Shading reduces exploitation of soil nitrate and phosphate by *Agropyron desertorum* and *Artemisia tridentata* from soils with patchy and uniform nutrient distributions. Oecologia 109:177–183
Dittmer HJ (1949) Root hair variation in plant species. Am J Bot 36:152–155
Downes GM, Alexander IJ, Cairney JWG (1992) A study of ageing of spruce [*Picea sitchensis* (Bong.) Carr.] ectomycorrhizas. I. Morphological and cellular changes in mycorrhizas formed by *Tylospora fibrillosa* (Burt.) Donk and *Paxillus involutus* (Batsch. ex Fr.) Fr. New Phytol 122:141–152
Dunbabin V, Rengel Z, Diggle A (2001) *Lupinus angustifolius* has a plastic uptake response to heterogeneously supplied nitrate while *Lupinus pilosus* does not. Aust J Agric Res 52:505–512
Eissenstat DM (1991) On the relationship of specific root length and the rate of root proliferation: a field study using citrus rootstocks. New Phytol 69:870–873
Eissenstat DM (1992) The costs and benefits of constructing roots of small diameter. J Plant Nutr 15:163–782
Eissenstat DM, Achor DS (1999) Anatomical characteristics of roots of citrus rootstocks that vary in specific root length. New Phytol 141:309–321
Eissenstat DM, Caldwell MM (1988) Competitive ability is linked to rates of water extraction: a field study of two aridland tussock grasses. Oecologia 75:1–7

Eissenstat DM, Yanai RD (1997) The ecology of root lifespan. Adv Ecol Res 27:1–62
Eissenstat DM, Yanai RD (2002) Root lifespan, turnover and efficiency. In: Waisel Y, Eshel A, Kafkafi U (eds) Plant roots: the hidden half, 3rd edn. Marcel Dekker, New York, pp 221–238
Eissenstat DM, Graham JH, Syvertsen JP, Drouillard DL (1993) Carbon economy of sour orange in relation to mycorrhizal colonization and phosphorus stress. Ann Bot 71:1–10
Eissenstat DM, Whaley EL, Volder A, Wells CE (1999) Recovery of citrus surface roots following prolonged exposure to dry soil. J Exp Bot 50:1845–1854
Eissenstat DM, Wells CE, Yanai RD, Whitbeck JL (2000) Building roots in a changing environment: implications for root longevity. New Phytol 147:33–42
Eissenstat DM, Wells CE, Wang L (2001) Root efficiency and mineral nutrition in apple. Acta Hortic 564:165–184
Enstone DE, Peterson CA, Hallgren SW (2001) Anatomy of seedling tap roots of loblolly pine (*Pinus taeda* L.) Trees Struct Funct 15:98–111
Espeleta JF, Eissenstat DM (1998) Response of citrus fine roots to localized soil drying: a comparison of seedlings with adult fruiting trees. Tree Physiol 18:113–119
Espeleta JF, Eissenstat DM, Graham JH (1999) Citrus root responses to localized drying soil: a new approach to studying mycorrhizal effects on the roots of mature trees. Plant Soil 206:1–10
Fahey TJ, Hughes JW (1994) Fine root dynamics in a northern hardwood forest ecosystem: Hubbard Brook Experimental Forest, NH. J Ecol 82:533–548
Fisher MCT, Eissenstat DM, Lynch JP (2002) Lack of evidence for programmed root senescence in common bean (*Phaseolus vulgaris* L.) at different levels of phosphorus supply. New Phytol 153:63–71
Fitter AH (1985) Functional significance of root morphology and root system architecture. In: Fitter AH, Atkinson D, Read DJ, Usher MB (eds) Ecological interaction in soil: plant, microbes and animals. Blackwell, London, pp 87–106
Fitter AH (1986a) The topology and geometry of plant root systems: influence of watering rate on the root system topology in *Trifolium pratense*. Ann Bot 57:81–101
Fitter AH (1986b) Effect of benomyl on leaf phosphorus concentration in alpine grasslands: a test of mycorrhizal benefit. New Phytol 103:767–776
Fitter AH (1994) Architecture and biomass allocation as components of the plastic response of root systems to soil heterogeneity. In: Caldwell MC, Pearcy RC (eds) Exploitation of environmental heterogeneity by plants. Academic Press, San Diego, pp 305–323
Fitter AH (1996) Characteristics and functions of root systems. In: Waisel Y, Eshel A, Kafkafi U (eds) Plant roots: the hidden half, 2nd edn. Marcel Dekker, New York, pp 1–20
Fitter AH, Graves JD, Wolfenden J, Self GK, Brown TK, Bogie D, Mansfield TA (1997) Root production and turnover and carbon budgets of two contrasting grasslands under ambient and elevated atmospheric carbon dioxide concentrations. New Phytol 137:247–255
Fusseder A (1987) The longevity and activity of the primary root of maize. Plant Soil 101:257–265
Gahoonia TS, Nielsen NE (1998) Direct evidence on participation of root hairs in phosphorus (^{32}P) uptake from soil. Plant Soil 198:147–152
Gao S, Pan WL, Koenig RT (1998) Integrated root system age in relation to plant nutrient uptake activity. Agron J 90:505–510
Gavito ME, Curtis PS, Jakobsen I (2001) Neither mycorrhizal inoculation nor atmospheric CO_2 concentration has strong effects on pea root production and root loss. New Phytol 149:283–290

Gill RA, Jackson RB (2000) Global patterns of rot turnover for terrestrial ecosystems. New Phytol 147:13–31
Gilroy S, Jones DL (2000) Through form and function: root hair development and nutrient uptake. Trees 5:56–60
Graham JH, Eissenstat DM (1998) Field evidence for the carbon cost of citrus mycorrhizas. New Phytol 140:103–110
Grayston SJ, Vaughan D, Jones D (1997) Rhizosphere carbon flow in trees, in comparison with annual plants: the importance of root exudation and its impact on microbial activity and nutrient availability. Appl Soil Ecol 5:29–56
Hayes DC, Seastedt TR (1987) Root dynamics of tallgrass prairie in wet and dry years. Can J Bot 65:787–791
Head GC (1967) Effects of seasonal changes in shoot growth on the amount of unsuberized root on apple and plum trees. J Hortic Sci 42:169–180
Hendrick RL, Pregitzer KS (1992) The demography of fine roots in a northern hardwood forest. Ecology 73:1094–1104
Henry CM, Deacon JW (1981) Natural (non-pathogenic) death of the cortex of wheat and barley seminal roots, as evidenced by nuclear staining with acridine orange. Plant Soil 60:255–274
Henry A, Kosola K (1999) Root age and phosphorus effects on colonization of *Andropogon gerardii* by mycorrhizal fungi. Soil Biol Biochem 31:1657–1660
Hodge A, Grayston SJ, Ord BG (1996) A novel method for characterisation and quantification of plant root exudates. Plant Soil 184:97–104
Hodge A, Stewart J, Robinson D, Griffiths BS, Fitter AH (1998) Root proliferation, soil fauna and plant nitrogen capture from nutrient-rich patches of soil. New Phytol 139:479–494
Hodge A, Robinson D, Griffiths BS, Fitter AH (1999) Why plants bother: root proliferation results in increased nitrogen capture from an organic patch when two grasses compete. Plant Cell Environ 22:811–820
Hoffland E, Findenegg GR, Nelemans JA (1989) Solubilization of rock phosphate by rape. 2. Local root exudation of organic acids as a response to P starvation. Plant Soil 113:161–165
Holden J (1975) Use of nuclear staining to assess rates of cell death in cortices of cereal roots. Soil Biol Biochem 7:333–334
Huang B, Eissenstat DM (2000) Root plasticity in exploiting water and nutrient heterogeneity. In: Wilkinson RE (ed) Plant–environment interactions. Marcel Dekker, New York, pp 111–132
Huang B, Fry JD (1998) Root anatomical, physiological and morphological responses to drought stress for tall fescue cultivars. Crop Sci 38:1017–1022
Huang B, Nobel PS (1993) Hydraulic conductivity and anatomy along lateral roots of cacti: changes with soil water status. New Phytol 123:499–507
Itoh S, Barber SA (1993a) Phosphorus uptake by six plant species as related to root hairs. Agron J 75:457–461
Itoh S, Barber SA (1993b) A numerical solution of whole plant nutrient uptake for soil-root systems with root hairs. Plant Soil 70:403–413
Jackson RB, Caldwell MM (1992) Shading and the capture of localized soil nutrients: nutrient contents, carbohydrates, and root uptake kinetics of a perennial tussock grass. Oecologia 91:457–462
Johansson G (1992) Release of organic-C from growing roots of meadow fescue (*Festuca pratensis* L.). Soil Biol Biochem 24:427–433
Johnson NC, Graham JH, Smith FA (1997) Functioning of mycorrhizal associations along the mutualism–parasitism continuum. New Phytol 135:575–585

Johnson MG, Phillips DL, Tingey DT, Storm MJ (2000) Effects of elevated CO_2, N-fertilization, and season on survival of ponderosa pine fine roots. Can J For Res 30:220–228

Jones MD, Durall DM, Tinker PB (1990) Phosphorus relationships and production of extramatrical hyphae by two types of willow ecto-mycorrhizas at different soil phosphorus levels. New Phytol 115:259–267

Jupp AP, Newman EI (1987) Morphological and anatomical effects of severe drought on the roots of *Lolium perenne* L. New Phytol 105:393–402

Kelly JM, Barber SA, Edwards GS (1992) Modeling magnesium, phosphorus, and potassium uptake by loblolly-pine seedlings using a Barber-Cushman approach. Plant Soil 139:209–218

Koide RT, Elliot G (1989) Cost, benefit and efficiency of the vesicular-arbuscular mycorrhizal symbiosis. Funct Ecol 3:252–255

Koide RT, Dickie IA, Goff MD (1999) Phosphorus deficiency, plant growth and the phosphorus efficiency index. Funct Ecol 13:733–736

Kosola KR, Eissenstat DM (1994) The fate of surface roots of citrus seedlings in dry soil. J Exp Bot 45:1639–1645

Kramer PJ, Bullock HC (1966) Seasonal variations in the proportions of suberised and unsuberized roots of trees in relation to the absorption of water. Am J Bot 53:200–204

Lambers H (1987) Growth, respiration, exudation and symbiotic associations: the fate of carbon translocated to the roots. In: Gregory PJ, Lake JV, Rose DA (eds) Root development and function. Cambridge University Press, Cambridge, pp 125–146

Lambrecht M, Okon Y, Van de Broek A, Vanderleyden J (2000) Indole-3-acetic acid: a reciprocal signalling molecule in bacteria-plant interactions. Trends Microbiol 8:298–300

Lauenroth WK, Sala OE, Milchunas DG, Lathrop RW (1987) Root dynamics of *Bouteloua gracilis* during short-term recovery from drought. Funct Ecol 1:117–124

Lemon CW, Considine JA (1993) Anatomy and histochemistry of the root system of the kiwifruit vine, *Actinidia deliciosa* var. *deliciosa*. Ann Bot 71:117–129

Mackay AD, Barber SA (1985) Effect of soil moisture and phosphate level on root hair growth of corn roots. Plant Soil 86:321–331

Mapfumo E, Aspinall D, Hancock TW (1994) Growth and development of roots of grapevine (*Vitis vinifera* L.) in relation to water uptake from the soil. Ann Bot 74:75–85

Maron JL (1998) Insect herbivory above- and belowground: individual and joint effects on plant fitness. Ecology 79:1281–1293

Marschner H (1998) Role of root growth, arbuscular mycorrhizae, and root exudates for the efficiency in nutrient acquisition. Field Crops Res 56:203–207

Matzner SL, Richards JH (1996) Sagebrush (*Artemisia tridentate* Nutt) roots maintain nutrient uptake capacity under water stress. J Exp Bot 47:1045–1056

McCrady RL, Comerford NB (1998) Morphological and anatomical relationships of loblolly pine fine roots. Trees 12:431–437

McCully ME (1999) Roots in soil: unearthing the complexities of roots and their rhizospheres. Annu Rev Plant Physiol Plant Mol Biol 50:695–718

McGonigle TP, Fitter AH (1988) Ecological consequences of arthropod grazing on VA mycorrhizal fungi. Proc R Soc Edinb B94:25–32

McKenzie BE, Peterson CA (1995) Root browning in *Pinus banksiana* Lamb. and *Eucalyptus pilularis* Sm. 1. Anatomy and permeability of the white and tannin zones. Bot Acta 108:127–137

Merryweather J, Fitter A (1996) Phosphorus nutrition of an obligately mycorrhizal plant treated with the fungicide benomyl in the field. New Phytol 132:307–311

Miller RM, Reinhardt DR, Jastrow JD (1995) External hyphal production of vesicular-arbuscular mycorrhizal fungi in pasture and tallgrass prairie communities. Oecologia 103:17–23

Okano K, Omae H (1996) Quantitative estimation of physiological functions of various roots with different diameters in the root system of the tea tree. Jpn J Crop Sci 65:605–611

Paterson E, Sim A (2000) Effect of nitrogen supply and defoliation on loss of organic compounds from roots of *Festuca rubra*. J Exp Bot 349:1449–1457

Peng S, Eissenstat DM, Graham JH, Williams K, Hodge NC (1993) Growth depression of mycorrhizal citrus at high phosphorus supply: analysis of carbon costs. Plant Physiol 101:1063–1071

Penning de Vries FWT, Brunsting AHM, van Laar HH (1974) Products, requirements and efficiency of biosynthesis: a quantitative approach. J Theor Biol 45:339–377

Peterson CA, Enstone DE, Taylor JH (1999) Pine root structure and its potential significance for root function. Plant Soil 217:205–213

Poorter H, Bergkotte M (1992) Chemical composition of 24 wild species differing in relative growth rate. Plant Cell Environ 15:221–229

Poorter H, Van Der Werf A, Atkin OK, Lambers H (1991) Respiratory energy requirements of roots vary with the potential growth rate of a species. Physiol Plant 83:469–475

Pregitzer KS, Hendrick RL, Fogel R (1993) The demography of fine roots in response to patches of water and nitrogen. New Phytol 125:575–580

Pregitzer KS, DeForest JL, Burton AJ, Allen MF, Ruess RW, Hendrick RL (2002) Fine root architecture of nine North American trees. Ecol Monogr 72:293–309

Pritchard SG, Rogers HH, Davis MA, van Santen E, Prior SA, Schlesinger WH (2001) The influence of elevated atmospheric CO_2 on fine root dynamics in an intact temperate forest. Global Change Biol 7:829–837

Reid JB, Sorensen I, Petrie RA (1993) Root demography of kiwifruit (*Actinidia deliciosa*). Plant Cell Environ 16:949–957

Richards D, Considine JA (1981) Suberization and browning of grapevine roots. In: Brouwer (ed) Structure and function of plant roots. Dr. W. Junk, London, pp 111–115

Robinson D (1994) The responses of plants to non-uniform supplies of nutrients. Tansley Rev no 73. New Phytol 127:635–674

Robinson D, Hodge A, Griffiths BS, Fitter AH (1999) Plant root proliferation in nitrogen-rich patches confers competitive advantage. Proc R Soc Lond B 266:431–435

Rousseau JVD, Sylvia DM, Fox AJ (1994) Contribution of ectomycorrhiza to the potential nutrient-absorbing surface of pine. New Phytol 128:639–644

Russel RS, Sanderson J (1967) Nutrient uptake rates by different parts of the intact roots of plants. J Exp Bot 18:491–508

Rygiewicz PT, Anderson CP (1994) Mycorrhizae alter quality and quantity of carbon allocated below ground. Nature 369:58–60

Ryser P (1996) The importance of tissue density for growth and life span of leaves and roots: a comparison of five ecologically contrasting grasses. Funct Ecol 10:717–723

Sanders FE, Tinker PB (1971) Mechanism of absorption of phosphate from soil by *Endogone* mycorrhizas. Nature 232:278–279

Sands R, Fiscus EL, Reid CPP (1982) Hydraulic properties of pine and bean roots with varying degrees of suberisation, vascular differentiation and mycorrhizal infection. Aust J Plant Physiol 9:559–569

Scheurwater I (1999) Variation in specific respiratory costs in the roots of fast- and slow-growing grass species. PhD Thesis, Utrecht University, The Netherlands

Scheurwater I, Cornelissen C, Dictus F, Welschen R, Lambers H (1998) Why do fast- and slow-growing grass species differ so little in their rate of root respiration, considering the large differences in rate of growth and ion uptake. Plant Cell Environ 21:995–1005

Scheurwater I, Clarkson DT, Purves JV, Van Rijt G, Saker LR, Welschen R, Lambers H (1999) Relatively large nitrate efflux can account for the high specific respiratory costs for nitrate transport in slow-growing grass species. Plant Soil 215:123–134

Shone MGT, Flood AV (1983) Effects of periods of localized water stress on subsequent nutrient uptake by barley roots and their adaptation by osmotic adjustment. New Phytol 94:561–572

Silberbush M, Barber SA (1983) Sensitivity of simulated phosphorus uptake to parameters used by a mechanistic-mathematical model. Plant Soil 74:93–100

Spaeth SC, Cortes PM (1995) Root cortex death and subsequent initiation and growth of lateral roots from bare steles of chickpeas. Can J Bot 73:253–261

Stasovski E, Peterson CA (1991) The effects of drought and subsequent rehydration on the structure and vitality of *Zea mays* seedling roots. Can J Bot 69:1170–1178

Stasovski E, Peterson CA (1993) Effects of drought and subsequent rehydration on the structure, vitality, and permeability of *Allium cepa* adventitious roots. Can J Bot 71:700–707

Swinnen J, Van Veen JA, Merckx R (1995) Carbon fluxes in the rhizosphere of winter-wheat and spring barley with conventional vs integrated farming. Soil Biol Biochem 27:811–820

Tarawaya K, Hashimoto K, Wagatsuma T (1998) Effect of root exudate fractions from P-deficient and P-sufficient onion plants on root colonisation by the arbuscular mycorrhizal fungus *Gigaspora margarita*. Mycorrhiza 8:67–70

Thomas H, Sadras VO (2001) The capture and gratuitous disposal of resources by plants. Funct Ecol 15:3–12

Thomas SM, Whithead D, Reid JB, Cook FJ, Adams JA, Leckie AC (1999) Growth, loss, and vertical distribution of *Pinus radiata* fine roots growing at ambient and elevated CO_2. Global Change Biol 5:107–121

Thornley JHM (1972) A balanced quantitative model for root:shoot ratios in vegetative plants. Ann Bot 36:431–441

Tierney GL, Fahey TJ (2001) Evaluating minirhizotron estimates of fine root longevity and production in the forest floor of a temperate broadleaf forest. Plant Soil 229:167–176

Tinker PB, Nye PH (2000) Solute movement in the rhizosphere. Oxford University Press, New York

van Rees KCJ, Comerford NB (1990) The role of woody roots of slash pine seedlings in water and potassium absorption. Can J For Res 20:1183–1191

Veen BW (1981) Relation between root respiration and root activity. Plant Soil 63:73–76

Vermeer J, McCully ME (1982) The rhizosphere in *Zea*: new insight into its structure and development. Planta 156:45–61

Visser EJW, Blom CWPM, Voesenek LACJ (1996) Flooding-induced adventitious rooting in *Rumex*: morphology and development in an ecological perspective. Acta Bot Neerl 45:17–28

Volder A, Smart DR, Bloom AJ, Eissenstat DM (2005) Rapid decline in nitrate uptake and respiration with age in fine lateral roots of grape: implications for root efficiency and competitive effectiveness. New Phytol 165:493–502

Wallace LL (1987) Mycorrhizas in grasslands: interactions of ungulates, fungi and drought. New Phytol 105:619–632

Watson CA, Ross JM, Bagnaresi U, Minotta GF, Roffi F, Atkinson D, Black KE, Hooker JE (2000) Environment-induced modifications to root longevity in *Lolium perenne* and *Trifolium repens*. Ann Bot 85:397–401

Wells CE, Eissenstat DM (2001) Marked differences in survivorship among apple roots of different diameters. Ecology 82:882–892

Wells CE, Glenn DM, Eissenstat DM (2002) Soil insects alter fine root demography in peach (*Prunus persica*). Plant Cell Environ 25:431–439

Williams M, Yanai RD (1996) Multi-dimensional sensitivity analysis and ecological implications of a nutrient uptake model. Plant Soil 180:311–324

Wright IJ, Westoby M (1999) Differences in seedling growth behaviour among species: trait correlations across species, and trait shifts along nutrient compared to rainfall gradients. J Ecol 87:85–97

Wullschleger SD, Norby RJ, Love JC, Runck C (1997) Energetic costs of tissue construction in yellow-poplar and white oak trees exposed to long-term CO_2 enrichment. Ann Bot 80:289–297

Yanai RD (1994) A steady-state model of nutrient uptake accounting for newly grown roots. Soil Sci Soc Am J 58:1562–1571

Yanai RD, Fahey TJ, Miller SL (1995) Efficiency of nutrient acquisition by fine roots and mycorrhizae. In: Smith WK, Hinckley TM (eds) Resource physiology of conifers. Academic Press, New York, pp 75–103

Yoder JI (2001) Host-plant recognition by parasitic *Scrophulariaceae*. Curr Opin Plant Biol 4:359–365

Zak DR, Pregitzer KS, King JS, Holmes WE (2000) Elevated atmospheric CO_2, fine roots and the response of soil microorganisms: a review and hypothesis. New Phytol 147:201–222

9 Action and Interaction in the Mycorrhizal Hyphosphere – a Re-evaluation of the Role of Mycorrhizas in Nutrient Acquisition and Plant Ecology

R. FINLAY

9.1 Introduction

The vast majority of terrestrial plants form symbiotic mycorrhizal associations with fungi, supporting the view that, in many ecosystems, mycorrhizas, not roots, are the chief organs of nutrient uptake. The underground organs of a wide range of extant lower land plants also form associations with fungi (Read et al. 2000), forming structures analogous to those found in higher plants. These observations, together with palaeobotanical (Nicolson 1975; Stubblefield et al. 1987a–c; Remy et al. 1994; Redecker et al. 2000) and molecular (Simon et al. 1993) evidence, suggest that the first land plants, which had no true roots, were colonised by mycorrhizal fungi belonging to the phylum Glomeromycota (Schüßler et al. 2001). Colonisation of land, 400×10^6 years ago, was thus probably by symbiotic organisms with fungal hyphal connections with the soil which could assist in the absorption of soil-derived nutrients. Mycorrhizal symbiosis thus occupies a central position in any consideration of plant nutrient acquisition from an ecological perspective.

At one end of the theoretical spectrum, the mycorrhizal mycelium has been seen as having a purely *quantitative* effect on the uptake of nutrients already present in the soil solution, the mycelium augmenting the surface area of the root system and providing an additional, larger and more finely distributed interface across which plant nutrients can be absorbed. The role of mycorrhizal mycelia as physical extensions of root systems is discussed in Section 9.2, together with a brief consideration of factors controlling root colonisation and hyphal growth. Since the water status of soil can have a profound effect on nutrient availability, the role of mycorrhizal mycelia in water uptake is also discussed in this section. The subject is dealt with rather superficially for reasons of space, and because there have been numerous

Ecological Studies, Vol. 181
H. BassiriRad (Ed.) Nutrient Acquisition by Plants
An Ecological Perspective

reviews and articles published on this traditional area of mycorrhizal research. Although some aspects of the internal, structural and physiological regulation of mycorrhizal roots are considered, the main emphasis of the chapter is on nutrient acquisition rather than assimilation, and on the functioning of the extraradical hyphae and their interactions with the biotic and abiotic environment.

Qualitative effects of mycorrhizal symbiosis on the supply of compounds from substrates not normally available to non-mycorrhizal roots are dealt with in Section 9.3. These include the capacity to access organic N and P compounds, as well as the likely ability of some mycorrhizal fungi to release nutrients by weathering of minerals. The possible roles which mycorrhizal fungi may play to *decrease* uptake of toxic elements are also discussed in this section. Interactions of mycorrhizal fungi with the biotic environment are dealt with in Section 9.4. These include interactions with saprotrophic, pathogenic and other mycorrhizal fungi, bacteria and soil animals, and the effects of these interactions on nutrient uptake and cycling. Although the main emphasis of this chapter is on nutritional roles of mycorrhizal symbiosis, non-nutritional roles of the symbiosis are also dealt with in Section 9.5 of this chapter, since they may also have an indirect effect on plant nutrition. This section emphasises the multiple roles that mycorrhizal fungi may have in plant communities as well as evolutionary aspects of the symbiosis and the phylogenetic relationships of mycorrhizal fungi to other groups of fungi with different trophic strategies. In Section 9.6 the implications of mycorrhizosphere effects for theories of nutrient cycling and plant ecology are considered. Many of these theories have been strongly influenced by a predominantly agronomic perspective which may not be relevant in natural plant communities. The need for both phytocentric and mycocentric perspectives (and *awareness* of these perspectives) in plant ecology is also discussed. Practical implications of the active and interactive effects of mycorrhizal fungi and other organisms and possible application areas such as biocontrol, biofertilisation and bioremediation are briefly outlined in Section 9.7. The different types of effect which mycorrhizal fungi may have are summarised in Fig. 9.1. Improved knowledge of these interactions and of the entire range and variation of activity of the many hitherto unexamined mycorrhizal fungal species are important prerequisites for improved, sustainable management of plant communities within agriculture and forestry, and for nature conservation.

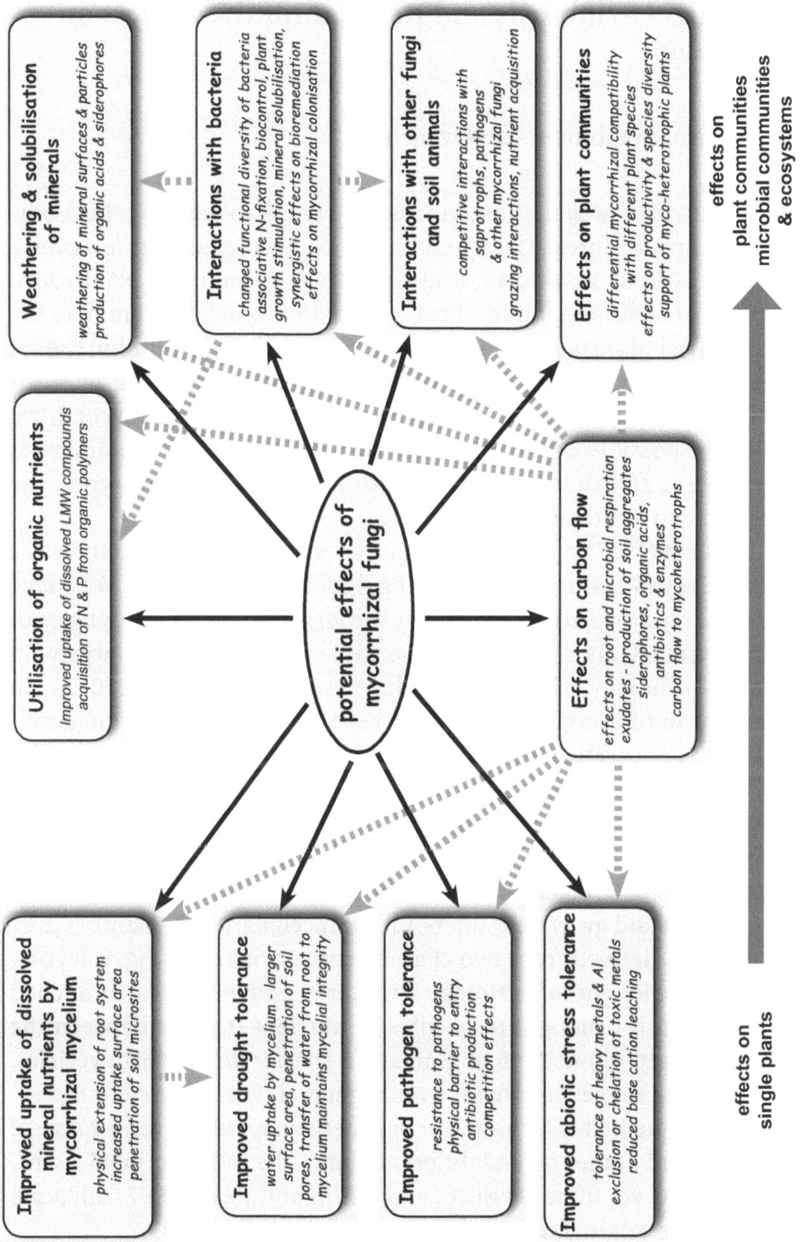

Fig. 9.1. Schematic diagram of the different types of effects which mycorrhizal mycelia have in nutrient acquisition and plant ecology

9.2 Mycorrhizal Symbiosis and Its Quantitative Effects on Plant Nutrition

9.2.1 Mycorrhizal Symbiosis – an Overview

The term "mykorrhiza" was first used in 1885 by Frank to describe the modified root structures of forest trees, and has since been extended to cover a range of mutualistic, symbiotic associations between fungi and plant roots. Different types of symbiotic mycorrhizal association typically dominate the different terrestrial biomes of the world, and their structural and functional adaptations to the environments they occupy have been reviewed by Read (1984, 1991). A complete description of the different types of mycorrhizal symbiosis is not possible here for reasons of space, but a brief summary is given below. For a full description of the different mycorrhizal types, the reader is referred to Smith and Read (1997).

Smith and Read (1997) distinguish seven types of mycorrhiza on the basis of the morphological features and the fungal associates involved. *Vesicular-arbuscular* (VA) *mycorrhizas* are formed by the largely aseptate endophytes of the Glomeromycota. Only 150 species have so far been described, largely on the basis of morphology. The fungi produce asexual spores 40–250 μm in diameter, and taxonomy is currently largely based on the structure and development of the spore walls (Morton and Benny 1990), although molecular analyses will undoubtedly play an increasingly important role in taxonomy. This type of mycorrhiza is the oldest (400×10^6 years B.P.) and most widespread of the various types, with plants of all phyla represented and an estimated 250,000 species of plants capable of forming the association. The fungi penetrate the roots and grow both inter- and intracellularly. The name "vesicular-arbuscular" is derived from two characteristic structures (Fig. 9.2a, b) – *vesicles*, which are thought to be storage structures, and *arbuscules*, branched intracellular structures thought to be sites of nutrient transfer between the fungus and the plant. About 20% of glomalean species do not form vesicles, which has led to the suggestion that the more general term *arbuscular mycorrhiza* (AM) be adopted. This term is now in general use, but a diverse range of structures is formed in the roots of different plants, the full functional significance of which has yet to be revealed, and Smith and Read (1997) advocate the retention of the original term "vesicular-arbuscular".

The septate fungi forming the remaining six types of mycorrhiza include members from almost all orders of the basidiomycetes and many ascomycetes. *Ectomycorrhizas* are characteristically formed by long-lived woody perennial plants in cool temperate, boreal or montaine regions. As many as 6,000 fungal species may form ectomycorrhizas (Molina et al. 1992) and, although this type of symbiosis is restricted to less than 5% of terrestrial plant species, its global importance is significant, since trees of the Pinaceae, Betulaceae, Salicaceae,

a

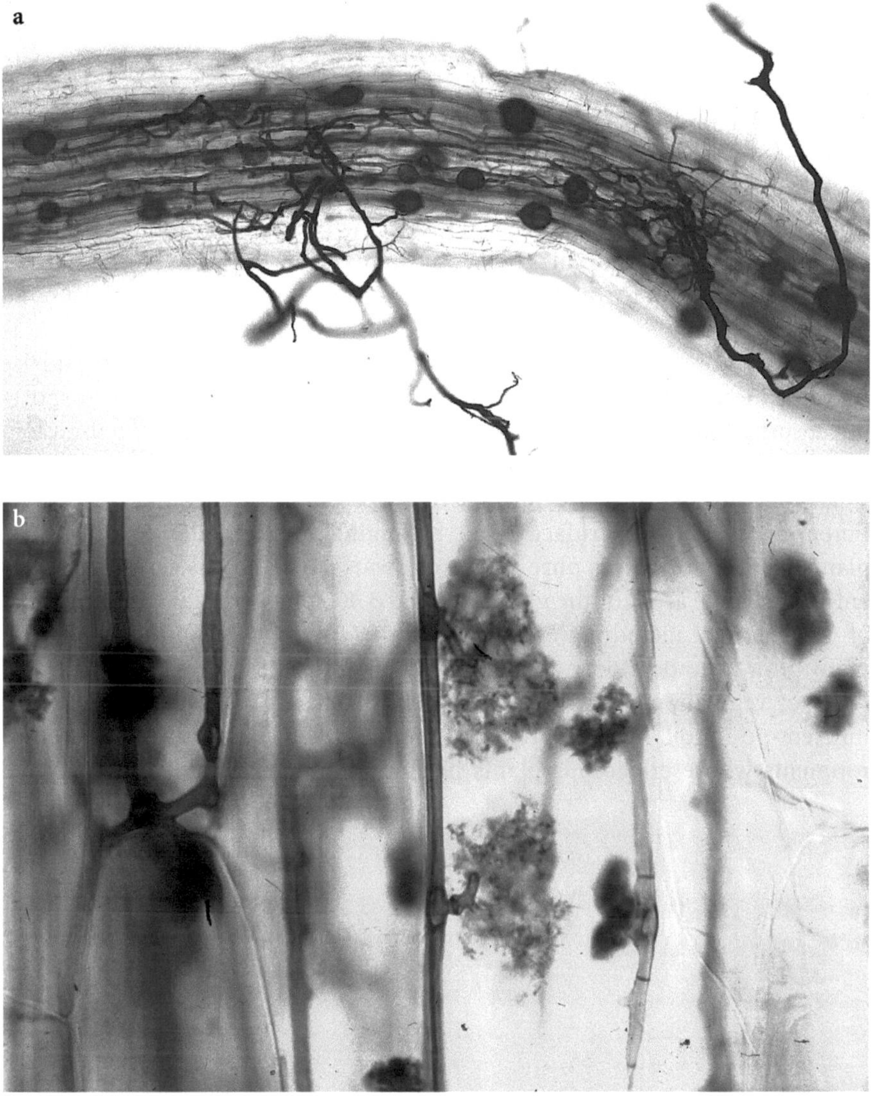

Fig.9.2. a Vesicles and external hyphae of the vesicular arbuscular mycorrhizal fungus *Glomus caledonium* colonising roots of *Lactuca sativa* (Photo: Roger Finlay). **b** Arbuscules of an unidentified *Glomus* vesicular arbuscular mycorrhizal fungus colonising roots of onion, *Allium cepa*. (Photo: Erland Bååth)

Myrtaceae and Fagaceae are predominately ectomycorrhizal and form vast areas of forest in both the northern and southern hemispheres. Ectomycorrhizas are also formed in arctic–alpine dwarf shrub communities (e.g. *Salix*, *Dryas*), Mediterranean/chaparral vegetation (*Cistus*, *Quercus*) and in tropical forests, where they are common in the Dipterocarpaceae and in leguminous species such as the Caesalpinoideae. Some herbaceous species, such as *Kobresia myosuroides* and *Polygonum viviparum*, also form ectomycorrhizas. The symbiosis is ectotrophic, without fungal penetration of host cells, and characterised by the formation of a more or less well-developed fungal *sheath* or *mantle* enveloping the short roots (Fig. 9.3a). An intercellular network of fungal hyphae penetrating between the epidermal and cortical cells, the *Hartig net* (Fig. 9.3a), forms the interface across which nutrients and carbon are exchanged. Like the arbuscules in VA mycorrhizas, the Hartig net represents an effective means of maximising the area of contact between the two symbiotic partners. Depending on the fungal species involved, the mantle is often connected to a more or less well-developed extraradical mycelium which, like the VA mycelium, provides an effective means of exploiting the soil and transporting absorbed nutrients to the root (Fig. 9.3b). In some species, the ectomycorrhizal mycelium may extend many centimetres from the mantle.

Ericoid mycorrhizas are formed in three plant families, the Ericaceae, Empetraceae and Epacridaceae, all belonging to the order Ericales. The plants form more or less pure stands of dwarf shrub heaths, covering nutrient-poor, fringe areas of the globe. Around 3,400 plant species form this type of association, including *Erica*, *Calluna*, *Vaccinium* and *Rhododendron*; many grow as understorey vegetation in boreal forests and in Mediterranean types of vegetation where they are often associated with strongly leached, nutrient-poor soil. Accumulation of organic matter is a feature of the environments where ericaceous plants become dominant. A single ascomycete

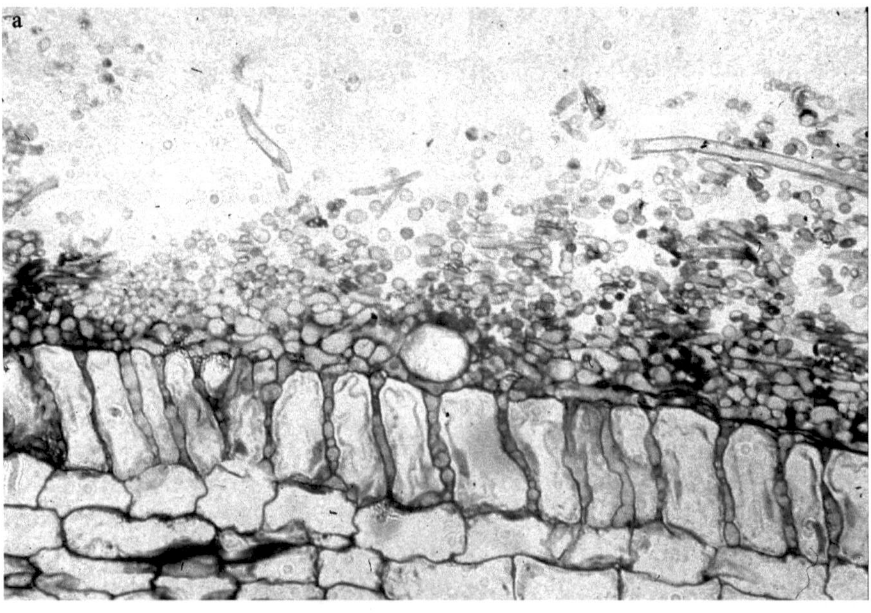

b

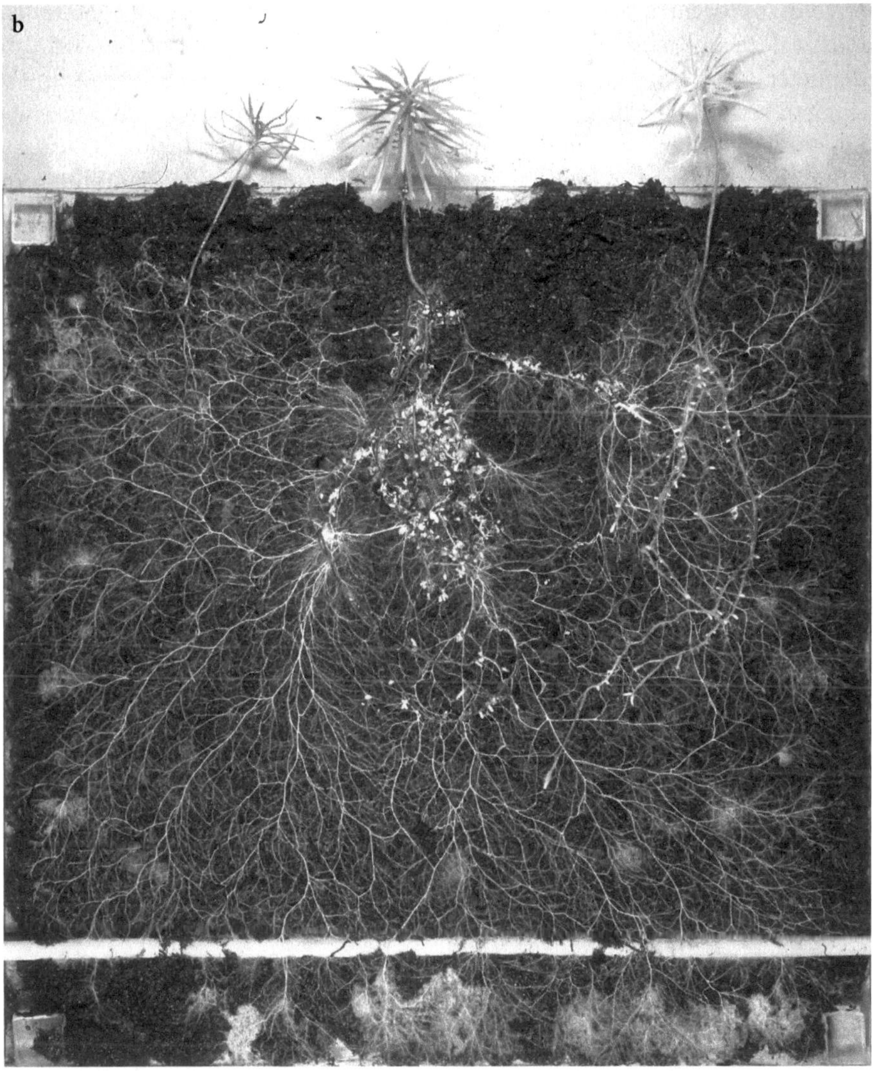

Fig. 9.3. a Longitudinal section of a *Betula pendula* short root colonised by the ectomycorrhizal fungus *Paxillus involutus* (Photo: Roger Finlay). **b** Microcosm containing *Larix eurolepis* and pine seedlings colonised by the fungus *Boletinus cavipes*. The fungal mycelium is clearly differentiated and forms extensive rhizomorphs, colonising the entire microcosm. (Photo: Roger Finlay)

species, *Hymenoscyphus ericae*, is most often involved in this type of symbiosis. Other species of the genus *Oidiodendron* have been isolated from ericoid mycorrhizas but their mycorrhizal status is unclear. Although *Oidiodendron griseum* can form ericoid mycorrhizas under axenic conditions (Dalpé 1986), its significance under field conditions is not clear. Like VA mycorrhiza, ericoid mycorrhiza are endomycorrhizal, penetrating the host cell walls. Most ericaceous plants possess "hair roots" which are very fine (40–100 μm diameter), consisting of a simple stele, surrounded by an endodermis and a single layer of cortical cells. The fungi grow along the root surface, penetrating the epidermal cells and producing coiled structures inside (Fig. 9.4). Often the infection can be very extensive, with a high proportion of the root volume occupied by hyphae (Read and Stribley 1975). The fungus may form a more or less dense covering of hyphae on the root but, in contrast to both ectomycorrhizas and VA mycorrhizas (which can produce extensive mycelia), the hyphae extend for only a few millimetres from the root surface. The architecture of the root system itself is such that the roots are densely packed into a mat between the litter above and the humified material below, and the production of an extensive mycelium would do little to enhance the exploitation of this substrate.

An important feature of most types of mycorrhizal symbiosis is that the autotrophic plant partner provides the fungus with photosynthetically derived carbon compounds. However, two kinds of mycorrhiza are formed with plants which are totally achlorophyllous for all or part of their lives. Plants in the Orchidaceae (about 17,500 species from 795 genera) are usually achlorophyllous at first, and have very small seeds with little or no reserves (Leake 1994). Orchids form *orchid mycorrhizas* with a range of basidiomycete fungi which were originally all thought to be highly effective saprotrophs or parasites of other plants, and thus capable of providing the orchids with the organic C compounds and nutrients needed for germination and growth. In 1997 Taylor and Bruns demonstrated, using molecular methods, that the fun-

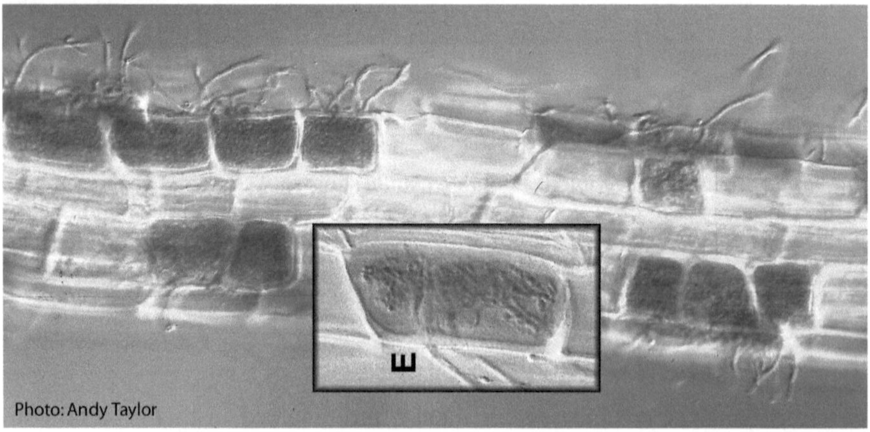

Fig. 9.4. Hair roots of *Vaccinium vitis idea* showing colonisation by the ericoid endophyte *Hymenoscyphus ericae*. Hyphal coils are seen filling most of the epidermal cells, and penetration of the cell wall by the endophyte is seen in the *enlarged inset*. (Photo: Andy Taylor)

gal symbionts of two orchid species, *Cephalanthera austinae* and *Corallorhiza maculata*, were members of the ectomycorrhizal fungal families Thelephoraceae and Russulaceae. There is now also increasing evidence that some orchids are dependent on fungi which form ectomycorrhizal associations with other, autotrophic plants, obtaining C from them as well as mineral nutrients derived from the soil (McKendrick et al. 2000a, b). Plants in the family Monotropaceae are all achlorophyllous and form *monotropoid mycorrhizas*. Like ectomycorrhiza, these normally have a well-developed fungal mantle but the Hartig net is restricted to the outer layer of epidermal cells and does not grow into the underlying cortex. Single hyphae grow from the Hartig net into the epidermal cells, forming structures known as fungal pegs. The achlorophyllous monotropoid plants are wholly dependent on the fungi for reduced carbon and soil nutrients, and it appears that the carbon comes indirectly from autotrophic host plants which are also attached to the same ectomycorrhizal mycelium (Leake et al. 2004). The monotropoid plants must therefore always occur in the vicinity of autotrophic host plants colonised by ectomycorrhizal fungi.

Many fungi, normally forming ectomycorrhiza on other host plants, also colonise plants in the genera *Arbutus* and *Arctostaphylos* and the family Pyrolaceae, forming *arbutoid mycorrhizas*. The structures formed usually have a well-developed mantle and Hartig net but, in contrast to ectomycorrhizas, extensive intracellular proliferation of the fungus occurs. In this type of association, as in ectomycorrhiza, the fungus receives reduced carbon from the plant host and supplies the plant with soil nutrients. A final category of mycorrhizal associations, *ectendomycorrhizas*, has been distinguished to describe mycorrhizal roots displaying some of the structural characteristics of both ectomycorrhizas and endomycorrhizas. The conifer ectendomycorrhizas are often called E-type mycorrhizas, and the fungi are ascomyctes which form ectendomycorrhizas on *Pinus* and *Larix* but normal ectomycorrhizas with other tree species. Although much work has been performed on ectendo–arbutoid and monotropid mycorrhizas, they can still be considered rather special cases of the general mycorrhizal habit. Achlorophyllous orchid mycorrhizas, although represented by a diverse group of host species, are a special case in that the carbon balance between the plant symbiont and fungus is reversed. For these reasons, I will confine the rest of this chapter to discussion of the three main mycorrhizal types – VA mycorrhiza, ericoid mycorrhiza and ectomycorrhiza.

9.2.2 Mycorrhizal Mycelia as Physical Extensions of Root Systems

The multiple roles of the mycelium and the role of mycorrhiza in plant communities have been the subject of many recent investigations, but the importance of the mycorrhizal mycelium as an extension of plant root systems has been appreciated for a long time. Peyronel described the extensive development of the VA mycelium as early as 1923. Nicolson (1959) studied the extraradical mycelium associated with grass roots. More recent studies of the architecture of VA mycelial systems (Read et al. 1985; Friese and Allen 1991) have demonstrated the existence of hyphal bridges between root systems. These structures are important in the rapid colonisation of seedlings, and have possible effects at the community level which are discussed below. Environmental factors also regulate root colonisation levels, and it is known that both low irradiance (Smith and Gianinazzi-Pearson 1990) and high P (Bruce et al. 1994) reduce colonisation, but the mechanisms involved are so far unclear. The importance of the mycelium as an extended nutrient absorbing surface was demonstrated by Rhodes and Gerdemann (1975) who injected radioactive tracer elements at different distances from a membrane which was permeable to hyphae but not to plant roots. Translocation occurred over distances of up to 7 cm, and the extensive mycelial systems growing from roots extend beyond the zones of soil depleted of relatively immobile nutrients such as P by rapid root uptake. Sanders and Tinker (1973) demonstrated increased inflow of P into mycorrhizal roots, and these early studies were followed by many hundreds of experiments investigating the uptake of different nutrients (e.g. Cooper and Tinker 1978; Ames et al. 1983; Pearson and Jakobsen 1993; Johansen et al. 1996). It seems clear that VA mycorrhizal hyphae can absorb non-mobile nutrients such as P, Zn and Cu from soil and translocate them rapidly to plants, avoiding problems of depletion. Hyphae are also able to penetrate soil pores inaccessible to roots, and may be able to compete effectively with micro-organisms for recently mineralised nutrients. Much emphasis has been placed upon the effects on P uptake, although Fitter (1985) suggested that a range of other factors would be likely to be more important under field conditions. Augé (2001) recently provided an extensive review of the effects of VA fungi on the water balance of plants. Although water relations are undoubtedly modified by VA mycorrhizal interactions, the exact mechanisms are difficult to determine and are more likely to be due to nutritional effects, rather than to direct transport of water via hyphae or alteration of hydraulic properties. Smith and Read (1997) have pointed out that, as soil dries, nutrients become less and less available due to the increasing tortuosity of the diffusion path, and the hyphal contribution to nutrient uptake becomes increasingly important. Querejeta et al. (2003) have also presented evidence suggesting that water obtained by hydraulic lift may be transferred from roots to mycorrhizal hyphae, maintaining the activity of the mycorrhizal mycelium during periods of drought.

Effects of ectomycorrhizal mycelial systems on nutrient uptake are also well established, and there is a long history of experimentation. Frank (1894) first described the modified fungus–root structures "mykorhiza" of forest trees, and realised their significance in nutrient acquisition. Many of the key, pioneering experiments demonstrating the importance of the extraradical mycelium in nutrient uptake were performed by Melin and Nilsson (e.g. 1950, 1953), and these have been reviewed by Finlay (1993). Different species of ectomycorrhizal fungi build more or less differentiated mycelia, the function of which has been studied in flat microcosm systems (Finlay and Read 1986a, b) demonstrating translocation of phosphorus, nitrogen (Finlay et al. 1988, 1989), K^+ (^{86}Rb) and Ca^{2+} (Finlay 1993). Evidence of the ability of ectomycorrhizal mycelium to absorb and translocate water has also been provided from laboratory experiments (Duddridge et al. 1980; Read and Boyd 1986; Lamhamedi et al. 1992). Some ectomycorrhizal genera have been studied more extensively than others (see Cairney and Chambers 1999), and it is becoming increasingly clear that the functional diversity exhibited by the 5,000–6,000 fungal species forming ectomycorrhizas is high. The relative efficiency with which different species transfer absorbed nutrients to their host plants is known to vary strongly (e.g. Abuzinadah and Read 1989a; Gorissen and Kuyper 2000), and there are clear inter-specific differences in both the extent of development of the extraradical mycelium and the range of compounds which can be absorbed. Increased levels of mineral N may decrease the development of the extraradical mycelium of some species (Wallander and Nylund 1992; Arnebrant 1994), and may explain the decline in ectomycorrhizal diversity along gradients of anthropogenic N deposition in Europe (Taylor et al. 2000). However, there is clearly considerable spatial heterogeneity in the expression of different enzymes by the extraradical mycelium (Timonen and Sen 1998), and studies by Brandes et al. (1998) and Jentschke et al. (2000) have shown localised increases in hyphal density in response to nutrients in situations where the plant roots themselves are not exposed to increased levels of nutrients.

Although ericoid mycorrhizal associations have been less widely studied than VA and ectomycorrhizal associations, the effects of the symbiosis have been well documented (Stribley and Read 1980; Bajwa and Read 1985, 1986; Read 1991) and include improved N and P nutrition. The role of the extraradical mycelium as a physical extension of the root system is less clear, however, since the hyphae appear only to extend a few millimetres from the root system. As pointed out by Read (1991), the fine, ericoid hair roots frequently form a dense mat between the overlying litter and the underlying humic material, and structurally are thus well adapted to nutrient capture themselves. The soils colonised by plants with ericoid mycorrhizas are typically strongly leached and poor in mineral nutrients. In these environments, essential nutrients such as N and P are bound in recalcitrant organic forms, and the main effect of the ericoid endophyte *Hymenoscyphus ericae* appears to be in making these nutrients available to their plant hosts by mobilising them from organic substrates. In ericoid mycorrhizas, at least, the effect of the mycelium on plant nutrition is more qualitative than quantitative, and different *qualitative* influences of mycorrhizas on plant nutrition are discussed in the following section.

9.3 Qualitative Effects of Interactions with the Abiotic Environment

There is now increasing recognition that, in addition to providing an increased interception area for uptake of nutrients in solution, mycorrhizal fungi play an active role in influencing the availability of different types of nutrients to plants. Recent research has concentrated on the utilisation of organic forms of N and P which may not be directly available for uptake by roots (Read and Perez-Moreno 2003; Table 9.1), as well as on the possibility that ectomycorrhizal fungi may release inorganic nutrients from mineral particles and rock surfaces through weathering (see Gadd 2005; Table 9.2).

9.3.1 Uptake and Utilisation of Organic Nutrients

The possibility that ectomycorrhizal fungi play a direct role in the acquisition of N from organic substrates was first suggested by Frank as early as 1894, and Frank's organic nitrogen theory has been discussed by Read (1987). More than 50 years ago, Melin and Nilsson (1953) demonstrated uptake of labelled nitrogen from glutamic acid by ectomycorrhizal mycelium, but the view that mycorrhizal fungi were unimportant in the mobilisation of nutrients from organic sources persisted until the late 1980s when experiments by Read and his coworkers first demonstrated a significant capacity of certain mycorrhizal fungi to utilise amino acids and proteins (Abuzinadah and Read 1986a, b, 1989a, b; Finlay et al. 1992). Subsequent studies have continued to investigate this important qualitative modification of plant nutrient uptake capacity, and utilisation of a range of substrates has been demonstrated in laboratory experiments. Proliferation of ectomycorrhizal hyphae in patches of organic material collected from the FH horizon (humic layer) of forest soil and introduced into peat-containing microcosms (Read 1991) suggests that the mycelium is able to forage for nutrients contained in organic residues, and this is further supported by studies of Bending and Read (1995a) showing depletion of N, P and K from similar organic patches. Although the role played by saprotrophs in this nutrient mobilisation was not clear, Bending and Read (1995b) were able to measure elevated activities of nutrient-mobilising enzymes such as protease, polyphenol oxidase and phosphomonoesterase in substrates intensively colonised by ectomycorrhizal fungi. A comparable study conducted under field conditions (Entry et al. 1991) also investigated decomposition and nutrient release from *Pseudotsuga menziesii* litter in mesh bags incubated in soil containing mats of mycorrhizal mycelium, and reported losses of similar amounts of nutrients, consistent with similar increases in the activity of key nutrient-mobilising enzymes (Griffiths and Caldwell 1992). Experiments by Chalot et al. (1994a, b, 1995a, b) using ^{14}C- and ^{15}N-labelled tracers and enzyme inhibitors have investigated amino acid uptake and metabolism by the fungus *Paxillus involutus*, and the physiology of organic nitrogen acquisition by ectomycorrhizal fungi has been reviewed by Chalot and Brun (1998).

There is no evidence that mycorrhizal fungi play a major role in mobilising nutrients from highly recalcitrant forms of humified material, but mobilisation from less recalcitrant organic compounds may be substantial. It is known that all of the major hydrolytic enzymes associated with the mobilisation of N and P from organic compounds by saprotrophs have also been recorded in ericoid and ectomycorrhizal fungi (see Leake and Read 1997). Proteinase activity in ericoid endophytes has been demonstrated by Leake and Read (1990a). Chitin, the major polymer of fungal cell walls and insect exoskeletons, is readily used by ericoid mycorrhizal fungi and some ectomycorrhizal fungi (Leake and Read 1990b), and can contain more than 20 % of the total nitrogen in the litter horizon of a mor-humus heath land and as much as 30 % in the fermentation horizon. Chitin degradation by the ericoid endophyte *Hymenoscyphus ericae* and the transfer of chitin-derived N to its host plants is significant, as has been shown by Kerley and Read (1995, 1997, 1998). Lindahl and Taylor (2004) demonstrated the widespread occurrence of *N*-acetylhexosaminidase-encoding genes in ectomycorrhizal basidiomycetes, indicating that these fungi have the potential to exploit polymers of amino sugars as a source of nitrogen.

Partial degradation of lignin by mycorrhizal fungi observed by Trojanowski et al. (1984) and Haselwandter et al. (1990) was initially interpreted as suggesting lignin peroxidase production (Cairney and Burke 1994), but has subsequently been attributed to the production of hydroxyl radicals (Cairney and Burke 1998). These artefacts can be avoided by measuring lignin peroxidase activity (Bending et al. 1996) but, although genes for ligninolytic enzymes such as lignin peroxidase and manganese peroxidase, normally associated with white rot fungi, were originally reported to be widespread in a broad taxonomic range of ectomycorrhizal fungi (Chen et al. 2001), it now appears that there is no evidence for lignin peroxidase genes in ectomycorrhizal fungi (Cairney et al. 2003). Although the catalytic activity of mycorrhizal fungi may be lower than that of non-mycorrhizal fungi when measured in pure culture (Bending and Read 1997), Leake and Read (1997) urge caution in making direct comparisons, since differences in activity may be offset in the field by high ectomycorrhizal mycelial biomass and provision of energy-rich carbon compounds from plant hosts. Furthermore, apparently weak expression of some enzymes such as cellulase by fungi such as the ericoid endophyte may be unimportant, since the fungus readily penetrates the cell walls of the host root (see Fig. 9.4) and may use the same *localised* capacity to penetrate plant litter and gain access to the nutrients contained within.

Table 9.1. Qualitative effects of mycorrhizal utilisation of organic nutrient sources

Details	Study (see reference list)
Studies of organic N or P uptake by ectomycorrhizal systems	
Use of peptides by mycorrhizal fungi	Abuzinadah and Read (1986a)
Protein utilisation by *Betula*, *Picea* and *Pinus* colonised by *Hebeloma crustuliniforme*	Abuzinadah and Read (1986b)
Utilisation of peptides by ectomycorrhizal *Betula pendula* plants	Abuzinadah and Read (1989a)
Differences in N transfer from organic sources to ectomycorrhizal *Betula pendula*	Abuzinadah and Read (1989b)
Protein utilisation/discussion of short-circuiting conventional decomposition pathways	Abuzinadah et al. (1986)
Foraging behaviour and translocation of nutrients from litter patches by ectomycorrhizal fungi	Bending and Read (1995a)
Activities of nutrient-mobilising enzymes in birch litter colonised by *Paxillus involutus*	Bending and Read (1995b)
Uptake and translocation of glutamate	Melin and Nilsson (1953)
N acquisition patterns in relation to successional stage	Finlay et al. (1992)
Studies of organic nitrogen uptake and metabolism	
Metabolism of glutamate by *Paxillus involutus*	Chalot et al. (1994a)
Respiration of alanine in the ectomycorrhizal fungus *Paxillus involutus*	Chalot et al. (1994b)
Metabolism of alanine in the ectomycorrhizal fungus *Paxillus involutus*	Chalot et al. (1995a)
Factors affecting amino acid uptake by *Paxillus involutus*	Chalot et al. (1995b)
Nutrient acquisition from different sources	
Nutrient acquisition from bacteria	Barron (1988)
Acquisition of organic N from micro-arthropods by *Laccaria bicolor*	Klironomos and Hart (2001)
Ectomycorrhizal mycelial acquisition of P from saprotrophic mycelium	Lindahl et al. (1999)
Mycorrhiza–saprotroph interactions – effect of resource size on direction of P transfer	Lindahl et al. (2001)
Studies of VA mycorrhizal systems	
Uptake and transport of organic N by arbuscular mycorrhiza (*G. mosseae*)	Hawkins et al. (2000)

Studies of ericoid mycorrhizal systems	
Analysis of free amino acids in relation to time and depth in soil profile	Abuarghub and Read (1988)
Utilisation of mineral and organic N by *H. ericae* and mycorrhizal *Vaccinium* plants	Bajwa and Read (1985, 1986)
Chitin degradation by *Hymenoscyphus ericae* – transfer of derived N to host plant	Kerley and Read (1995)
Fungal mycelium as an N source for *Hymenoscyphus ericae* and its host plants	Kerley and Read (1997)
Plant and mycorrhizal necromass as nitrogenous substrates for *Hymenoscyphus ericae*	Kerley and Read (1998)
Proteinase activity in mycorrhizal fungi – effects of pH, and N sources on production and activity	Leake and Read (1990a)
Chitin as an N source for mycorrhizal fungi	Leake and Read (1990b)
Field experiments	
Influence of ectomycorrhizal mats on lignin and cellulose degradation	Entry et al. (1991)
Boreal forest plants take up organic N	Näsholm et al. (1998)
Decomposition of complex substrates	
Decomposition of lignin, holocellulose and lignocellulose	Trojanowski et al. (1984)
Degradation of ^{14}C-labelled lignin by ericoid and ectomycorrhizal fungi	Haselwandter et al. (1990)
Genes for lignin and manganese peroxidases in ectomycorrhizal fungi	Chen et al. (2001)
No evidence for lignin peroxidase genes in ectomycorrhizal fungi	Cairney et al. (2003)
Occurrence of *N*-acetylhexosaminidase genes in ectomycorrhizal basidiomycetes	Lindahl and Taylor (2004)
Fungal enzymes degrading plant cell walls	Cairney and Burke (1994)
N mobilisation from protein–polyphenol complex by ericoid and ectomycorrhizal fungi	Bending and Read (1996)
Lignin and soluble phenolic degradation by ericoid and ectomycorrhizal fungi	Bending and Read (1997)
Reviews of mycorrhizal uptake of organic compounds	
In support of Frank's organic N theory	Read (1987)
Review of exploitation of organic N and P sources by mycorrhiza	Leake and Read (1997)
Review of organic N uptake by ectomycorrhiza	Chalot and Brun (1998)
Role of mycorrhizas in nutrient cycling in ecosystems	Read and Perez-Moreno (2003)
Enzymatic activities of mycorrhizal mycelia	Lindahl et al. (2005)

Table 9.2. Studies of weathering interactions involving ectomycorrhizal fungi

Details	Study (see reference list)
Production and significance of organic acids	
Production of oxalic acid by *Paxillus involutus*	Lapeyrie et al. (1987)
Production of oxalic acid by ectomycorrhiza – chelation of Al and Cu	Ahonen Jonnarth et al. (2000)
Root production of organic acids in the rhizosphere – fate of organic acids	Jones (1998)
Fungal production of citric and oxalic acid and its importance in biogeochemical processes, including corrosion of stone and building materials	Gadd (1999)
Calcium oxalate – occurrence and effect on nutrient and geochemical cycles	Graustein et al. (1977)
Exudates of ectomycorrhizal roots enhance dissolution rates of minerals	Ochs (1996)
Altered chemistry of ectomycorrhizosphere soil due to organic acid production	Arocena et al. (1999)
Mycorrhizal siderophore production	
Identification and characterisation of siderophores of mycorrhizal fungi	Haselwandter and Winkelmann (1998)
Identification of ferrichrome and ferricrocin production by ectomycorrhizal fungi	Holmström et al. (2004)
Production of hydroxamate siderophore iron chelators by ectomycorrhizal fungi	Szaniszlo et al. (1981)
Soil chemistry of ectomycorrhizal mats	
Calcium oxalate accumulation and soil weathering in mats of *Hysterangium crassum*	Cromack et al. (1979)
Soil solution chemistry of ectomycorrhizal mats in forest soil	Griffiths et al. (1994)
Solubilisation of minerals	
Solubilisation of phosphate	Lapeyrie et al. (1991)
Ash solubilisation by ectomycorrhizal fungi	Mahmood et al. (2001)
Fungal weathering/etching of minerals by mycorrhizal and other fungi	
Application of the lichen model to understanding mineral weathering in the rhizosphere	Banfield et al. (1999)
Biogenic etching in amorphous and crystalline silicates	Callot et al. (1987)
Pore formation in rocks and mineral particles	van Breemen et al. (2000a)

Occurrence of pores in minerals	Jongmans et al. (1997)
Contribution of mineral tunnelling to total feldspar weathering	Smits et al. (2005)
Weathering of ammonium- or calcium-saturated phyllosilicates by ectomycorrhizal fungi in vitro	Paris et al. (1995a)
Weathering of phlogopite by ectomycorrhizal fungi – effect of K^+ and Mg^{2+} deficiency	Paris et al. (1995b)
Phlogopite weathering by ectomycorrhizal fungi – effect of K^+ and Mg^{2+} deficiency and N sources on accumulation of oxalate	Paris et al. (1996)
Microbial interactions and weathering	
Interactions between ectomycorrhizal fungi and bacteria	Olsson and Wallander (1998)
Hyphal exudation/re-absorption in ectomycorrhizal fungi and its significance for hyphal interactions with soil and soil organisms	Sun et al. (1999)
Relative contribution of mycorrhizal fungi and bacteria to weathering	Leyval and Berthelin (1991, 1993)
C utilisation profiles of mycorrhizosphere bacterial communities in the illuvial mineral horizon	Heinonsalo et al. (2001)
Release and uptake of nutrients from minerals by ectomycorrhizal plants	
Ectomycorrhiza–bactera–beech root interactions – influence on release of P, K, Mg and Fe	Leyval and Berthelin (1989)
Biotite and microcline as sources of potassium to ectomycorrhizal seedlings	Wallander and Wickman (1999)
Uptake of P from apatite by ectomycorrhizal seedlings	Wallander (2000a)
Weathering of biotite and uptake of K by ectomycorrhizal seedlings	Wallander (2000b)
Acquisition of P from apatite by ectomycorrhizal seedlings	Wallander et al. (1997)
Reviews of mycorrhizal weathering interactions and related phenomena	
Fungi in biogeochemical cycles (and articles therein)	Gadd (2005)
Role of ectomycorrhiza in weathering minerals – review	Landeweert et al. (2001)
Discussion of evidence that mycorrhiza may drive podzolization	van Breemen et al. (2000b)
Role of microbial loops in nutrient cycling	Sun et al. (1999)
Role of cation uptake, Al tolerance on mediation of soil acidification effects	Finlay (1995)
Reduced leaching of base cations due to mycorrhizal mycelium	Ahonen-Jonnarth et al. (2003)
Vertical distribution of ectomycorrhizal fungal species	Rosling et al. (2003)
Ectomycorrhizal carbon allocation to mineral substrates	Rosling et al. (2004a)
Mycelial growth and substrate acidification in response to different minerals	Rosling et al. (2004b)

It now seems clear that mycorrhizal roots and the fungi colonising them play a potentially important role in the direct recycling of organic nutrients, and Abuzinadah et al. (1986) pointed out that this would restrict immobilisation of nutrients in tissues of decomposer organisms and enable tighter nutrient cycling than conventionally proposed pathways, facilitating the direct return of N to the host trees. Some of the implications of these "short circuits" in nutrient cycling are discussed below in Section 9.6.

The inability to grow the fungi forming arbuscular mycorrhizal associations in pure culture has contributed to the notion that these fungi completely lack saprotrophic capabilities. This is not correct, and it should by now be clear that the fungi produce plant cell wall-degrading enzymes enabling penetration of, and growth within, their host plant roots (Garcia-Garrido et al. 1992). St. John et al. (1983) showed that arbuscular mycorrhizal fungi are capable of proliferating preferentially in decomposing organic materials. Several studies (see Leake and Read 1997) have investigated the possible utilisation of organic forms of P, but the potential for mobilising organic forms of N has received much less attention. Hawkins et al. (2000) examined uptake and transport of organic N by arbuscular mycorrhizal fungi but concluded that in their experiments the amounts of N involved did not make a large contribution to the nutrition of the host plants. Recent experiments by Hodge et al. (2001) have shown that the arbuscular mycorrhizal fungus *Glomus hoi* was able to enhance decomposition and increase plant N capture from grass leaves. However, further research is still needed to distinguish between the direct capacity of AM fungi to mobilise organic substrates and their possible, indirect effects on decomposition and plant nutrient uptake, caused by stimulation of decomposers and subsequent uptake of their decomposition products by mycorrhizal hyphae.

Leake and Read (1997) emphasise that, while the nutrient-mobilising potential of some ecto- and ericoid mycorrhizal fungi has clearly been established, available studies include only a restricted range of fungi, and there is a need to investigate the extent to which these processes occur in a wider range of fungi under field conditions. Although uptake of simple amino compounds has been shown under field conditions (Näsholm et al. 1998), further experiments need to be conducted in which mycelial uptake and root uptake can be partitioned. Further experiments are still needed to determine the relative importance of different organic nutrient sources, and to examine inter-specific variation in the efficiency of nutrient transfer across the fungal–host interface. Analyses of the true enzymatic potential of mycorrhizal fungi, using in situ measurements in the field, will require the continued development of molecular tools, as well as their application under ecologically realistic conditions (Lindahl et al. 2005). Different possible sources of organic nutrients, such as Collembola (Klironomos and Hart 2001), nematodes (Perez-Moreno and Read 2001a), pollen (Perez-Moreno and Read 2001b) and bacteria (Barron 1988), are discussed below in relation to the interactions of mycorrhizal fungi with their biotic environment.

9.3.2 Weathering and Acquisition of Poorly Soluble Minerals

Although it is well accepted that mycorrhiza assist in acquisition of mineral nutrients already in the soil solution, there has recently been increasing interest in the idea that mycorrhizal mycelia, either by themselves or in association with bacteria or other fungi, may actively release nutrients from mineral particles and rock surfaces through weathering (Landeweert et al. 2001). Some recent studies are listed in Table 9.2. Low molecular weight (LMW) organic acids are considered to play an important role in the weathering process (Berthelin 1983; Gadd 1999), and there have been frequent reports of their production by mycorrhizal fungi. Oxalic acid production has been reported (Lapeyrie et al. 1987; Ahonen-Jonnarth et al. 2000) to play a role in solubilisation of phosphate from insoluble sources such as calcium phosphate (Lapeyrie et al. 1991) or hardened wood ash (Mahmood et al. 2001). Production of LMW organic acids such as oxalic acid has been demonstrated in forest soils (Graustein et al. 1977), and has been discussed in relation to its effects on nutrient and geochemical cycles as well as on the soil chemistry of ectomycorrhizal mycelial mats (Cromack et al. 1979; Entry et al. 1992; Griffiths et al. 1994). Results of laboratory experiments by Ochs (1996) suggested that exudation of simple organic ligands by mycorrhizal roots and fungi enhanced dissolution of minerals whereas exudates from non-mycorrhizal roots and humic substances did not significantly affect weathering. This interpretation was confirmed by Arocena et al. (1999) who compared the chemistry of soil solution from ectomycorrhizosphere and non-ectomycorrhizosphere soils under the subalpine fir *Abies lasiocarpa*. In soil associated with mycorrhizal roots and fungi, concentrations of the base cations Ca^{2+}, K^{+} and Mg^{2+} were higher than in soil devoid of roots and hyphal rhizomorphs. This was attributed to a reduced pH due to exudation of organic acids, as found by Cromack et al. (1979) who reported that the pH of soil colonised by mats of the ectomycorrhizal fungus *Hysterangium crassum* was reduced to 4.9, compared with uncolonised soil which had a pH of 6.1.

In 1997 Jongmans et al. published evidence suggesting that weathering-prone minerals in podzol surface soils and shallow granitic rock under European coniferous forests were commonly crisscrossed by networks of tubular pores 3–10 μm in diameter. Although their aetiology could not be established, the authors speculated that they might have been caused by hyphae of ectomycorrhizal fungi exuding organic acids at their tips. No conclusive proof of the physical weathering process, or of the direct or exclusive involvement of mycorrhizal fungi was provided by Jongmans et al. (1997) or van Breemen et al. (2000a), and it has now been shown that the contribution of these tunnels to total feldspar weathering is less than 1 % (Smits et al. 2005). However, it is well known that fungi have physical effects on minerals and rock surfaces. Callot et al. (1987) reported biogenic etching in amorphous and crystalline silicates, and the role of fungal organic acid production in the corrosion of stone monuments has been discussed by Hirsch et al. (1995). Callot et al. (1987) suggested that siderophores may be involved in fungal weathering. Production of siderophores by ecto- and ericoid mycorrhizal fungi has been demonstrated by Szaniszlo et al. (1981) and Haselwandter and Winkelmann (1998), and may play a role in the weathering of minerals. Weathering by lichens is well understood, and Banfield et al. (1999) have discussed the application of the "lichen model" to the understanding of mineral weathering in the rhizosphere. Suggested features of weathering by lichens, such as dissolution of minerals by exudation of organic acids derived from photosynthetic products of an autotrophic plant host, bear a striking similarity to the proposed mycorrhizal weathering, with the additional feature that ectomycorrhizal host trees should provide an even bigger source of assimilates whilst simultaneously constituting a large sink for mineral nutrients released by dissolution. Recent work by Rosling et al. (2003) has revealed that over half of the ectomycorrhizal species in a forest soil occurred exclusively in the underlying mineral horizons. Laboratory studies of ectomycorrhizal mycelia colonising different mineral substrates in vivo, or in plant microcosms, also suggest that patterns of mycelial growth, carbon allocation and substrate acidification may be influenced by mineral composition (Rosling et al. 2004a, b).

Other experimental evidence supporting the idea of mineral weathering by mycorrhizal hyphae comes from the X-ray diffraction studies of Paris et al. (1995a, b, 1996), showing that fungi such as *Paxillus involutus* and *Pisolithus tinctorius* are able to weather phlogopite mica and vermiculite, displacing interlayer K^+, Al^{3+} and/or Mg^{2+} from the interlattice spaces. Leyval and Berthelin (1989) demonstrated the influence of beech ectomycorrhiza–bacteria interactions on release of P, K^+, Mg^{2+} and Fe^{3+}, but suggested (Leyval and Berthelin 1991) that rhizosphere bacteria were mainly responsible for the solubilisation of mineral nutrients, and that the ability of mycorrhizal fungi to enhance nutrient uptake was solely due to the greater absorptive surface area of mycorrhizal roots. Wallander et al. (1997) demonstrated the acquisition of P from apatite by mycorrhizal seedlings but there was no clear relationship between exudation of organic acids and P uptake, presumably due to degradation of the organic acids by bacteria (Jones 1998). Wallander and Wickman (1999) found that pots containing pine seedlings colonised with *Suillus variegatus* and with access to biotite had soil solutions higher in citric and oxalic acids, and that citric acid concentration was positively correlated with fungal biomass in the soil as well as with foliar K^+ concentrations. More recent studies of mycorrhizal effects on weathering of apatite (Wallander 2000a) and biotite (Wallander 2000b) suggest that exudation of oxalic acid by mycorrhizal fungi can influence dissolution of apatite and increase P uptake by *P.*

sylvestris, but evidence of a direct mycorrhizal effect on K^+ uptake was less clear in the latter experiment. In addition to the above effects, ectomycorrhizal hyphae may also help restrict losses of base cations through leaching (Ahonen-Jonnarth et al. 2003).

The possible effects of other groups of mycorrhizal fungi on weathering are less clear. Ericoid mycorrhizal fungi have not been examined in detail but it is possible that they might contribute to the effects found in Podzols in situations where ericaceous plants form understorey vegetation. However, the extension of ericoid mycorrhizal hyphae from their host roots is limited. The effects of VA mycorrhizal fungi are less clear because pure culture studies are not possible, and evidence that they can improve P nutrition of plants by solubilising poorly soluble forms is conflicting.

In conclusion, it seems possible that ectomycorrhizal fungi may play an important role in mineral weathering but the relative contributions of different organism groups and the overall importance of biological weathering in relation to physical and chemical processes are still unclear. Sun et al. (1999) highlighted the possible importance of mycorrhizal mycelial exudation of water and organic acids for interactions with other micro-organisms at the soil–mycelium interface. The possibility that bacteria associated with mycorrhizal hyphae might be responsible for weathering was acknowledged by van Breemen et al. (2000a) but there have been few direct studies of such interactions in relation to weathering. Olsson and Wallander (1998) found that bacterial communities were differently influenced by one and the same mycorrhizal fungus, depending on the mineral sources the fungi were exposed to, and concluded that the effects could have been due to fungal production of different organic acids. Interactions between mycorrhizal mycelia and other organisms may have consequences for plant nutrition and nutrient cycling at a number of levels, and these are discussed below.

9.4 Biotic Interactions

Functioning of mycorrhizal root systems has increasingly been considered within the context of the ways in which mycorrhizal hyphae interact with other soil organisms. Many different types of interaction have been documented, with different effects on both the nutrition of individual plants and the cycling of nutrients at the plant ecosystem level (see Tables 9.3 and 9.4). For convenience, these are divided below into interactions with bacteria (of which there are many studies; Table 9.3) and interactions with other organisms (Table 9.4) such as fungi (fewer studies) and soil animals, principally Collembola and nematodes.

Garbaye (1991) and Fitter and Garbaye (1994) have provided general reviews of the subject but, as mycorrhizal scientists increasingly acknowledge

Table 9.3. Mycorrhizal interactions with bacteria

Details	Authors (see reference list)
VA mycorrhiza–bacteria interactions	
Qualitative effects of AM taxa on bacteria are more important than quantitative ones	Andrade et al. (1997)
AM fungi selectively influence persistence of inoculated bacteria and may influence native bacteria	Andrade et al. (1998a)
Mycorrhiza influence water-stable soil aggregates, in turn affecting soil bacteria numbers	Andrade et al. (1998b)
G. intraradices-induced reduction in *P. fluorescens* through reduction in root C exudation	Marschner et al. (1997)
Synergistic effects of the PGPR *P. fluorescens* DF57 and *Glomus caledonium* on P uptake	Ravnskov and Jakobsen (1999)
Negative effects of *Glomus intraradices* on numbers of the PGPR *P. fluorescens* DF57 in hyphosphere soil	Ravnskov et al. (1999)
Co-inoculation of AM fungi, *Rhizobium* and PGPR bacteria – multiple microbial inoculation improved plant development, nutrient uptake and N fixation	Requena et al. (1997)
Management of indigenous plant–microbe symbioses and its use in restoration of desertified ecosystems	Requena et al. (2001)
AM fungi and PGPR *Pseudomonas* have positive effects independently but these are reduced in combination	Sastry et al. (2000)
Interactions between AM fungi, *Azospirillum* and *Rhizobium* and their effects on photosynthetic activity of alfalfa are both synergistic and antagonistic	Tsimilli et al. (2000)
Positive effects of dual inoculation with AM fungi and *Bacillus subtilis* on growth of micro-propagated potatoes	Votsáka and Gryndler (2000)
Interactions with biocontrol organisms	
Biocontrol *Pseudomonas* strains have no negative effect on *Glomus mosseae* – bacteria adhere to spores	Barea et al. (1998)
Phytostimulators (*Azospirillum*) and two biological control agents (*Pseudomonas* and *Trichoderma*) were tested for negative effects on two inoculated arbuscular mycorrhizal fungi as well as indigenous AMF – no negative effects on mycorrhizal activity were found. Mycorrhiza-induced qualitative effects on the bacterial flora	Vázquez et al. (2000)
Bacterial endosymbionts of mycorrhiza and interactions with N-fixers	
N-fixing *Bacillus* sp. associated with Douglas fir tuberculate mycorrhiza – possible contribution to N cycling	Li et al. (1992)
Ectomycorrhizal nutrient mobilisation from litter patches – N loss smaller than N gain by plants – discussion of possible associative N fixation	Perez-Moreno and Read (2000)
Discussion of possible associative bacterial N fixation in the ectomycorrhizal hyphosphere	Sen (2000)
Interrelations between *Azospirillum*, *Rhizobium* and AM fungi in rhizosphere of alfalfa	Biró et al. (2000)

Identification and characterisation of DNA region containing putative *nif* genes in *Burkholderia* growing endosymbiotically on *Gigaspora*	Minerdi et al. (2001)
Translocation of N between actinorhizal alder seedlings and pine seedlings linked by a common ECM mycelium	Arnebrant et al. (1993)
Detection of *Burkholderia* endosymbionts in arbuscular mycorrhizal fungi belonging to the Gigasporacea	Bianciotto et al. (2000a)
Endosymbiotic *Burkholderia* strain living inside the AM fungus *Gigaspora margarita* possesses the *vacB* gene involved in colonisation	Ruiz-Lozano and Bonfante (2000)
Bacterial biofilms – structure and regulation	
Regulation of bacterial biofilms	Costerton et al. (1999)
Acyl-homoserine lactone signalling and its possible functioning of biofilms – quorum sensing	Parsek and Greenberg (2000)
Mucoid mutants of *Pseudomonas fluorescens* show improved attachment to *G. intraradices* hyphae and roots	Bianciotto et al. (2001b)
Ectomycorrhiza interactions	
Stimulation of *P. fluorescens* biovar 1 and repression of biovar V and II by mycorrhizosphere of *Laccaria bicolor* and Douglas fir – preferential use of trehalose by biovar 1	Frey et al. (1997)
Decline of inoculated *P. fluorescens* but mycorrhiza helper bacterium enhances pre-symbiotic growth of fungus	Frey Klett et al. (1997)
Dose effects – helper bacterium BBc6R8 acts at low population densities – lowest dose increased mycorrhizal colonisation more than highest dose	Frey Klett et al. (1999)
Effects of *Laccaria bicolor* and *P. fluorescens* BBc6 on rooting of de-rooted shoot hypocotyls of *P. abies*	Karabaghli et al. (1998)
Abundant α- and β-proteobacteria found using FISH, and confocal microscopy implications of non-culturability	Mogge et al. (2000)
Ectomycorrhizospheres of *Paxillus involutus* and *Suillus bovinus* on pine have different effects on bacteria	Nurmiaho-Lassila et al. (1997)
+ve and −ve effects of mycorrhizal fungi on bacterial activity and numbers, depending upon soil substrate	Olsson and Wallander (1998)
Reductions in mycorrhizosphere-associated bacterial activity – greatest with *Paxillus involutus*	Olsson et al. (1996)
Significance of exudation–absorption at the mycelial–soil interface	Sun et al. (1999)
Ectomycorrhizosphere community structure (based on carbon source utilisation patterns) differs between nursery peat and forest soil, and between fungal symbionts (*Suillus bovinus* and *Paxillus involutus*)	Timonen et al. (1998)
Insensitivity of detrital food web micro-organisms to the presence of ectomycorrhiza – additional production of plant-assimilated C does not reach detrital food web	Setälä et al. (1999)
Ectomycorrhizal effects on functional diversity of bacteria	Frey-Klett et al. (2005)

Table 9.3. (*Continued*)

Details	Authors (see reference list)
Bioremediation interactions	
Microbial biofilms with catabolic bacteria at PHC (petroleum hydrocarbon)-contaminated soil–fungal interface	Sarand et al. (1998)
Tolerance and biodegradation of *m*-toluate by *Pinus sylvestris*, *Suillus bovinus* and *Pseudomonas fluorescens* harbouring TOL catabolic plasmids, individually and under associative conditions	Sarand et al. (1999)
Effects of inoculation of a TOL plasmid containing mycorrhizosphere bacterium on development of Scots pine seedlings, their mycorrhizosphere and microbial flora in *m*-toluate-amended soil – implications for bioremediation	Sarand et al. (2000)
Retardation of fluorene degradation in the ectomycorrhizosphere	Genney et al. (2004)
Reviews	
"Interactions between mycorrhizal fungi and other soil micro-organisms"	Fitter and Garbaye (1994)
"Helper bacteria: a new dimension to the mycorrhizal symbiosis"	Garbaye (1994)
"Ectomycorrhizas – extending the capabilities of rhizosphere remediation"	Meharg and Cairney (2000)
"Bacterial associations with mycorrhizal fungi, close and distant friends in the rhizosphere"	Perotto and Bonfante (1997)

the existence of other soil organisms, this research area has expanded rapidly since then. The concept of the "rhizosphere", first introduced by Hiltner in 1904 to describe the volume of soil around roots which is influenced by plant activity, has thus been extended to include the influence of the mycorrhizal hyphae colonising the roots. The term "mycorrhizosphere" was first used by Rambelli (1973) to describe the rhizosphere of roots colonised by mycorrhizal hyphae. The more specific term "mycorrhizal hyphosphere" refers only to the volume of soil influenced by the mycorrhizal hyphae themselves.

9.4.1 Interactions with Bacteria

Bowen and Theodorou (1979) demonstrated both positive and negative interactions between ectomycorrhizal fungi and bacteria. Garbaye and Bowen (1989) later showed that some bacteria closely associated with ectomycorrhizal fungi could stimulate mycorrhiza formation, and called these "mycorrhization helper bacteria" (MHB). Possible mechanisms underlying the effect of MHBs are discussed by Garbaye (1994) and include (1) enhancement of the susceptibility of roots to mycorrhizal infection, (2) enhancement of the root–fungus recognition process, (3) effects on the pre-symbiotic growth of the fungi, (4) modification of the rhizospheric soil, and (5) stimulation of the germination of fungal propagules. Subsequent studies have examined these possible effects in more detail in both arbuscular mycorrhizal and ectomycorrhizal systems.

Andrade et al. (1997) grew *Sorghum bicolor* with or without AM fungi *Glomus etunicatum*, *G. intraradices* or *G. mosseae*. Bacterial numbers were greater in the rhizosphere than in the hyphosphere. Isolates of *Bacillus* and *Arthrobacter* were frequent in the hyphosphere, and *Pseudomonas* in the rhizosphere. Development of AM mycelium in the soil had little influence on the composition of the microflora in the hyphosphere but AM root colonisation was positively related with bacterial numbers in the hyphosphere and with the presence of *Pseudomonas* in the rhizosphere. Andrade et al. (1998a) found that hyphosphere-specific *Alcaligenes eutrophus* depended upon the presence of *Glomus mosseae* but that the non-specific *Arthrobacter globiformis* did not. Their experiments suggested that the mycorrhizal status of soils may selectively influence the persistence of bacterial inoculants as well as affecting the numbers of other native bacteria. Another study by Andrade et al. (1998b) showed that root and fungal components of mycorrhiza affect soil bacteria indirectly by influencing water-stable soil aggregates and providing a favourable, protective environment for the organisms. Other positive effects of co-inoculation of bacteria with AM fungi have been reported in micropropagated potatoes (Vosátka and Gryndler 2000) and cucumber plants (Ravnskov and Jakobsen 1999) where *Glomus caledonium* and *Pseudomonas fluorescens* had a synergistic effect on P uptake. Positive effects of multiple

Table 9.4. Interactions of mycorrhiza with fungi and other organisms

Details	Study (see reference list)
Mycorrhizal interactions with other fungi (including mycorrhizal fungi)	
Ectomycorrhizal acquisition of P from a saprotroph *Hypholoma fasciculare*	Lindahl et al. (1999)
Influence of resource size on transfer of P between saprotrophic and ectomycorrhizal fungi	Lindahl et al. (2001)
Reduced carbon flux to ectomycorrhizal mycelium interacting with *Phanerochaetae velutina*	Leake et al. (2001)
Ectomycorrhizal root colonisation on pine suppressed by saprotrophic fungi	Shaw et al. (1995)
Interactions between tree roots, mycorrhizal and a saprotrophic fungus *Mycena galopus* give synergistic and antagonistic effects on decomposition	Dighton et al. (1987)
Inhibitory effect of *Trichoderma* on mycorrhiza formation by *Laccaria bicolor* in vitro	Summerbell (1987)
Micro-fungi associated with the mantle and mycorrhizosphere of black spruce	Summerbell (1989)
Inhibition of phytopathogenic fungi in vitro by cell-free culture media of ectomycorrhizal fungi	Kope and Fortin (1989)
Effects of temperature and incubation time on the ability of three ectomycorrhizal fungi to colonise *P. sylvestris* roots	Erland and Finlay (1992)
Competitive interactions between ectomycorrhizal fungi under field conditions	McAffee and Fortin (1986)
Competition between ectomycorrhizal fungi colonising *Pinus densiflora*	Wu et al. (1999)
Influence of ectomycorrhizal fungi on resistance to pathogenic fungi and soil bacteria	Marx (1969)
Ectomycorrhizae as biological deterrents to pathogenic root infections	Marx (1972)
Disease suppression and antibiosis by the ectomycorrhizal fungus *Paxillus involutus*	Duchesne et al. (1989)
Probable mycorrhizal suppression of decomposition demonstrated by trenching experiments	Gadgil and Gadgil (1971)
Fungal mycelium and mycorrhizal necromass as N sources for *Hymenoscyphus ericae* and its host plants	Kerley and Read (1997, 1998)
Mycorrhizal interactions with other organisms	
Grazing interactions between Collembola and vesicular-arbuscular mycorrhiza (+ve and –ve effects)	Finlay (1985)
Grazing and influence of *Folsomia candida* on growth and nutrient uptake of mycorrhizal leeks	Warnock et al. (1982)
Reduction of mycorrhizal fungal growth by a nematode *Aphelenchoides saprophilus*	Ruess and Dighton (1996)
No effect of collembolan grazing on functioning of three arbuscular mycorrhizal fungi	Larsen and Jakobsen (1996)
Ectomycorrhizal associations with *Laccaria bicolor* obtain N from Collembola	Klironomos and Hart (2001)

Reductions of ectomycorrhizal soil mycelia by faunal grazing – but effects related to soil N status	Setälä et al. (1997)
Non-nutritional interactions between protozoa and ectomycorrhiza – plant growth increased by protozoa, protozoan abundance increased by mycorrhizal colonisation	Jentschke et al. (1995)
Collembola prefer saprobic fungi to mycorrhizal ones – previous estimates of negative grazing effects may be exaggerated	Klironomos et al. (1999)
VA mycorrhizal root symbiont modifies plant resistance to the insect herbivore *Epilachna varivestis*	Borowicz (1997)
Grazing by Collembola likely to have only a small impact on effectiveness of mycorrhizal fungi in promotion of nutrient uptake	McGonigle (1995)

microbial inoculants including AM fungi and rhizobacteria have also been reported by Requena et al. (1997, 2001) who recommend management of indigenous plant–microbe symbioses as a tool in the restoration of ecosystems affected by desertification.

Both synergistic and antagonistic effects of co-inoculated arbuscular mycorrhizal fungi and nitrogen-fixing bacteria were reported by Tsimilli-Michael et al. (2000). Sastry et al. (2000) also documented inhibitory effects of co-inoculation of AM fungi and plant growth-promoting rhizobacteria (PGPR), suggesting that microbial components with stimulatory effects when inoculated independently may not always have a positive, synergistic effect on plant growth when co-inoculated. Negative effects of *Glomus intraradices* on populations of the PGPR *P. fluorescens* DF57 were shown by Ravnskov et al. (1999) who suggested that competition for inorganic nutrients might explain the effect, since the mechanism did not require cell-to-cell contact. Marschner et al. (1997) suggested that similar negative effects of *G. intraradices* on *P. fluorescens* 2-79RL might be due to mycorrhiza-induced decreases in root exudation, affecting the composition of the rhizosphere soil solution. Possible antagonistic effects between mycorrhizal fungi and biocontrol organisms have obvious applied importance but have so far not been studied extensively. Barea et al. (1998) investigated the effects of *Pseudomonas* strains used for biocontrol of soil-borne fungal plant pathogens on arbuscular mycorrhiza formation. The wild-type strain and a mutant strain not producing the antifungal compound DAPG (2,4-diacetylphlorogluconol) stimulated spore production and mycorrhizal root infection by *Glomus mosseae*. A DAPG overproducer did not stimulate *G. mosseae* but had no negative effect on the mycorrhizal fungus. Adhesion of the bacteria to mycorrhizal spores appeared to be unrelated to the ability to produce DAPG. In a more recent study, Vázquez et al. (2000) examined effects of phytostimulators (*Azospirillum*) and biological control agents (*Pseudomonas* and *Trichoderma*) on arbuscular mycorrhizal fungi, and found no negative effects on mycorrhizal activity. Mycorrhizal colonisation by different fungi, on the other hand, induced species-dependent, qualitative changes in bacterial community structure and rhizosphere enzyme activity.

Perotto and Bonfante (1997) reviewed various mycorrhiza–bacteria interactions and discussed their possible significance in rhizosphere ecology, distinguishing different levels of cellular integration ranging from simple association, through surface attachment, to obligatory, endosymbiotic association. The possible interactive effects of obligate and associative diazotrophs and different mycorrhizal fungi have been investigated in a number of studies. Biró et al. (2000) examined interactions between the AM fungus *Glomus fasciculatum* and *Azospirillum* and *Rhizobium* N-fixers and found synergistic effects of multilevel inoculation, compared with single mycorrhizal or dual (*Azospirillum*-mycorrhiza) inoculation. Ruiz-Lozano and Bonfante (2000) demonstrated endosymbiotic *Burkholderia* bacteria inside the arbuscular mycorrhizal fungus *Gigaspora margarita*, and Bianciotto et al. (2000) documented *Burkholderia* endosymbionts in spores of 11 fungal isolates from 5 fungal species from different geographic areas, suggesting that intracellular bacteria are a common feature of the family Gigasporaceae. Minerdi et al. (2001) have since demonstrated that *Burkholderia* bacteria in *G. margarita* contain the nitrogen-fixing genes *nifHDK* which are expressed at least during spore germination.

Nitrogen-fixing bacteria have also been suggested to be associated with ectomycorrhizal fungi (Chanway and Holl 1991), and Li et al. (1992) demonstrated N-fixing *Bacillus* sp. associated with tuberculate mycorrhizal roots of Douglas fir, as well as low levels of nitrogenase activity which could contribute to the long-term nitrogen dynamics of N-limited Douglas fir forests. Nitrogen budgets of ectomycorrhizal birch microcosm systems (Perez-Moreno and Read 2000) suggested that associative N fixation in intensively colonised litter patches may have contributed to the N nutrition of seedlings, and this idea has been further discussed by Sen (2000). Vigorous mycelial growth might be expected to lead to the low oxygen tension and high availability of carbon substrates necessary for micro-aerophilic N fixation. Indirect interactions between ectomycorrhizal plants and actinorhizal trees forming tripartite symbioses with both ectomycorrhiza and *Frankia* (Arnebrant et al. 1993) may also contribute to N dynamics of plant communities.

Clearly, the full complexity of bacterial interactions with different mycorrhizal associations has yet to be revealed, and recent advances in biofilm research (Costerton et al. 1999) involving bacterial signalling and quorum sensing (Parsek and Greenberg 2000) may be important in understanding the structure, maintenance and functional heterogeneity of bacterial communities colonising hyphal surfaces. Advances in the use of techniques such as denaturing gradient gel electrophoresis (DGGE; Kozdrój and van Elsas 2000) and confocal laser scanning microscopy (Palmer and Sternberg 1999; Mogge et al. 2000) should improve the resolution with which we can study these bacterial communities, and their functional interactions with different mycorrhizal symbioses. Extracellular polysaccharides play an important role in the formation of biofilms, and Bianciotto et al. (2001) demonstrated that a wild-type non-mucoid strain of *Pseudomonas fluorescens* had a much lower ability to anchor to extraradical hyphae of *Glomus intraradices* compared with two mucoid mutants with increased production of acidic extracellular polysaccharides.

Interactions between ectomycorrhizal fungi and bacteria have also been studied more intensively during the past decade. Earlier studies by Garbaye and coworkers concentrating on the role of bacteria in mycorrhizal formation in nurseries (see Garbaye 1994) have been supplemented by assessments of the roles of bacterial interactions in forest soils. In the mycorrhizosphere of Douglas fir, Frey et al. (1997) demonstrated total culturable bacterial densities and fluorescent pseudomonad populations which were 14 and 4 times greater respectively than the values recorded in the adjacent bulk soil. Similar effects were also observed in the mycorrhizosphere of hybrid larch, Sitka spruce and sycamore by Grayston et al. (1994). In another study, Frey-Klett et al. (1997) found that *Pseudomonas fluorescens* BBc6R8 consistently promoted *Laccaria bicolor*–Douglas fir mycorrhiza formation, even at low doses, although the BBc6R8 bacteria declined in all experiments, and disappeared below detection level within 19 weeks. This was attributed to an early effect of the bacterium on the pre-symbiotic growth of the fungus, stimulating fungal growth and increasing the chance of encounter between the root and the fungus. Later studies (Frey-Klett et al. 1999) showed that BBc6R8 acted at very low population densities, well below the inoculation dose associated with most PGPR bacteria. Karabaghli et al. (1998) found that *Laccaria bicolor* stimulated rooting of *P. abies* shoot hypocotyls, which was also stimulated by indole-3-acetic acid (IAA), but that the bacterium *P. fluorescens* BBc6 did not. Both the fungus and the bacterium produced IAA, suggesting that IAA was not the only substance required to stimulate rooting, or that the bacterium also produces other compounds which counteract the stimulatory effects. Mogge et al. (2000) used fluorescence in situ hybridisation (FISH), with fluorescent oligonucleotide probes targeting phylogenetic sequences of the16S and 23S rRNA, to examine bacterial community structure and colonisation patterns on mycorrhizal beech roots. *Lactarius subdulcis, L. vellerus* and *Laccaria amethystina* were found to be common mycorrhizal fungi colonising beech roots. Confocal laser scanning microscopy, in conjunction with FISH, revealed abundant α, β and γ subclass proteobacteria on the mantle surfaces. The subclasses α and β contain numerous N-fixing and plant growth-promoting rhizobacteria, such as members of the genera *Burkholderia*, *Azospirillum* and *Acetobacter*. Other studies (e.g. Frey et al. 1997) using culture-dependent methods have concentrated on the γ class of proteobacteria, including *Pseudomonas fluorescens/putidans*, but culture-free methods may ultimately provide less biased estimates of community structure in mycorrhiza associated bacteria.

Nurmiaho-Lassila et al. (1997) compared the ectomycorrhizosphere of *Pinus sylvestris* seedlings colonised by *Paxillus involutus* or *Suillus bovinus.* Bacterial colonisation patterns were different both within and between these systems. Bacteria were detected on the mantle surfaces and at inter- and intracellular locations in the mantle and Hartig net of *S. bovinus*-colonised roots. Fungal rhizomorphs had few bacteria but the outermost fine hyphae were colonised by extensive monolayers of bacteria. Mycorrhizas of *Paxillus* were almost devoid of bacteria but the external mycelium had both colonies and solitary bacteria. Further studies of bacterial community structure based on carbon source utilisation patterns (Timonen et al. 1998) showed divergent species composition and significant differences between communities from two different fungal mycorrhizospheres, and even clearer differences between forest and nursery mycorrhizospheres and between mycorrhizosphere and non-mycorrhizosphere soils. Earlier observations of reduction of bacterial activity (thymidine and leucine incorporation) by mycelium of *Paxillus involutus* (Olsson et al. 1996) also support the results of Nurmiaho-Lassila et al. (1997). Olsson and Wallander (1998) showed that the same fungus may both stimulate and decrease bacteria under different conditions. In microcline-amended and unamended soils, both *Suillus variegatus* and *Paxillus involutus* reduced bacterial activity measured as thymidine incorporation. *Suillus variegatus* grew best in the biotite soil where it increased both bacterial activity and biomass (counts and PLFAs). *Suillus variegatus* modified the community structure based on bacteria-specific PLFA composition. A tenfold increase in citric acid was observed, and it is possible that increases in bacterial biomass and changes in community structure may have been due to organic substances produced by fungus. In the apatite-amended soil, two isolates of *S. variegatus* and an unidentified species all tended to stimulate bacterial activity.

Clearly, the mycorrhizal mycelium can play a role in conditioning its biotic environment and acting as a support surface for bacteria (Sun et al. 1999). Further research is required to understand the ways in which these bacteria modify nutrient supply at the mycorrhizal mycelial surface. Carbon supply via mycorrhizal fungi is evidently of potential significance, although results of Setäla et al. (1999) suggest that direct entry of plant-derived photosynthates into the detrital food web via ectomycorrhizal fungi is probably not large. Recent experiments by Frey-Klett et al. (2005) suggest that the ectomycorrhizal fungus *Laccaria bicolor* may significantly influence the functional diversity of bacteria in the mycorrhizosphere of *Pseudostuga menziesii* seedlings in ways which are potentially beneficial to the symbiosis and the plant host.

The metabolic diversity of the interactions involved may have important implications for future applications in biocontrol, biofertilisation and bioremediation (Meharg and Cairney 2000) – some of these are discussed in Section 9.7. Possible effects include modifying access to organic N and P compounds, effects on weathering of minerals, bioremediation of organic pollutants (Sarand et al. 1998, 1999, 2000), and stimulation of biocontrol organisms.

9.4.2 Interactions with Fungi

Interactions of mycorrhizal fungi with other fungi have in general been much less well studied than those with bacteria. Much of the initial interest was in the possible role of mycorrhizal symbiosis in disease suppression (Marx 1969). Ectomycorrhizal fungi such as *Pisolithus tinctorius* and *Thelephora terrestris* have been shown to provide their host plants with increased resistance to attack by the root pathogen *Phytophthora cinnamomi*, and it is know that many ectomycorrhizal fungi produce antibiotic substances as well as acting as a physical barrier to penetration by pathogens (Marx 1972). In some interactions of *P. cinnamomi* with VA mycorrhizal plants, *prior* inoculation with the mycorrhizal fungus is necessary to achieve protection (Davis and Menge 1980), and it has often been difficult to separate indirect effects of mycorrhizal inoculation, arising from improved nutritional status, from direct antibiotic effects. Evidence of direct effects was provided by Duchesne et al. (1988a, b) who observed reduced pathogenicity of *Fusarium oxysporum* in *Pinus resinosa* seedlings inoculated with *Paxillus involutus* before mycorrhizal colonisation took place. Increased seedling survival was associated with a six-fold reduction in sporulation of *F. oxysporum* in the rhizosphere (Duchesne et al. 1987), and ethanol extracts of the rhizosphere indicated fungitoxic effects within 3 days of inoculation with the mycorrhizal fungus (Duchesne et al. 1989). Antibiotic activity of cell-free culture media was also demonstrated by Kope and Fortin (1989) who isolated anti-fungal compounds produced by *Pisolithus tinctorius*.

Competition between different mycorrhizal fungi also has the potential to influence plant nutrient acquisition but so far there are hardly any studies of these interactions (McAffee and Fortin 1986; Erland and Finlay 1992; Wu et al. 1999). Studies of mycorrhizal interactions with saprotrophic fungi are also infrequent but suggest that there may be both direct and indirect effects on plant nutrition. Ectomycorrhizal root colonisation may be suppressed by competition with saprotrophic fungi (Shaw et al. 1995), and *Trichoderma* species have also been shown to inhibit mycorrhiza formation by *Laccaria bicolor* in vitro (Summerbell 1987). Dighton et al. (1987) reported both synergistic and antagonistic effects of interactions between *Mycena galopus* and an ectomycorrhizal fungus on decomposition. Gadgil and Gadgil (1971) observed increased decomposition rates in response to trenching of plots to sever ectomycorrhizal connections with their tree roots, suggesting that ectomycorrhizal fungi may have an inhibitory effect on decomposition. This negative impact was interpreted in terms of competition for nitrogen from ectomycorrhizal fungi (Abuzinadah et al. 1986). However, direct transfer of labelled ^{32}P from the saprotrophic mycelium of *Hypholoma fasciculare* to mycorrhizal plants has also been demonstrated by Lindahl et al. (1999), suggesting that direct interactions may alter the nutrient status of mycorrhizal plants. The balance of competition between ectomycorrhizal and saprotrophic fungi, and direction of nutrient transfer resulting from antagonistic interactions are likely to be strongly influenced by resource availability (Lindahl et al. 2001). Experiments by Leake et al. (2001) have also demonstrated inhibitory effects of interactions with the mycelium of the saprotrophic fungus *Phanerochaete velutina*, reducing C allocation to the ectomycorrhizal mycelium. Clearly, the existence of such direct interactions has important implications for models of nutrient uptake, nutrient cycling, carbon flow and forest ecology – these are discussed in Section 9.6 below.

9.4.3 Interactions with Other Soil Organisms

Studies of interactions between mycorrhizal mycelia and other organisms have been mostly restricted to Collembola and nematodes. Warnock et al. (1982) demonstrated negative impacts on nutrient uptake by mycorrhizal leek plants in response to mycelial grazing by the collembolan *Folsomia candida*. In later field and laboratory experiments using VA mycorrhizal plants, Finlay (1985) demonstrated both negative and positive effects of mycelial grazing, depending upon the density of Collembola. McGonigle (1995) discussed the significance of the location of grazing, and suggested that preferential grazing of thinner hyphae, further away from the root, would have only a small impact on the growth-promoting effect of mycorrhizal fungi on nutrient uptake. Larsen and Jakobsen (1996) concluded that grazing by Collembola seemed to have little effect on uptake by mycorrhizal *Trifolium subterraneum* plants. Klironomos et al. (1999) make the point that most experiments on mycorrhizal grazing exclude other, non-mycorrhizal fungi which could be possible alternative food sources, and that effects of animal grazing may thus have been exaggerated. Further experiments examining the influence of grazing under more natural conditions are clearly necessary. Experiments involving ectomycorrhizal plants are less frequent but studies with a mixed soil fauna (Setäla 1995) showed no detrimental effect on plant growth, despite reductions in the amount of ectomycorrhizal fungal biomass. More recent work by Setäla et al. (1997) has demonstrated that grazing effects are dependent on soil nutrient status. Ectomycorrhizal fungi have also been shown to influence population development of the fungal-feeding nematode *Aphelenchoides saprophilus* which, in turn, may influence mycorrhizal development and fungal competition (Ruess and Dighton 1996). Few studies have been conducted on the effects of soil protozoa but experiments by Jentschke et al. (1995) suggest that the abundance of naked amoebae at the rhizoplane of *Picea abies* can be significantly increased by mycorrhizal colonisation with the ectomycorrhizal fungus *Lactarius rufus*, and that protozoa can increase seedling growth without affecting concentrations of nutrients in the plant needles. Although the predominant focus of grazing studies has been the effects of animal grazing on hyphae, and thus on plant nutrient uptake, one recent study by Klironomos and Hart (2001) showed that the ectomycorrhizal fungus *Laccaria bicolor* can kill the soil micro-arthropod *Folsomia candida*, take up nitrogen from the animals, and transfer the derived N to *Pinus strobus* host plants. If this predatory activity (see also Lindahl et al. 1999) is widespread in ectomycorrhizal fungi, then nutrient cycling may be more complicated and tight than previously believed. Recent experiments by Perez-Moreno and Read (2001a, b) support this idea, and have demonstrated the ability of the ectomycorrhizal fungus *Paxillus involutus* to recycle nutrients from sources such as dead nematodes and pollen.

9.5 Multiplicity of Function – Non-nutritional Roles

Since the first land plants were rootless, it has been argued that mycorrhizal fungal hyphae were required for the colonisation of land because of the need of a system to scavenge for immobile ions such as phosphate (Pirozynski and Malloch 1975). Plant root systems have evolved to become efficient at extracting nutrients but this has not led to loss of the symbiosis – non-mycorrhizal plants are unexpectedly rare (as few as 5% of plant species; Fitter and Moyerson 1996; Daniell et al. 1999). The symbiosis has apparently remained ubiquitous because it has evolved other roles, including defence against pathogens and drought resistance. The symbiosis is thus *multifunctional* (Newsham et al. 1995). Mycorrhizal interactions with different components of the abiotic and biotic environment can influence plant nutrition in a multitude of ways. Some of these are summarised in Table 9.5. Division of mycorrhizal effects into "nutritional" and "non-nutritional" categories is somewhat arbitrary, since any processes affecting plant growth at the population and community level will ultimately have an impact on nutrient cycling. Instead, different possible interactions are listed according to whether their main effects are at the level of the plant individual, the plant community, or at the ecosystem level.

Possible effects on individual plants include improved acquisition of various organic and inorganic nutrients, through increased absorptive surface area, enzymatic activity or weathering, improvements in drought tolerance, and hormonal effects on root growth (Gay et al. 1994). Mycorrhizal hyphae enable penetration and exploitation of soil micro-sites as well as stabilising soil aggregates (Tisdall and Oades 1979). Mediation of the negative effects of heavy metals and aluminium may also result from reduced uptake or binding to compounds such as metallothioneins or oxalic acid (Finlay 1995; Ahonen-Jonnarth et al. 2000). Disease suppression and resistance to pathogens may be brought about by both physical and chemical (antibiotic) means, although interpretation of these effects is often complicated by changes in nutritional status of the host plant, and the reproduction of individual plants may also be influenced (Koide et al. 1994). At the level of the plant community, there may be effects on competition between different species (Fitter 1977; Perry et al. 1989). Experiments in laboratory mesocosms (Grime et al. 1987) suggest that mycorrhizal symbiosis may promote seedling survivorship and species richness of grassland communities, and more recent studies in mesocosms and field plots (van der Heijden et al. 1998) suggest that the belowground diversity of arbuscular mycorrhizal fungi may influence both the diversity and productivity of plant communities. These results emphasise the potential contribution of arbuscular mycorrhizal fungi to the maintenance of plant biodiversity and ecosystem functioning, and the need to consider these "invisible" symbionts in sustainable management practices designed to minimise impacts on (plant) species diversity. Possible ecosystem-level effects include the supply of carbon to the soil ecosystem and its subsequent release to the atmosphere. Recent tree-girdling experiments in forest ecosystems dominated by ectomycorrhizas (Högberg et al. 2001) have shown that soil respiration is strongly driven by the flux of current assimilates through roots and mycorrhizal fungi, confirming earlier laboratory experiments by Soderström and Read (1987). In situ $^{13}CO_2$ pulse-labelling experiments in an upland grassland have also recently demonstrated the rapid flux of carbon through arbuscular mycorrhizal fungi, showing that the fixation by, and loss from the mycorrhizal mycelium was at least as large as that from fine roots (Johnson et al. 2002). This supply of carbon to mycorrhizal fungi has been suggested to allow them to compete efficiently with saprotrophic organisms for nutrients, and may have an inhibitory effect on decomposition (Abuzinadah et al. 1986). Other possible effects on the soil ecosystem include the idea that mycorrhizal fungi may even drive podzolisation (van Breemen et al. 2000b).

Molecular methods have greatly improved our ability to distinguish different mycorrhizal fungal species in the field. As the resolution with which we can study communities of ectomycorrhizal fungi improves, it will become increasingly important to improve our knowledge of the functional properties of these different species. It is already apparent that there are large inter- and intraspecific differences in functional properties such as carbon use efficiency, and the efficiency with which different fungi take up or transfer nutrients to their host plants or retain them in their mycelium (Colpaert et al. 1996; Colpaert and Verstuyft 1999; Gorissen and Kuyper 2000; Smith et al. 2000). Further studies of the functional significance of different levels of host–fungus compatibility (Finlay 1989) are needed. Data from a phylogenetic analysis of rDNA sequences of a wide range of basidiomycetes (Hibbett et al. 2000) suggest that ectomycorrhizal fungi with diverse hosts may have evolved repeatedly from saprotrophic precursors, but also that there have been multiple reversals to a free-living condition and that ectomycorrhizas are unstable, evolutionarily dynamic associations. Within this context, variation in function and efficiency would not be surprising. Parniske (2000) has discussed evidence for the presence of a common developmental program for symbiotic and pathogenic micro-organisms and their interactions with plants. Further development and application of modern tools within functional and environmental genomics will help to reveal the full spectrum of possible interactions (Martin 2001; Voiblet et al. 2001), but advances in the resolution with which we can distinguish fungal species and in the in situ spatial analysis of gene expression within individual mycelia will also be necessary to improve our understanding of the multiplicity of functions which mycorrhizal fungi are involved in.

Table 9.5. Multiplicity of mycorrhizal effects at the plant, community and ecosystem levels

Details	Authors (see reference list)
Effects on individual plants	
Uptake of mineral nutrients	Smith and Read (1997)
Mycelial effects on soil aggregation	Tisdall (1974); Tisdall and Oades (1979)
Modified water relations (VA)	Augé (2001)
Water uptake (ecto)	Boyd et al. (1986)
Relationship between mycorrhizal rhizomorph diameter and xylem water potential	Lamhamedi et al. (1992)
Water transfer from roots to mycorrhizal mycelia during soil drying maintains mycelial integrity	Querejeta et al. (2003)
Disease suppression and resistance to pathogens	Marx (1972), Duchesne et al. (1989)
Uptake and utilisation of organic nutrients	Table 9.1, Leake and Read (1997)
Effects of mycorrhiza on plant fecundity	Koide et al. (1994)
Ectomycorrhizal uptake of P from saprotrophic mycelium	Lindahl et al. (1999)
Ectomycorrhizal acquisition of N from Collembola	Klironomos and Hart (2001)
Heavy metal tolerance of ericoid plants	Bradley et al. (1982)
Ericoid mycorrhizal resistance to arsenate	Sharples et al. (2000)
Ectomycorrhizal mediation of heavy metal effects	Wilkins (1991)
Ectomycorrhizal mediation of acidification effects and responses to elevated Al	Finlay (1995)
Metal-binding capacity of arbuscular mycorrhizal mycelium	Joner et al. (2000)
Ectomycorrhizal effects on weathering of minerals	Table 9.2, Landeweert et al. (2001)
Ectomycorrhizal production of oxalic acid in response to elevated Al levels	Ahonen-Jonnarth et al. (2000)
Interactions with mycorrhiza helper bacteria and endosymbiotic N-fixers	Garbaye (1994); Minerdi et al. (2001)
Effects on plant communities	
Effects on the balance of plant competition	Fitter (1977); Perry et al. (1989)
Mycorrhizal links between plants – their ecological significance	Newman (1988)
Mycorrhizal effects on plant survivorship and diversity in microcosms	Grime et al. (1987)
Carbon transfer between *Betula papyrifera* and *Pseudotsuga menziesii*	Simard et al. (1997a, b)
– but see critical comments of Robinson and Fitter	Robinson and Fitter (1999)

Transfer of C from ectomycorrhizal plants to orchids through shared hyphal connections	McKendrick et al. (2000a, b)
Germination and development of *Monotropa hypopitys* dependent on C transfer from ectomycorrhizal hosts	Leake et al. (2004)
Natural ^{13}C abundance reveals host origin of carbon in mycorrhizal fungi in mixed forests	Högberg et al. (1999)
Ecosystem-level effects	
Potential for mycorrhiza to contribute to below ground C sequestration	Treseder and Allen (2000)
- but see also comments by Schlesinger and Lichter	Schlesinger and Lichter (2001)
Tighter nutrient cycling – direct links restrict losses to immobilisation	Abuzinadah et al. (1986)
Tighter nutrient cycling via nutrient micro-loops at the mycelium–soil microbe–soil interface	Sun et al. (1999)
Supply of carbon to belowground ecosystem	Finlay and Söderström (1992)
Coupling of current photosynthesis to mycorrhizal activity and soil respiration	Högberg et al. (2001)
Belowground mycorrhizal diversity promotes aboveground diversity and productivity	van der Heijden et al. (1998)
Possible effects of ectomycorrhizal fungi on podzolisation	van Breemen et al. (2000b)

9.6 Implications for Nutrient Cycling and Plant Ecology

Several general features emerge from consideration of the above interactions. These are in contrast to traditional ideas of nutrient cycling. One idea is that nutrient cycling may take place on different spatial scales, and that cycling at very small scales – nutrient micro-loops – may contribute to overall cycling at larger scales. Many "short-circuits" in traditional pathways of nutrient cycling are emerging, and the significance of these needs to be more closely investigated. These have been discussed by Abuzinadah et al. (1986) in relation to direct recycling of organic nutrients back to plants, by Lindahl et al. (1999) in relation to interactions between saprotrophic and ectomycorrhizal fungi, and by Jongmans et al. (1997) and van Breemen et al. (2000a) in relation to release and uptake of base cations from inside mineral particles. In the latter case, sites of base cation uptake could be isolated from the bulk soil solution, and critical loads of acid deposition calculated from base cation/Al ratios in the bulk soil solution would be irrelevant, since base cations released by weathering inside minerals would already be absorbed by hyphae and transported directly to plants, without entering the bulk solution. In a similar way, competitive interactions between nutrient-conservative saprotrophic and mycorrhizal fungi might result in rapid movement from one biological pool to another. In these types of interaction, with intense competition for organic nutrients, the concept of mineral nutrient uptake from a soil solution continually replenished by mineralisation may not be realistic (Lindahl et al. 1999, 2001; Schimel and Bennet 2004).

Many of the theories and methods used in studies of nutrient cycling have been developed for agricultural soils, which are very different from the boreal forest ecosystems dominated by ectomycorrhizal fungi. Many of the more traditional models developed within this framework are characterised by a "phytocentric" approach in which micro-organisms are seen merely as processors of mineral nutrients required by plants. Little attention has been paid to the filamentous nature of the fungi, their ability to translocate carbon and nutrients between different substrates, and the capacity of ectomycorrhizal fungi to utilise organic nutrients. In nutrient-limited systems, there is a need to develop new models emphasising competition for organic nutrients between decomposer organisms and plants, as well as antagonistic interactions involving nutrient transfer between decomposer and mycorrhizal fungi (Lindahl et al. 2002).

Mycorrhizal mycelia may play a more active role in nutrient uptake than hitherto supposed. In addition to the uptake of simple organic and inorganic compounds already in solution, it now seems likely that many mycorrhizal fungi may have the capacity to modify otherwise unavailable substrates through the production of enzymes or organic acids. Energy for this is made available by the supply of energy-rich carbon compounds from their autotrophic hosts, and recent measurements of this process in arbuscular mycorrhizas (Johnson et al. 2002) and ectomycorrhizal trees (Högberg et al. 2001) underline the quantitative significance of the amounts of carbon involved. There has been intensive discussion and controversy about the ultimate fate of carbon reaching the mycorrhizal mycelium, and the possible significance of any transfer between plants connected by shared hyphal networks. Some authors (Francis and Read 1984; Simard et al. 1997a, b) contend that the amounts of C transferred can be ecologically significant whereas other authors (Newman 1988; Robinson and Fitter 1999) question the significance of the process or advocate a more mycocentric viewpoint (Fitter et al. 2000). Finlay and Söderström (1992) pointed out that connection of plants to a common mycelial network could be of significance without needing to invoke any arguments concerning net interplant transfer of carbon, since seedlings could gain access to a large absorptive network of hyphae with minimal investment of carbon. The results of Högberg et al. (1999) are consistent with the above argument, since these authors found that promiscuous ectomycorrhizal fungi forming associations with several tree species had $\delta\,^{13}C$ values closer to those of overstorey trees, suggesting that the overstorey trees partly or wholly support the carbon demands of the nutrient-absorbing mycelia of their alleged competitors, the understorey trees. Another situation in which net carbon transfer via hyphal links seems less equivocal is where achlorophyllous monotropoid plants or orchids receive carbon fixed by ectomycorrhizal hosts sharing one and the same mycelium. Whether this can be dismissed as a special case is doubtful, since the Orchidaceae contain many species.

9.7 Practical Implications of Mycorrhizosphere Effects

Improved understanding of mycorrhizosphere interactions may have implications in a number of practical applications within soil biology, such as biocontrol, biofertilisation and bioremediation. The role of mycorrhizal symbiosis in suppression of disease and pathogenic organisms has so far mostly been investigated with respect to the fungi themselves, and the role of mycorrhizal-associated bacteria in controlling other infections has not yet been adequately evaluated. Another applied aspect of research concerns possible effects of biocontrol organisms on mycorrhizal fungi. Barea et al. (1998) and Vázquez et al. (2000) found no negative effects of different biocontrol strains of *Pseudomonas* and *Trichoderma* on arbuscular mycorrhizal fungi, but clearly this is an important future area of investigation. Microbial interactions in the mycorrhizosphere may also have significance for sustainable agriculture (Johansson et al. 2004).

There is still a need for improved understanding of the ways in which different groups of organisms interact to contribute to soil fertility and plant growth. Possible interactions between bacteria and mycorrhizal fungi may be important in biological weathering, and need further investigation (Landeweert et al. 2001). Mycorrhizal fungi may play a central role in sequestering nutrients from ash applications following removal of forest residues during biofuel harvesting (Mahmood et al. 2001).

Possible applications of mycorrhizosphere interactions include both the use of mycorrhizal fungi themselves and the use of mycorrhiza-associated micro-organisms. Mycorrhizal fungi can play an important role in ameliorating problems of toxic metals (Colpaert and Van Assche 1987, 1992; Sharples et al. 2000). Meharg and Cairney (2000) have reviewed possible ways in which rhizosphere mediation of persistent organic pollutants (POPs) can be extended by using ectomycorrhizal fungi. Donnelly et al. (1993) demonstrated the ability of ericoid and ectomycorrhizal fungi to degrade two chlorinated aromatic herbicides (2,4-dichlorophenoxyacetic acid and atrazine), and found that total degradation increased as the concentration of N increased. In these experiments, the ericoid fungus *Hymenoscyphus ericae* incorporated the highest level of atrazine carbon into its own tissue. Meharg et al. (1997a) demonstrated the ability of the two ectomycorrhizal fungi *Paxillus involutus* and *Suillus variegatus* to degrade 2,4-dichlorophenol, and that this was higher when the fungi were growing in symbiosis with *Pinus sylvestris* than when they were in pure culture. In other experiments (Meharg et al. 1997b), *Suillus variegatus* has been show to be effective in degrading 2,4,6-trinitrotoluene (TNT). One potential advantage of ectomycorrhizal basidiomycetes is that they receive carbon directly from their hosts.

Attempts to introduce genetically modified micro-organisms often fail because the micro-organisms introduced fail to establish themselves. Mycorrhizal hyphae may facilitate the establishment of associated bacteria by providing a support surface with a supply of energy-rich carbon compounds, particularly around hyphal tips (Sun et al. 1999). Research in this area is still in its infancy but initial experiments by Sarand et al. (1998) suggest that mycorrhizal hyphae are able to support microbial biofilms of catabolic plasmid harbouring bacteria which could be active in bioremediation of petroleum-contaminated soil. In further experiments, the number of Tol^+ bacteria was higher in mycorrhizospheric soil compared with bulk soil, and inoculation with bacteria had a positive effect on plant and fungal development (Sarand et al. 2000). The presence of easily available plant-derived carbon sources did not impede the degradation of the *m*-toluate by the bacteria (Sarand et al. 1999). However, in other experiments Genney et al. (2004) found that degradation of the polycyclic aromatic hydrocarbon fluorene was retarded in a Scots pine ectomycorrhizosphere.

In general, the role of arbuscular mycorrhizal fungi is less well investigated with respect to bioremediation but recent experiments by Joner et al. (2000) suggest that dissipation of polycyclic aromatic hydrocarbons (PAHs) may be enhanced in the presence of arbuscular mycorrhizas, and that changes in the composition of the mycorrhiza-associated microflora may be responsible for the observed decreases in PAH concentrations.

9.8 Concluding Remarks

Over the past two decades, experiments investigating the effects of mycorrhizal fungi on single plants have gradually given way to experiments designed to investigate the effects of the symbiosis on plant communities and ecosystems. The complexity of the context within which these symbiotic associations are considered has increased. The mycorrhizal mycelium has conventionally been considered as a passive extension of plant root systems, and the view many ecologists have had is largely phytocentric. However, there has been a shift in focus during recent years from simple inorganic substrates to the role mycorrhizal fungi play in accessing more complex organic substrates, coupled with an increasing awareness of the probable importance of interactions with other organisms. More emphasis has been placed upon the multifunctional roles the symbiosis plays. The picture emerging from this new research is one in which the mycorrhizal mycelium actively conditions and modifies its own biotic and abiotic environment, and takes part in hitherto only superficially described interactions with other micro-organisms. It seems these small-scale interactions have the potential to make a significant contribution to nutrient cycling at the ecosystem level. Improved knowledge of actions and interactions in the mycorrhizosphere is essential not only for a more complete understanding of natural ecosystem functioning, but also when we need to apply this knowledge to achieve sustainable management of plant communities.

9.9 Summary

This chapter synthesises recent developments in our understanding of mycorrhizal symbiosis, and the multiple roles which mycorrhizal associations can play in nutrient acquisition and plant ecology. The aim is to describe a conceptual progression which has taken place from theories which regard mycorrhizal mycelia as more or less passive, physical extensions of the root system, taking up nutrients which are already in solution, to theories which consider the ways in which mycorrhizal fungi actively condition and modify their own biotic and abiotic environment, regulating nutrient acquisition. During the past 20 years, a parallel development has also taken place in the complexity of the context within which mycorrhizal symbiosis is considered. Early experiments investigating the effects of mycorrhizal fungi on individual plants have been complemented by more recent studies in which the effects of the symbiosis are considered at the plant community or ecosystem levels. Recent studies have also increasingly considered the possible interactions of mycorrhizal fungi with other soil organisms. The increasing emphasis on the *active* role of mycorrhizal fungi in nutrient acquisition, and their *interaction* with the biotic and abiotic environment is reflected in the chapter title and the topics discussed. Qualitative as well as quantitative effects on nutrient uptake are discussed, together with non-nutritional effects which indirectly influence plant nutrition. Mycorrhizal fungi can play a significant role in acquisition of N and P from organic polymers, and more research is necessary to investigate their potential to influence release of nutrients through weathering of minerals. Small-scale interactions of mycorrhizal hyphae with plant roots and with organic and inorganic substrates influence not only the nutrient status of individual plants but also nutrient and carbon cycling, and ultimately plant community and ecosystem dynamics. The biotic interactions which take place in the mycorrhizal hyphosphere between mycorrhizal hyphae and bacteria, soil animals and saprophytic or pathogenic fungi may have important implications in application areas such as biocontrol, biofertilisation and bioremediation. Improved understanding of these interactions is likely to play an important future role in the sustainable management of plant communities and ecosystems.

Acknowledgements. Thanks are due to Torgny Unestam, Andy Taylor, Björn Lindahl, Les Paul, Petra Fransson, Anna Rosling and Jonathan Leake for stimulating discussions and valuable comments. Financial support from the Swedish Natural Sciences Research Council (VR), the Swedish Research Council for Environment, Agricultural Sciences and Spatial Planning (FORMAS), the European Commission, the Swedish Environmental Protection Agency (SNV) and the Swedish National Energy Administration (STEM) is gratefully acknowledged.

References

Abuarghub SM, Read DJ (1988) The biology of mycorrhiza in the Ericaceae. XII. Quantitative analysis of individual free amino acids in relation to time and depth in the soil profile. New Phytol 108:433–441

Abuzinadah RA, Read DJ (1986a) The role of proteins in the nitrogen nutrition of ectomycorrhizal plants. I. Utilization of peptides and proteins by ectomycorrhizal fungi. New Phytol 103:481–493

Abuzinadah RA, Read DJ (1986b) The role of proteins in the nitrogen nutrition of ectomycorrhizal plants. III. Protein utilization by *Betula*, *Picea* and *Pinus* in mycorrhizal association with *Hebeloma crustuliniforme*. New Phytol 103:507–514

Abuzinadah RA, Read DJ (1989a) The role of proteins in the nitrogen nutrition of ectomycorrhizal plants. V. Nitrogen transfer in birch (*Betula pendula* L.) grown in association with mycorrhizal and non-mycorrhizal fungi. New Phytol 112:61–68

Abuzinadah RA, Read DJ (1989b) The role of proteins in the nitrogen nutrition of ectomycorrhizal plants. IV. The utilization of peptides by birch (*Betula pendula* Roth.) infected with different mycorrhizal fungi. New Phytol 112:55–60

Abuzinadah RA, Finlay RD, Read DJ (1986) The role of proteins in the nitrogen nutrition of ectomycorrhizal plants. II. Utilization of protein by mycorrhizal plants of *Pinus contorta*. New Phytol 103:495–506

Ahonen-Jonnarth U, Van Hees PAW, Lundström US, Finlay RD (2000) Production of organic acids by mycorrhizal and non-mycorrhizal *Pinus sylvestris* L. seedlings exposed to elevated concentrations of aluminium and heavy metals. New Phytol 146:557–567

Ahonen-Jonnarth U, Göransson A, Finlay RD (2000) Growth and nutrient uptake of ectomycorrhizal *Pinus sylvestris* seedlings treated with elevated Al concentrations. Tree Physiol 23:157–167

Ames RN, Reid CPP, Porter L, Cambardella C (1983) Hyphal uptake and transport of nitrogen from two ^{15}N-labelled sources by *Glomus mosseae*, a vesicular-arbuscular mycorrhizal fungus. New Phytol 95:381–396

Andrade G, Mihara KL, Linderman RG, Bethlenfalvay GJ (1997) Bacteria from rhizosphere and hyphosphere soils of different arbuscular-mycorrhizal fungi. Plant Soil 192:71–79

Andrade G, Linderman RG, Bethlenfalvay GJ (1998a) Bacterial associations with the mycorrhizosphere and hyphosphere of the arbuscular mycorrhizal fungus *Glomus mosseae*. Plant Soil 202:79–87

Andrade G, Mihara KL, Linderman RG, Bethlenfalvay GJ (1998b) Soil aggregation status and rhizobacteria in the mycorrhizosphere. Plant Soil 202:89–96

Arnebrant K (1994) Nitrogen amendments reduce the growth of extramatrical mycelium. Mycorrhiza 5:7–15

Arnebrant K, Ek H, Finlay RD, Söderström B (1993) Translocation of nitrogen between *Alnus glutinosa* seedlings inoculated with *Frankia* sp. and *Pinus contorta* seedlings connected by a common ectomycorrhizal fungus. New Phytol 124:231–242

Arocena JM, Glowa KR, Massicotte HB, Lavkulich L (1999) Chemical and mineral composition of ectomycorrhizosphere soils of subalpine fir (*Abies lasiocarpa* (Hook.) Nutt.) in the Ae horizon of a luvisol. Can J Soil Sci 79:25–35

Augé RM (2001) Water relations, drought and vesicular arbuscular mycorrhizal symbiosis. Mycorrhiza 11:3–42

Bajwa R, Read DJ (1985) The biology of mycorrhiza in the Ericaceae. IX. Peptides as nitrogen sources for the ericoid endophyte and for mycorrhizal and non-mycorrhizal plants. New Phytol 101:459–467

Bajwa R, Read DJ (1986) Utilisation of mineral and amino N sources by the ericoid mycorrhizal endophyte *Hymenoscyphus ericae* and by mycorrhizal and non-mycorrhizal seedlings of *Vaccinium*. Trans Br Mycol Soc 87:269–277

Banfield JF, Barker WW, Welch SA, Taunton A (1999) Biological impact on mineral dissolution: application of the lichen model to understanding mineral weathering in the rhizosphere. Proc Natl Acad Sci USA96:3404–3411

Barea JM, Andrade G, Bianciotto V, Dowling D, Lohrke S, Bonfante P, O'Gara F, Azcon-Aguilar C (1998) Impact on arbuscular mycorrhiza formation of *Pseudomonas* strains used as inoculants for biocontrol of soil-borne fungal plant pathogens. Appl Environ Microbiol 64:2304–2307

Barron GL (1988) Microcolonies of bacteria as a nutrient source for lignicolous and other fungi. Can J Bot 66:2505–2510

Bending GD, Read DJ (1995a) The structure and function of the vegetative mycelium of ectomycorrhizal plants. V. Foraging behaviour and translocation of nutrients from exploited litter. New Phytol 130:401–409

Bending GD, Read DJ (1995b) The structure and function of the vegetative mycelium of ectomycorrhizal plants. VI. Activities of nutrient mobilizing enzymes in birch litter colonized by *Paxillus involutus* (Fr.) Fr. New Phytol 130:411–417

Bending GD, Read DJ (1996) Nitrogen mobilization from protein-polyphenol complex by ericoid and ectomycorrhizal fungi. Soil Biol Biochem 28:1603–1612

Bending GD, Read DJ (1997) Lignin and soluble phenolic degradation by ectomycorrhizal and ericoid mycorrhizal fungi. Mycol Res 101:1348–1354

Berthelin J (1983) Microbial weathering processes. In: Krumbein WE (ed) Microbial geochemistry. Blackwell, Oxford, pp 223–263

Bianciotto V, Lumini E, Lanfranco L, Minerdi D, Bonfante P, Perotto S (2000) Detection and identification of bacterial endosymbionts in arbuscular mycorrhizal fungi belonging to the family Gigasporaceae. Appl Environ Microbiol 66:4503–4509

Bianciotto V, Andreotti S, Balestrini R, Bonfante P, Perotto S (2001) Mucoid mutants of the biocontrol strain *Pseudomonas fluorescens* CHA0 show increased ability in biofilm formation on mycorrhizal and non-mycorrhizal carrot roots. Mol Plant-Microbe Interact 14:255–260

Biró B, Köves-Péchy K, Vörös I, Takács T, Eggenberger P, Strasser RJ (2000) Interrelations between *Azospirillum* and *Rhizobium* nitrogen fixers and arbuscular mycorrhizal fungi in the rhizosphere of alfalfa in sterile AMF-free or normal conditions. Appl Soil Ecol 15:159–168

Borowicz VA (1997) A fungal root symbiont modifies plant resistance to an insect herbivore. Oecologia 112:534–542

Bowen GD, Theodorou C (1979) Interactions between bacteria and ectomycorrhizal fungi. Soil Biol Biochem 11:119–126

Boyd R, Furbank RT, Read DJ (1986) Ectomycorrhiza and water relations of trees. In: Gianinazzi-Pearson V, Gianinazzi S (eds) Physiological and genetical aspects of mycorrhizae. INRA, Paris, pp 689–693

Bradley R, Burt AJ, Read DJ (1982) The biology of mycorrhiza in the Ericaceae. VIII. The role of mycorrhizal infection in heavy metal resistance. New Phytol 91:197–201

Brandes B, Godbold DL, Kuhn AJ, Jentschke G (1998) Nitrogen and phosphorus acquisition by the mycelium of the ectomycorrhizal fungus *Paxillus involutus* and its effect on host nutrition. New Phytol 140:735–743

Bruce A, Smith SE, Tester M (1994) The development of mycorrhizal infection in cucumber: effects of P supply on root growth, formation of entry points and growth of infection units. New Phytol 127:507–514

Cairney JWG, Burke RM (1994) Fungal enzymes degrading plant cell walls: their possible significance in the ectomycorrhizal symbiosis. Mycol Res 98:1345–1356

Cairney JWG, Burke RM (1998) Do ecto- and ericoid mycorrhizal fungi produce peroxidase activity? Mycorrhiza 8:61–65

Cairney JWG, Chambers SM (eds) (1999) Ectomycorrhizal fungi: key genera in profile. Springer, Berlin Heidelberg New York

Cairney JWG, Taylor AFS, Burke RM (2003) No evidence for lignin peroxidase genes in ectomycorrhizal fungi. New Phytol 160:461–462

Callot G, Maurette M, Pottier L, Dubois A (1987) Biogenic etching of microfeatures in amorphous and crystalline silicates. Nature 328:147–149

Chalot M, Brun A (1998) Physiology of organic nitrogen acquisition by ectomycorrhizal fungi and ectomycorrhizas. FEMS Microbiol Rev 22:21–44

Chalot M, Brun A, Finlay RD, Söderström B (1994a) Metabolism of [^{14}C]glutamate and [^{14}C]glutamine by the ectomycorrhizal fungus *Paxillus involutus*. Microbiology 140:1641–1649

Chalot M, Brun A, Finlay RD, Söderström B (1994b) Respiration of [^{14}C] alanine by the ectomycorrhizal fungus *Paxillus involutus*. FEMS Microbiol Lett 121:87–92

Chalot M, Kytöviita MM, Finlay RD, Söderström B (1995a) Factors affecting amino acid uptake in the ectomycorrhizal fungus *Paxillus involutus*. Mycol Res 99:1131–1138
Chalot M, Finlay RD, Ek H, Söderström B (1995b) Metabolism of [^{15}N] alanine by the ectomycorrhizal fungus *Paxillus involutus*. Exp Mycol 19:297–304
Chanway CP, Holl FB (1991) Biomass increase and associative nitrogen fixation of mycorrhizal *Pinus contorta* seedlings inoculated with a plant growth promoting *Bacillus* strain. Can J Bot 69:507–511
Chen DM, Taylor AFS, Burke RM, Cairney JWG (2001) Identification of genes for lignin peroxidases and manganese peroxidases in ectomycorrhizal fungi using PCR. New Phytol 152:151–158
Colpaert JV, Van Assche JA (1987) Heavy metal tolerance in some ectomycorrhizal fungi. Funct Ecol 1:415–421
Colpaert JV, Van Assche JA (1992) The effects of cadmium and the cadmium-zinc interaction on the axenic growth of ectomycorrhizal fungi. Plant Soil 145:237–243
Colpaert JV, Verstuyft I (1999) The Ingestad concept in ectomycorrhizal research: possibilities and limitations. Physiol Plant 105:233–238
Colpaert JV, van Laere A, van Assche JA (1996) Carbon and nitrogen allocation in ectomycorrhizal and non-mycorrhizal *Pinus sylvestris* L. seedlings. Tree Physiol 16:787–793
Cooper KM, Tinker PB (1978) Translocation and transfer of nutrients in vesicular-arbuscular mycorrhizas. II. Uptake and translocation of phosphorus, zinc and sulphur. New Phytol 81:43–52
Costerton JW, Stewart PS, Greenberg EP (1999) Bacterial biofilms: a common cause of persistent infections. Science 284:1318–1322
Cromack K Jr, Sollins P, Graustein WC, Speidel K, Todd AW, Spycher G, Li CY, Todd RL (1979) Calcium oxalate accumulations and soil weathering in mats of the hypogeous fungus *Hysterangium crassum*. Soil Biol Biochem 11:463–468
Dalpé Y (1986) Axenic synthesis of ericoid mycorrhiza in *Vaccinium angustifolium* Ait. by *Oidiodendron* species. New Phytol 103:391–396
Daniell TJ, Hodge A, Young JPW, Fitter AH (1999) How many fungi does it take to change a plant community? Trends Plant Sci 4:81–82
Davis RM, Menge JA (1980) Influence of *Glomus fasciculatus* and soil phosphorus on *Phytophthora* root rot of citrus. Phytopathology 70:447–452
Dighton J, Thomas ED, Latter PM (1987) Interactions between tree roots, mycorrhizas, a saprotrophic fungus and the decomposition of organic substrates in a microcosm. Biol Fertil Soil 4:145–150
Donnelly PK, Entry JA, Crawford DL (1993) Degradation of Atrazine and 2.4-dichlorophenoxyacetic acid by mycorrhizal fungi at 3 nitrogen concentrations in-vitro. Appl Environ Microbiol 59:2642–2647
Duchesne LC, Peterson RL, Ellis BE (1987) The accumulation of plant-produced antimicrobial compounds in response to ectomycorrhizal fungi: a review. Phytoprotection 68:17–27
Duchesne LC, Peterson RL, Ellis BE (1988a) Interaction between the ectomycorrhizal fungus *Paxillus involutus* and *Pinus resinosa* induces resistance to *Fusarium oxysporum*. Can J Bot 66:558–562
Duchesne LC, Peterson RL, Ellis BE (1988b) Pine root exudate stimulates antibiotic synthesis by the ectomycorrhizal fungus *Paxillus involutus*. New Phytol 108:471–476
Duchesne LC, Peterson RL, Ellis BE (1989) The time course of disease suppression by the ectomycorrhizal fungus *Paxillus involutus*. New Phytol 111:693–698
Duddridge JA, Malibari A, Read DJ (1980) Structure and function of mycorrhizal rhizomorphs with special reference to their role in water transport. Nature 287: 834–836

Entry JA, Donnelly PK, Cromack K (1991) Influence of ectomycorrhizal mat soils on lignin and cellulose degradation. Biol Fertil Soils 11:75–78

Entry JA, Rose CL, Cromack K (1992) Microbial biomass and nutrient concentrations in hyphal mats of the ectomycorrhizal fungus *Hysterangium setchellii* in a coniferous forest soil. Soil Biol Biochem 24:447–453

Erland SE, Finlay RD (1992) Effects of temperature and incubation time on the ability of three ectomycorrhizal fungi to colonize *P. sylvestris* roots. Mycol Res 96:270–272

Finlay RD (1985) Interactions between soil micro-arthropods and endomycorrhizal associations of higher plants. In: Fitter AH (ed) Ecological interactions in soil: plants, microbes and animals. Blackwell, Oxford, pp 319–331

Finlay RD (1989) Functional aspects of phosphorus uptake and carbon translocation in incompatible ectomycorrhizal associations between *Pinus sylvestris* and *Suillus grevillei* and *Boletinus cavipes*. New Phytol 112:185–192

Finlay RD (1993) Uptake and mycelial translocation of nutrients by ectomycorrhizal fungi. In: Read DJ, Lewis DH, Fitter AH, Alexander IJ (eds) Mycorrhiza in ecosystems. Proc 3rd Eur Symp Mycorrhizas. CAB International, Wallingford, pp 91–97

Finlay RD (1995) Interactions between soil acidification, plant growth and nutrient uptake in ectomycorrhizal associations of forest trees. Ecol Bull 44:197–214

Finlay RD, Read DJ (1986a) The structure and function of the vegetative mycelium of ectomycorrhizal plants. I. Translocation of ^{14}C-labelled carbon between plants interconnected by a common mycelium. New Phytol 103:143–156

Finlay RD, Read DJ (1986b) The structure and function of the vegetative mycelium of ectomycorrhizal plants. II. The uptake and distribution of phosphorus by mycelial strands interconnecting plants. New Phytol 103:157–165

Finlay R, Söderström B (1992) Mycorrhiza and carbon flow to the soil. In: Allen MF (ed) Mycorrhizal functioning. Chapman and Hall, New York

Finlay RD, Ek H, Odham G, Söderström B (1988) Mycelial uptake, translocation and assimilation of nitrogen from ^{15}N-labelled ammonium by *Pinus sylvestris* plants infected with four different ectomycorrhizal fungi. New Phytol 110:59–66

Finlay RD, Ek H, Odham G, Söderström B (1989) Uptake, translocation and assimilation of nitrogen from ^{15}N-labelled ammonium and nitrate sources by intact ectomycorrhizal systems of *Fagus sylvatica* infected with *Paxillus involutus*. New Phytol 113:47–55

Finlay RD, Frostegård Å, Sonnerfeldt A-M (1992) Utilization of organic and inorganic nitrogen sources by ectomycorrhizal fungi in pure culture and in symbiosis with *Pinus contorta* Dougl. ex Loud. New Phytol 120:105–115

Fitter AH (1977) Influence of mycorrhizal infection on competition for phosphorus and potassium by two grasses. New Phytol 79:119–125

Fitter AH (1985) Functioning of vesicular-arbuscular mycorrhizas under field conditions. New Phytol 99:257–265

Fitter AH, Garbaye J (1994) Interactions between mycorrhizal fungi and other soil organisms. Plant Soil 159:123–132

Fitter AH, Moyerson B (1996) Evolutionary trends in root-microbe symbioses. Philos Trans R Soc Lond Ser B 351:1367–1375

Fitter AH, Heinemeyer A, Staddon PL (2000) The impact of elevated CO_2 and global climate change on arbuscular mycorrhizas: a mycocentric approach. New Phytol 147:179–187

Francis R, Read DJ (1984) Direct transfer of carbon between plants connected by vesicular-arbuscular mycorrhizal mycelium. Nature 307:53–56

Frank AB (1885) Über die auf Wurzelsymbiose beruhende Ernährung gewisser Bäume durch unterirdische Pilze. Ber Dtsch Bot Ges 3:128–145

Frank AB (1894) Die Bedeutung der Mykorrhizapilze für die gemeine Kiefer. Forstwissen Centralbl 16:1852–1890

Frey P, Frey-Klett P, Garbaye J, Berge O, Heulin T (1997) Metabolic and genotypic fingerprinting of fluorescent pseudomonads associated with the Douglas fir *Laccaria bicolor* mycorrhizosphere. Appl Environ Microbiol 63:1852–1860

Frey-Klett P, Pierrat J-C, Garbaye J (1997) Location and survival of mycorrhiza helper *Pseudomonas fluorescens* during the establishment of ectomycorrhizal symbiosis between *Laccaria bicolor* and Douglas Fir. Appl Environ Microbiol 63:139–144

Frey-Klett P, Churin J-L, Pierrat J-C, Garbaye J (1999) Dose effect in the dual inoculation of an ectomycorrhizal fungus and a mycorrhizal helper bacterium in two forest nurseries. Appl Environ Microbiol 63:139–144

Frey-Klett P, Chavatte M, Clausse M-L, Courrier S, Le Roux C, Raaijmakers J, Martinotti MG, Pierrat J-C, Garbaye J (2005) Ectomycorrhizal symbiosis affects functional diversity of rhizosphere fluorescent pseudomonads. New Phytol 165:317–328

Friese CF, Allen MF (1991) The spread of VA mycorrhizal fungal hyphae in the soil: inoculum types and external hyphal architecture. Mycologia 83:409–418

Gadd GM (1999) Fungal production of citric and oxalic acid: importance in metal speciation, physiology and biogeochemical processes. Adv Microb Physiol 41:47–92

Gadd GM (ed) (2005) Fungi in biogeochemical cycles. Cambridge University Press, Cambridge

Gadgil RL, Gadgil PD (1971) Mycorrhiza and litter decomposition. Nature 233:133

Garbaye J (1991) Biological interactions in the mycorrhizosphere. Experientia 47:370–375

Garbaye J (1994) Helper bacteria – a new dimension to the mycorrhizal symbiosis. New Phytol 128:197–210

Garbaye J, Bowen GD (1989) Stimulation of ectomycorrhizal infection of *Pinus radiata* by some microorganisms associated with the mantle of ectomycorrhizas. New Phytol 112:383–388

Garcia-Garrido JM, Garcia-Romera I, Ocampo JA (1992) Cellulase production by the vesicular-arbuscular mycorrhizal fungus *Glomus mosseae* (Nichol. and Gerd.) Gerd. and Trappe. New Phytol 121:221–226

Gay G, Normand L, Marmeisse R, Sotta B, Debaud JC (1994) Auxin overproducer mutants of *Hebeloma cylindrosporum* Romagnesi have increased mycorrhizal activity. New Phytol 128:645–657

Genney DR, Alexander IJ, Killham K, Meharg AA (2004) Degradation of the polycyclic aromatic hydrocarbon (PAH) fluorene is retarded in a Scots pine ectomycorrhizosphere. New Phytol 163:641–649

Gorissen A, Kuyper ThW (2000) Fungal species-specific responses of ectomycorrhizal Scots pine (*Pinus sylvestris*) to elevated [CO_2]. New Phytol 146:163–168

Graustein WC, Cromack K Jr, Sollins P (1977) Calcium oxalate: occurrence in soils and effect on nutrient and geochemical cycles. Science 198:1252–1254

Grayston SJ, Campbell CD, Vaughan D (1994) Microbial diversity in the rhizospheres of different tree species. In: Pankhurst CE (ed) Soil biota: management in sustainable farming systems. CSIRO, Adelaide

Griffiths RP, Caldwell BA (1992) Mycorrhizal mat communities in forest soils. In: Read DJ, Lewis DH, Fitter AH, Alexander IJ (eds) Mycorrhiza in ecosystems. Proc 3rd Eur Symp Mycorrhizas. CAB International, Wallingford, pp 98–105

Griffiths RP, Baham JE, Caldwell BA (1994) Soil solution chemistry of ectomycorrhizal mats in forest soil. Soil Biol Biochem 26:331–337

Grime JP, Mackey JML, Hillier SH, Read DJ (1987) Floristic diversity in a model system using experimental microcosms. Nature 328:420–422

Haselwandter K, Winkelmann G (1998) Identification and characterisation of siderophores of mycorrhizal fungi. In: Varma A (ed) Mycorrhiza manual. Springer, Berlin Heidelberg New York, pp 243–254

Haselwandter K, Bobleter O, Read DJ (1990) Degradation of ^{14}C-labelled lignin and dehydropolymer of coniferyl alcohol by ericoid and ectomycorrhizal fungi. Arch Microbiol 153:352–354

Hawkins H-J, Johansen A, George E (2000) Uptake and transport of organic and inorganic nitrogen by arbuscular mycorrhizal fungi. Plant Soil 226:275–285

Heinonsalo J, Jorgensen KS, Sen R (2001) Microcosm-based analyses of Scots pine seedling growth, ectomycorrhizal fungal community growth and bacterial carbon utilization profiles in boreal forest humus and underlying illuvial mineral horizons. FEMS Microbiol Ecol 36:73–84

Hibbett DS, Gilbert L-B, Donaghue MJ (2000) Evolutionary instability of ectomycorrhizal symbioses in basidiomycetes. Nature 407:506–508

Hiltner L (1904) Über neuere Erfahrungen und Probleme auf dem Gebiet der Bodenbakteriologie und unter besonderer Berücksichtigung der Gründüngung und Brache. Arb Dtsch Landwirtsch Ges Berlin 98:59–78

Hirsch P, Eckhardt FEW, Palmer RJ (1995) Fungi active in weathering of rock and stone monuments. Can J Bot 73:S1384-S1390

Hodge A, Campbell CD, Fitter AH (2001) An arbuscular mycorrhizal fungus accelerates decomposition and acquires nitrogen directly from organic material. Nature 413:297–299

Högberg P, Plamboeck AH, Taylor AFS, Fransson PMA (1999) Natural ^{13}C abundance reveals trophic status of fungi and host-origin of carbon in mycorrhizal fungi in mixed forests. Proc Natl Acad Sci USA 96:8534–8539

Högberg P, Nordgren A, Buchmann N, Taylor AFS, Ekblad A, Högberg MN, Nyberg G, Ottosson-Löfvenius M, Read DJ (2001) Large-scale forest girdling shows that current photosynthesis drives soil respiration. Nature 411:789–792

Holmström SJM, Lundström US, Finlay RD, van Hees PAW (2004) Siderophores in forest soil solution. Biogeochemistry 71:247–258

Jentschke G, Bonkowski M, Godbold DL, Scheu S (1995) Soil protozoa and forest tree growth – non-nutritional effects and interactions with mycorrhizae. Biol Fertil Soils 20:263–269

Jentschke G, Brandes B, Kuhn AJ, Schröder WH, Becker JS, Godbold DL (2000) The mycorrhizal fungus *Paxillus involutus* transports magnesium to Norway spruce seedlings. Evidence from stable isotope labeling. Plant Soil 220:243–246

Johansen A, Finlay RD, Olsson P-A (1996) N assimilation in the external hyphae of the arbuscular mycorrhizal fungus *Glomus intraradices*. New Phytol 133:705–712

Johansson J, Paul LR, Finlay RD (2004) Microbial interactions in the mycorrhizosphere and their significance for sustainable agriculture. FEMS Microbiol Ecol 48:1–12

Johnson D, Leake JR, Ostle N, Ineson P, Read DJ (2002) In-situ $^{13}CO_2$ pulse-labelling of upland grassland demonstrates that a rapid pathway of carbon flux from arbuscular mycorrhizal mycelia to the soil. New Phytol 153:327–334

Joner EJ, Briones R, Leyval C (2000) Metal-binding capacity of arbuscular mycorrhizal mycelium. Plant Soil 226:227–234

Jones DL (1998) Organic acids in the rhizosphere – a critical review. Plant Soil 205:25–44

Jongmans AG, van Breemen N, Lundström U, Finlay RD, van Hees PAW, Giesler R, Melkerud P-A, Olsson M, Srinivasan M, Unestam T (1997) Rock-eating fungi: a true case of mineral plant nutrition? Nature 389:682–683

Karabaghli C, Frey-Klett P, Sotta B, Bonnet M, Le Tacon F (1998) In vitro effects of *Laccaria bicolor* S238 N and *Pseudomonas fluorescens* strain BBc6 on rooting of derooted shoot hypocotyls of Norway spruce. Tree Physiol 18:103–111

Kerley SJ, Read DJ (1995) The biology of mycorrhiza in the Ericaceae. XVIII. Chitin degradation by *Hymenoscyphus ericae* and transfer of chitin-nitrogen to the host plant. New Phytol 131:369–375

Kerley SJ, Read DJ (1997) The biology of mycorrhiza in the Ericaceae. XIX. Fungal mycelium as a nitrogen source for the ericoid mycorrhizal fungus *Hymenoscyphus ericae* and its host plants. New Phytol 136:691–701
Kerley SJ, Read DJ (1998) The biology of mycorrhiza in the Ericaceae. XX. Plant and mycorrhizal necromass as nitrogenous substrates for the ericoid mycorrhizal fungus *Hymenoscyphus ericae* and its host. New Phytol 139:353–360
Klironomos JN, Hart MM (2001) Food-web dynamics – animal nitrogen swap for plant carbon. Nature 410:651–652
Klironomos JN, Bednarczuk EM, Neville J (1999) Reproductive significance of saprobic and arbuscular mycorrhizal fungi by the collembolan, *Folsomia candida*. Funct Ecol 13:756–761
Koide R, Shumway DL, Mabon SA (1994) Mycorrhizal fungi and reproduction of field populations of *Abutilon theophrasti* Medic. (Malvaceae). New Phytol 126:123–130
Kope HH, Fortin JA (1989) Inhibition of phytopathogenic fungi in vitro by cell free culture media of ectomycorrhizal fungi. New Phytol 113:57–63
Kozdrój J, van Elsas JD (2000) Response of the bacterial community to root exudates in soil polluted with heavy metals assessed by molecular and cultural approaches. Soil Biol Biochem 32:1405–1417
Lamhamedi MS, Bernier PY, Fortin JA (1992) Hydraulic conductance and soil water potential at the soil root interface of *Pinus pinaster* seedlings inoculated with different dikaryons of *Pisolithus* sp. Tree Physiol 10:231–244
Landeweert R, Hoffland E, Finlay RD, Kuyper TW, van Breemen (2001) Linking plants to rocks: ectomycorrhizal fungi mobilize nutrients from minerals. Trends Ecol Evol 16:248–254
Lapeyrie F, Chilvers GA, Bhem CA (1987) Oxalic acid synthesis by the mycorrhizal fungus *Paxillus involutus* (Batsch ex Fr.) Fr. New Phytol 106:139–146
Lapeyrie F, Raager J, Vairelles D (1991) Phosphate-solubilizing activity of ectomycorrhizal fungi in vitro. Can J Bot 69:342–346
Larsen J, Jakobsen I (1996) Effects of a mycophagous Collembola on the symbioses between *Trifolium subterraneum* and three arbuscular mycorrhizal fungi. New Phytol 133:295–302
Leake JR (1994) The biology of myco-heterotrophic ("saprophytic") plants. Tansley Review no 69. New Phytol 127:171–216
Leake JR, Read DJ (1990a) Proteinase activity in mycorrhizal fungi. I. The effect of extracellular pH on the production and activity of proteinase by ericoid endophytes from soils of contrasted pH. New Phytol 115:243–250
Leake JR, Read DJ (1990b) Chitin as a nitrogen source for mycorrhizal fungi. Mycol Res 94:993–995
Leake JR, Read DJ (1997) Mycorrhizal fungi in terrestrial habitats. In: Wicklow DT, Söderström B (eds) The Mycota. IV. Environmental and microbial relationships. Springer, Berlin Heidelberg New York, pp 281–301
Leake JR, Donnelly DP, Saunders EM, Boddy L, Read DJ (2001) Rates and quantities of carbon flux to ectomycorrhizal mycelium following ^{14}C pulse labeling of *Pinus sylvestris* seedlings: effects of litter patches and interaction with a wood-decomposer fungus. Tree Physiol 21:71–82
Leake JR, McKendrick SL, Bidartondo M, Read DJ (2004) Symbiotic germination and development of the myco-heterotroph *Monotropa hypopitys* in nature and its requirement for locally distributed *Tricholoma* spp. New Phytol 163:405–423
Leyval C, Berthelin J (1989) Interactions between *Laccaria laccata-Agrobacterium radiobacter* and beech roots influence on phosphorus, potassium, magnesium and iron mobilisation from minerals and plant growth. Plant Soil 117:103–110

Leyval C, Berthelin J (1991) Weathering of a mica by roots and rhizospheric microorganisms of pine. Soil Sci Soc Am J 55:1009–1016

Leyval C, Berthelin J (1993) Rhirodeposition and net release of soluble organic-compounds by pine and beech seedlings inoculated with rhizobacteria and ectomycorrhizal fungi. Biol Fertil Soil 15:259–267

Li CY, Massicotte HB, Moore LVH (1992) Nitrogen-fixing *Bacillus* sp. Associated with Douglas-fir tuberculate ectomycorrhizae. Plant Soil 140:35–40

Lindahl B, Stenlid J, Olsson S, Finlay R (1999) Translocation of ^{32}P between interacting mycelia of a wood decomposing fungus and ectomycorrhizal fungi in microcosm systems. New Phytol 144:183–193

Lindahl B, Stenlid J, Finlay RD (2001) Effects of resource availability on mycelial interactions and ^{32}P transfer between a saprotrophic and an ectomycorrhizal fungus in soil microcosms. FEMS Microbiol Ecol 38:43–52

Lindahl B, Taylor AFS, Finlay RD (2002) Defining nutritional constraints on carbon cycling – towards a less "phytocentric" perspective. Plant Soil 242:123–135

Lindahl BD, Taylor AFS (2004) Occurrence of N-acetylhexosaminidase-encoding genes in ectomycorrhizal basidiomycetes. New Phytol 164:193–199

Lindahl BD, Finlay RD, Cairney JWG (2005) Enzymatic activities of mycelia in mycorrhizal fungal communities. In: Dighton J, Oudemans P, White J (eds) The fungal community: its organization and role in the ecosystem. Marcel Dekker, New York, pp 331–348

Mahmood S, Finlay RD, Erland S, Wallander H (2001) Solubilisation and colonisation of wood ash by ecto-mycorrhizal fungi isolated from a wood ash fertilised spruce forest. FEMS Microbiol Ecol 35:151–161

Marschner P, Crowley DE, Higashi RM (1997) Root exudation and physiological status of a root colonizing fluorescent pseudomonad in mycorrhizal and nonmycorrhizal pepper (*Capsicum annum* L.). Plant Soil 189:11–20

Martin F (2001) Frontiers in molecular mycorrhizal research – genes, loci, dots and spins. New Phytol 150:499–507

Marx DH (1969) The influence of ectotrophic mycorrhizal fungi on the resistance of pine roots to pathogenic infections. I. Antagonism of mycorrhizal fungi to root pathogenic fungi and soil bacteria. Phytopathology 59:153–163

Marx DH (1972) Ectomycorrhizae as biological deterrents to pathogenic root infections. Annu Rev Phytopathol 10:429–454

McAfee BJ, Fortin JA (1986) Competitive interactions of ectomycorrhizal mycobionts under field conditions. Can J Bot 64:848–852

McGonigle TP (1995) The significance of grazing on fungi in nutrient cycling. Can J Bot 73:1370–1376

McKendrick SL, Leake J, Taylor DL, Read DR (2000a) Symbiotic germination and development of myco-heterotrophic plants in nature: ontogeny of *Corallorhiza trifida* and characterization of its mycorrhizal fungi. New Phytol 145:523–537

McKendrick SL, Leake J, Read DR (2000b) Symbiotic germination and development of myco-heterotrophic plants in nature: transfer of carbon from ectomycorrhizal *Salix repens* and *Betula pendula* to the orchid *Corallorhiza trifida* through shared hyphal connections. New Phytol 145:539–548

Meharg AA, Cairney JWG (2000) Ectomycorrhizas – extending the capabilities of rhizosphere remediation? Soil Biol Biochem 32:1475–1484

Meharg AA, Cairney JWG, Maguire N (1997a) Mineralisation of 2,4-dichlorophenol by ectomycorrhizal fungi in axenic culture and in symbiosis with pine. Chemosphere 34:2495–2504

Meharg AA, Dennis GR, Cairney JWG (1997b) Biotransformation of 2,4,6-trinitrotoluene (TNT) by ectomycorrhizal basidiomycetes. Chemosphere 35:513–521

Melin E, Nilsson H (1950) Transfer of radioactive phosphorus to pine seedlings by means of mycorrhizal hyphae. Physiol Plant 3:88–92

Melin E, Nilsson H (1953) Transfer of labelled nitrogen from glutamic acid to pine seedlings through the mycelium of *Boletus variegatus*. Nature 171:134

Minerdi D, Fani R, Gallo R, Boarino A, Bonfante P (2001) Nitrogen fixation genes in an endosymbiotic *Burkholderia* strain. Appl Environ Microbiol 67:725–732

Mogge B, Loferer C, Hutzler P, Hartman A (2000) Bacterial community structure and colonization patterns of *Fagus sylvatica* L. Ectomycorhizospheres as determined by fluorescence in situ hybridization and confocal laser scanning microscopy. Mycorrhiza 9:271–278

Molina R, Massicotte H, Trappe JM (1992) Specificity phenomena in mycorrhizal symbiosis: community-ecological consequences and practical applications. In: Allen MF (ed) Mycorrhizal functioning. Chapman and Hall, London

Morton JB, Benny GL (1990) Revised classification of arbuscular mycorrhizal fungi (Zygomycetes): a new order, Glomales, two new suborders, Glomineae and Gigasporineae, and two new families, Acaulosporaceae and Gigasporaceae, with an emendation of Glomaceae. Mycotaxon 37:477–491

Näsholm T, Ekblad A, Nordin A, Giesler R, Högberg M, Högberg P (1998) Boreal forest plants take up organic nitrogen. Nature 392:914–916

Newman EI (1988) Mycorrhizal links between plants: their functioning and ecological significance. Adv Ecol Res 18:243–270

Newsham KK, Fitter AH, Watkinson AR. 1995. Multi-functionality and biodiversity in arbuscular mycorrhizas. Trends Ecol Evol 10:407–411

Nicolson TH (1959) Mycorrhiza in the Graminae. I. Vesicular-arbuscular endophytes, with special reference to the external phase. Trans Br Mycol Soc 42:421–438

Nicolson TH (1975) Evolution of vesicular-arbuscular mycorrhizas. In: Sanders FE, Mosse B, Tinker PB (eds) Endomycorrhizas. Academic Press, London, pp 25–34

Nurmiaho-Lassila EL, Timonen S, Haahtela K, Sen R (1997) Bacterial colonization patterns of intact *Pinus sylvestris* mycorrhizospheres in dry pine forest soil: an electron microscopy study. Can J Microbiol 43:1017–1035

Ochs M (1996) Influence of humified and non-humified natural organic compounds on mineral dissolution. Chem Geol 132:119–124

Olsson PA, Wallander H (1998) Interactions between ectomycorrhizal fungi and the bacterial community in soil amended with various primary minerals. FEMS Microbiol Ecol 27:195–205

Olsson PA, Chalot M, Bååth E, Finlay RD, Söderström B (1996) Reduced bacterial activity in a sandy soil with ectomycorrhizal mycelia growing with *Pinus contorta* seedlings. FEMS Microbiol Ecol 21:77–86

Palmer RJ Jr, Sternberg C (1999) Modern microscopy in biofilm research: confocal microscopy and other approaches. Curr Opin Biotechnol 10:263–268

Paris F, Bonnaud P, Ranger J, Robert M, Lapeyrie F (1995a) Weathering of ammonium- or calcium-saturated 2:1 phyllosilicates by ectomycorrhizal fungi in vitro. Soil Biol Biochem 27:1237–1244

Paris F, Bonnaud P, Ranger J, Lapeyrie F (1995b) In vitro weathering of phlogopite by ectomycorrhizal fungi. I. Effect of K^+ and Mg^{2+} deficiency on phyllosilicate evolution. Plant Soil 177:191–201

Paris F, Botton B, Lapeyrie F (1996) In vitro weathering of phlogopite by ectomycorrhizal fungi. 2. Effect of K^+ and Mg^{2+} deficiency and N sources on accumulation of oxalate and H^+. Plant Soil 179:141–150

Parniske M (2000) Intracellular accommodation of microbes by plants: a common developmental program for symbiosis and disease? Curr Opin Plant Biol 3:320–328

Parsek MR, Greenberg EP (2000) Acyl-homoserine lactone quorum sensing in Gram-negative bacteria: a signaling mechanism involved in associations with higher organisms. Proc Natl Acad Sci USA97:8789–8793

Pearson JN, Jakobsen I (1993) The relative contribution of hyphae and roots to phosphorus uptake by arbuscular mycorrhizal plants measured by dual labelling with ^{32}P and ^{33}P. New Phytol 124:489–494

Perez-Moreno J, Read DJ (2000) Mobilization and transfer of nutrients from litter to tree seedlings via vegetative mycelium of ectomycorrhizal plants. New Phytol 145:301–309

Perez-Moreno J, Read DJ (2001a) Nutrient transfer from soil nematodes to plants: a direct pathway provided by the mycorrhizal mycelial network. Plant Cell Environ 24:1219–1226

Perez-Moreno J, Read DJ (2001b) Exploitation of pollen by mycorrhizal mycelial systems with special reference to nutrient cycling in boreal forests. Proc R Soc Lond Ser B 268:1329–1335

Perotto S, Bonfante P (1997) Bacterial associations with mycorrhizal fungi: close and distant friends in the rhizosphere. Trends Microbiol 5:496–501

Perry DA, Margolis H, Choquette C, Molina R, Trappe JM (1989) Ectomycorrhizal mediation of competition between coniferous tree species. New Phytol 112:501–511

Peyronel B (1923) Fructification de l'endophyte à arbuscules et à vesicules des mycorhizes endotrophes. Bull Soc Mycol 39:119–126

Pirozynski KA, Malloch DW (1975) The origin of land plants: a matter of mycotrophism. Biosystems 6:153–164

Querejeta JI, Egerton-Warburton LM, Allen MF (2003) Direct nocturnal water transfer from oaks to their mycorrhizal symbionts during severe soil drying. Oecologia 134:55–64

Rambelli A (1973) The rhizosphere of mycorrhizae. In: Marks GL, Koslowski TT (eds) Ectomycorrhizae. Academic Press, New York, pp 299–343

Ravnskov S, Jakobsen I (1999) Effects of *Pseudomonas fluorescens* DF57 on growth and P uptake of two arbuscular mycorrhizal fungi in symbiosis with cucumber. Mycorrhiza 8:329–334

Ravnskov S, Nybroe O, Jakobsen I (1999) Influence of an arbuscular mycorrhizal fungus on *Pseudomonas fluorescens* DF57 in rhizosphere and hyphosphere soil. New Phytol 142:113–122

Read DJ (1984) The structure and function of the vegetative mycelium of mycorrhizal roots. In: Jennings DH, Rayner ADM (eds) The ecology and physiology of the fungal mycelium. Cambridge University Press, Cambridge, pp 215–240

Read DJ (1987) In support of Frank's organic nitrogen theory. Angew Bot 61:25–37

Read DJ (1991) Mycorrhizas in ecosystems. Experientia 47:376–391

Read DJ, Boyd R (1986) Water relations of mycorrhizal fungi and their host plants. In: Ayres P, Boddy L (eds) Water, fungi and plants. Cambridge University Press, Cambridge, pp 287–303

Read DJ, Perez-Moreno J (2003) Mycorrhizas and nutrient cycling in ecosystems – a journey towards relevance? New Phytol 157:475–492

Read DJ, Stribley DP (1975) Some mycological aspects of the biology of mycorrhiza in the Ericaceae. In: Sanders FE, Mosse B, Tinker PB (eds) Endomycorrhizae. Academic Press, London, pp 105–117

Read DJ, Francis R, Finlay RD (1985) Mycorrhizal mycelia and nutrient cycling in plant communities. In: Fitter AH, Atkinson D, Read DJ, Usher MB (eds) Ecological interactions in soil: plants, microbes and animals. Blackwell, Oxford, pp 193–217

Read DJ, Duckett JG, Francis R, Ligrone R, Russell A (2000) Symbiotic fungal associations in "lower" land plants. Philos Trans R Soc Lond B 355:815–831

Redecker D, Kodner R, Graham LE (2000) Glomalean fungi from the Ordovician. Science 289:1920–1921
Remy W, Taylor TN, Haas H, Kerp H (1994) Four hundred-million-year-old vesicular-arbuscular mycorrhizae. Proc Natl Acad Sci USA 91:11841–11843
Requena N, Jimenez I, Toro M, Barea JM (1997) Interactions between plant-growth promoting rhizobacteria (PGPR), arbuscular mycorrhizal fungi and *Rhizobium* spp. in the rhizosphere of *Anthyllis cytisoides*, a model legume for revegetation in Mediterranean semi-arid ecosystems. New Phytol 136:667–677
Requena N, Perez-Solis E, Azcón-Aguilar C, Jefferies P, Barea JM (2001) Management of indigenous plant-microbe symbioses aids restoration of desertified ecosystems. Appl Environ Microbiol 67:495–498
Rhodes LH, Gerdemann JW (1975) Phosphate uptake zones of mycorrhizal and non-mycorrhizal onions. New Phytol 75:555–561
Robinson D, Fitter AH (1999) The magnitude and control of carbon transfer between plants linked by a common mycorrhizal network. J Exp Bot 50:9–13
Rosling A, Landeweert R, Lindahl BD, Larsson K-H, Kuyper TW, Taylor AFS, Finlay RD (2003) Vertical distribution of ectomycorrhizal fungal taxa in a podzol profile determined by morphotyping and genetic verification. New Phytol 159:775–783
Rosling A, Lindahl BD, Finlay RD (2004a) Carbon allocation in intact mycorrhizal systems of *Pinus sylvestris* L. seedlings colonizing different mineral substrates. New Phytol 162:795–802
Rosling A, Lindahl BD, Taylor AFS, Finlay RD (2004b) Mycelial growth and substrate acidification of ectomycorrhizal fungi in response to different minerals. FEMS Microbiol Ecol 47:31–37
Ruess L, Dighton J (1996) Cultural studies on soil nematodes and their fungal hosts. Nematologica 42:330–346
Ruiz-Lozano JM, Bonfante P (2000) A *Burkholderia* strain living inside the arbuscular mycorrhizal fungus *Gigaspora margarita* possesses the vacB gene, which is involved in host cell colonization by bacteria. Microb Ecol 39:137–144
Sanders FE, Tinker PB (1973) Phosphate flow into mycorrhizal roots. Pest Sci 4:385–395
Sarand I, Timonen S, Nurmiaho-Lassila E-L, Koivula T, Haahtela K, Romantschuk M, Sen R (1998) Microbial biofilms and degradative catabolic plasmid harbouring fluorescent pseudomonads in Scots pine mycorrhizospheres developed on petroleum contaminated soil. FEMS Microbiol Ecol 27:115–126
Sarand I, Timonen S, Koivula T, Peltola R, Haahtela K, Sen R, Romantschuk M (1999) Tolerance and biodegradation of *m*-toluate by Scots pine, a mycorrhizal fungus and fluorescent pseudomonads individually and under associative conditions. J Appl Microbiol 86:817–826
Sarand I, Haario H, Jørgensen, KS, Romantschuk M (2000) Effect of inoculation of a TOL plasmid containing mycorrhizosphere bacterium on development of Scots pine seedlings, their mycorrhizosphere and the microbial flora in m-toluate-amended soil. FEMS Microbiol Ecol 31:127–141
Sastry MSR, Sharma AK, Johri BN (2000) Effect of an AM fungal consortium and *Pseudomonas* on the growth and nutrient uptake of *Eucalyptus hybrid*. Mycorrhiza 10:55–61
Schimel JP, Bennet J (2004) Nitrogen mineralization: challenges of a changing paradigm. Ecolgy 85(3):591–602
Schlesinger WH, Lichter J (2001) Limited carbon storage in soil and litter of experimental forest plots under increased atmospheric CO_2. Nature 411:466–469
Schüßler A, Schwarzott D, Walker C (2001) A new fungal phylum, the *Glomeromycota*: phylogeny and evolution. Mycol Res 105:1413–1421
Sen R (2000) Budgeting for the wood-wide web. New Phytol 145:161–165

Setälä H (1995) Growth of birch and pine seedlings in relation to grazing by soil fauna on ectomycorrhizal fungi. Ecology 76:1844–1851

Setälä H, Rissanen J, Markkola AM (1997) Conditional outcomes in the relationship between pine and ectomycorrhizal fungi in relation to biotic and abiotic environment. Oikos 80:112–122

Setälä H, Kulmala P, Mikola J, Markkola AM (1999) Influence of ectomycorrhiza on the structure of detrital food webs in pine rhizosphere. Oikos 87:113–122

Sharples JM, Meharg AA, Chambers SM, Cairney JWG (2000) Symbiotic solution to arsenic contamination. Nature 404:951–952

Shaw TM, Dighton T, Sanders F (1995) Interactions between ectomycorrhizal and saprotrophic fungi on agar and in association with seedlings of lodgepole pine (*Pinus contorta*). Mycol Res 99:159–165

Simard SW, Perry DA, Jones MD, Myrold DD, Durall DM, Molina R (1997a) Net transfer of carbon between ectomycorrhizal tree species in the field. Nature 388:579–582

Simard SW, Jones MD, Durall DM, Perry DA, Myrold DD, Molina R (1997b) Reciprocal transfer of carbon isotopes between ectomycorrhizal *Betula papyrifera* and *Pseudotsuga menziesii*. New Phytol 137:529–542

Simon L, Bousquet J, Levesque RC, Lalonde M (1993) Origin and diversification of endomycorrhizal fungi and coincidence with vascular land plants. Nature 363:67–69

Smith SE, Gianinazzi-Pearson V (1990) Phosphate uptake and arbuscular activity in mycorrhizal *Allium cepa* L.: effects of photon irradiance and phosphate nutrition. Aust J Plant Physiol 17:177–188

Smith SE, Read DJ (1997) Mycorrhizal symbiosis. Academic Press, London

Smith SE, Smith FA (1997) Structural diversity in (vesicular)-arbuscular mycorrhizal symbioses. Tansley Rev no 96. New Phytol 137:373–388

Smith FA, Jakobsen I, Smith SE (2000) Spatial differences in acquisition of soil phosphate between two arbuscular mycorrhizal fungi in symbiosis with *Medicago truncatula*. New Phytol 147:357–366

Smits MM, Hoffland E, Jongmans AG, van Breemen N (2005) Contribution of mineral tunneling to total feldspar weathering. Geoderma 125:59–69

Söderström B, Read DJ (1987) Respiratory activity of intact and excised ectomycorrhizal mycelial systems growing in unsterilised soil. Soil Biol Biochem 19:231–236

St. John TV, Coleman DC, Reid CPP (1983) Growth and spatial distribution of nutrient absorbing organs: selective exploitation of soil heterogeneity. Plant Soil 71:487–493

Stribley DP, Read DJ (1980) The biology of mycorrhiza in the Ericaceae. VII. The relationship between mycorrhizal infection and the capacity to utilise simple and complex organic nitrogen sources. New Phytol 86:365–371

Stubblefield SP, Taylor TN, Trappe JM (1987a) Fossil mycorrhizae: a case for symbiosis. Science 237:59–60

Stubblefield SP, Taylor TN, Trappe JM (1987b) Antarctic VAM fossils. Am J Bot 74:1904–1911

Stubblefield SP, Taylor TN, Seymour RL (1987 c) A possible endogonaceous fungus from the Triassic of Antarctica. Mycologia 79:905–906

Summerbell RC (1987) The inhibitory effect of *Trichoderma* species and other soil microfungi on mycorrhiza formation by *Laccaria bicolor* in vitro. New Phytol 105:437–448

Summerbell RC (1989) Microfungi associated with the mycorrhizal mantle and adjacent microhabitats within the rhizosphere of black spruce. Can J Bot 67:1085–1095

Sun Y-P, Unestam T, Lucas SD, Johanson KJ, Kenne L, Finlay RD (1999) Exudation-reabsorption in mycorrhizal fungi, the dynamic interface for interaction with soil and other microorganisms. Mycorrhiza 9:137–144

Szaniszlo PJ, Powell PE, Reid CPP, Cline GR (1981) Production of hydroxamate siderophore iron chelators by ectomycorrhizal fungi. Mycologia 73:1158–1174

Taylor DL, Bruns TD (1997) Independent, specialized invasions of ectomycorrhizal mutualism by two nonphotosynthetic orchids. Proc Natl Acad Sci USA 94:4510–4515

Taylor AFS, Martin F, Read DJ (2000) Fungal diversity in ectomycorrhizal communities of Norway spruce (*Picea abies* (L.) Karst.) and Beech (*Fagus sylvatica* L.) in forests along north-south transects in Europe. In: Schulze ED (ed) Carbon and nitrogen cycling in European forest ecosystems. Ecological studies, vol 142. Springer, Berlin Heidelberg New York, pp 343–365

Timonen S, Sen R (1998) Heterogeneity of fungal and plant enzyme expression in intact Scots pine-*Suillus bovinus* and -*Paxillus involutus* mycorrhizospheres developed in natural forest humus. New Phytol 138:355–366

Timonen S, Jørgensen K, Haahtela K, Sen R (1998) Bacterial community structure at defined locations of the *Pinus sylvestris-Suillus bovinus* and -*Paxillus involutus* mycorrhizospheres in dry forest humus and nursery peat. Can J Microbiol 44:499–513

Tisdall JM (1974) Possible role of soil microorganisms in aggregation of soils. Plant Soil 159:115–121

Tisdall JM, Oades JM (1979) Stabilisation of soil aggregates by the root systems of ryegrass. Aust J Soil Res 17:429–441

Treseder KK, Allen MF (2000) Mycorrhizal fungi have a potential role in soil carbon storage under elevated CO_2 and nitrogen deposition. New Phytol 147:189–200

Trojanowski J, Haider H, Hüttermann A (1984) Decomposition of ^{14}C-labelled lignin, holocellulose and lignocellulose by mycorrhizal fungi. Arch Microbiol 139:202–206

Tsimilli-Michael M, Eggenberg P, Biro B, Köves-Pechy K, Vörös I, Strasser RJ (2000) Synergistic and antagonistic effects of arbuscular mycorrhizal fungi and *Azospirillum* and *Rhizobium* nitrogen fixers on the photosynthetic activity of alfalfa, probed by the polyphasic chlorophyll a fluorescence transient O-J-I-P. Appl Soil Ecol 15:169–182

van Breemen N, Finlay RD, Lundström US, Jongmans AG, Giesler R, Olsson M (2000a) Mycorrhizal weathering: a true case of mineral plant nutrition? Biogeochemistry 49:53–67

van Breemen N, Lundström US, Jongmans AG (2000b) Do plants drive podzolization via rock-eating mycorrhizal fungi? Geoderma 94:163–171

van der Heijden MGA, Klironomos JN, Ursic M, Moutoglis P, Streitwolf-Engel R, Boller T, Wiemken A, Sanders IR (1998) Mycorrhizal fungal diversity determines plant biodiversity, ecosystem variability and productivity. Nature 396:69–72

Vázquez M Mar, César S, Azcón R, Barea JM (2000) Interactions between arbuscular mycorrhizal fungi and other microbial inoculants (*Azospirillum*, *Pseudomonas*, *Trichoderma*) and their effects on microbial population and enzyme activities in the rhizosphere of maize plants. Appl Soil Ecol 15:261–272

Voiblet C, Duplessis S, Encelot N, Martin F (2001) Identification of symbiosis-regulated genes in *Eucalyptus globulus-Pisolithus tinctorius* ectomycorrhiza by differential hybridization of arrayed cDNAs. Plant J 25:181–191

Vosátka M, Gryndler M (2000) Response of micropropagated potatoes transplanted to peat media to post-vitro inoculation with arbuscular mycorrhizal fungi and soil bacteria. Appl Soil Ecol 15:145–152

Wallander H (2000a) Uptake of P from apatite by *Pinus sylvestris* seedlings colonized by different ectomycorrhizal fungi. Plant Soil 218:249–256

Wallander H (2000b) Use of strontium isotopes and foliar K content to estimate weathering of biotite induced by pine seedlings colonised by ectomycorrhizal fungi from two different soils. Plant Soil 222:215–229

Wallander H, Nylund J-E (1992) Effects of excess nitrogen and phosphorus starvation on the extramatrical mycelium of ectomycorrhizas on *Pinus sylvestris* L. New Phytol 119:405–411

Wallander H, Wickman T (1999) Biotite and microcline as potassium sources in ectomycorrhizal and non- mycorrhizal *Pinus sylvestris* seedlings. Mycorrhiza 9:25–32

Wallander H, Wickman T, Jacks G (1997) Apatite as a P source in mycorrhizal and non-mycorrhizal *Pinus sylvestris* seedlings. Plant Soil 196:123–131

Warnock AJ, Fitter AH, Usher MB (1982) The influence of a springtail, *Folsomia candida* (Insecta, Collembola) on the mycorrhizal association of leek, *Allium porrum* and the vesicular-arbuscular endophyte, *Glomus fasciculatus.* New Phytol 90:283–292

Wilkins DA (1991) The influence of sheathing (ecto-)mycorrhizas of trees on the uptake and toxicity of metals. Agric Ecosys Environ 35:245–260

Wu B, Nara K, Hogetsu T (1999) Competition between ectomycorrhizal fungi colonizing *Pinus densiflora*. Mycorrhiza 9:151–159

10 Effects of Soil Temperature on Nutrient Uptake

K.S. Pregitzer and J.S. King

10.1 Introduction

More than a decade ago, Bowen (1991) underscored the curious fact that soil temperature had received relatively little attention from scientists, even though it is of fundamental importance to plant productivity. With the realization that soil temperatures may rise as a part of anthropogenic global change, there is renewed interest in studying the mechanisms and effects of terrestrial ecosystem responses to changes in soil temperature (Billings et al. 1982; Peterjohn et al. 1994; Rosenzweig and Hillel 2000). However, this area of investigation is still much underrepresented. In any consideration of soil temperature, we must be cognizant that it continually interacts with biotic and abiotic components of the soil–plant system in both space and time (Pregitzer et al. 2000). For example, not only does soil temperature influence rate constants of chemical reactions, water content, and nutrient transport in soil, but it simultaneously affects plant physiological aspects of ion uptake, root growth, and the composition and function of soil microbial communities. In fact, virtually all processes occurring in soil, from the weathering of primary minerals to plant nutrition and storage of organic carbon, are strongly influenced by soil temperature (Fig. 10.1; also Chap. 2, this Vol.). In particular, acquisition of mineral nutrients by plants is governed by myriad chemical, physical, and biological processes that are all sensitive to temperature. Soil temperature therefore exerts a major influence on primary productivity in terrestrial ecosystems around the globe.

In this chapter, we address how plant nutrient uptake is influenced by changes in soil temperature. There have been several excellent treatments of soil temperature effects on root growth and related processes (Gliński and Lipiec 1990; Bowen 1991; McMichael and Burke 1996, 1998; Pregitzer et al. 2000). Our contribution is a synthesis of information from a wide range of sources to provide insight on controls and global patterns of soil temperature, and how it affects the biotic and abiotic control points of nutrient uptake. Recent work has demonstrated the capacity of some plants to utilize organic

Ecological Studies, Vol. 181
H. BassiriRad (Ed.) Nutrient Acquisition by Plants
An Ecological Perspective

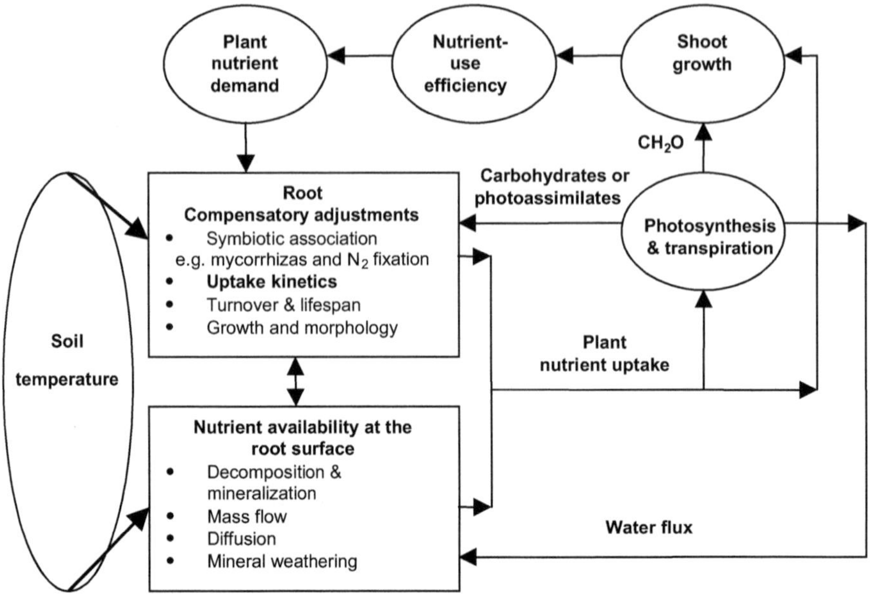

Fig. 10.1. A conceptual diagram depicting relationships between plant and soil factors that control the availability and uptake of mineral nutrients. Soil temperature affects nutrient uptake directly by altering root growth, morphology, and uptake kinetics. Indirect effects include altered rates of decomposition and nutrient mineralization, mineral weathering, and nutrient transport processes (mass flow and diffusion). (Modified from Fig. 1 in BassiriRad 2000, with kind permission)

forms of nutrients (Näsholm et al. 1998; Lipson and Näsholm 2001). The relative importance of essential elements acquired directly from organic matter is not well understood at this time (see Chap. 9, this Vol.), and it is difficult to speculate about how soil temperature might influence this process. Therefore, our review concentrates on how soil temperature influences uptake of inorganic ions. Nonetheless, we fully expect that soil temperature will directly influence the enzymatic processes that must control acquisition of nutrients from organic sources.

The chapter begins with a description of the soil energy budget, ranges of soil temperatures experienced at different latitudes, and covers and briefly explores how soil temperature is expected to change in the future. The second section discusses temperature effects on chemical and physical properties of soils, namely, rates of chemical reactions, soil moisture, mass flow, and diffusion. In the third section, we examine how soil temperature influences various aspects of root biology, including growth and morphology of root systems, root distributions in soil, and root physiology. The fourth section looks at effects of temperature on soil fauna and microbial communities, decomposition of organic matter and nitrogen transformations, nitrogen fixation, and mycorrhizae. We close with a summary and suggested areas of future research.

10.2 Dynamics of Soil Temperature

10.2.1 The Soil Energy Balance

The temperature of the soil is determined by the balance of energy input (short wave) and output (long wave), which changes continuously on a diurnal and seasonal basis. Formal treatments of the soil energy balance can be found in Hanks and Ashcroft (1980), Jury et al. (1991), and Hillel (1998). The main determinant of soil thermal regime is the net amount of radiation reaching the Earth's surface, termed insolation, which is a function of latitude, time of year, and cloud cover. Aspect significantly affects insolation, with south- and west-facing slopes generally receiving the most solar energy (Geiger 1965). Tajchman and Minton (1986) reported greater monthly mean soil temperature on south- and west-facing exposures relative to north-facing slopes in a forested Appalachian watershed. Altitude also significantly affects soil temperature, with lower-elevation soils warming more and earlier in the spring than those higher up (Woodward 1998).

How much of the incoming solar radiation is absorbed by the soil depends in large part on the surface albedo, which is a function of solar angle and the degree to which the soil is covered with vegetation, organic detritus, and snow. Plant cover generally reduces albedo, which can be extremely important to the energy balance of high-latitude ecosystems. Hall et al. (1996) found that jack pine-spruce forests in Saskatchewan and Manitoba had an average albedo of 0.08 compared to a value of 0.8 for bare snow. It has also been shown that the tussock growth form of *Eriophorum vaginatum* improves the soil thermal regime and nutrient cycling in Alaskan tundra (Chapin et al. 1979). Part of the insulating effect of vegetation is due to the buildup of a layer of organic detritus on the soil surface, which can decrease the amplitude of daily and seasonal temperature fluctuations. Breshears et al. (1998) reported that winter soil temperatures under patches of woody plants (*Pinus edulis* and *Juniperus monosperma*) in semi-arid woodland were warmer than interpatch areas due to the buildup of a litter layer. The insulating properties of litter stem from the relatively low thermal conductivity of organic matter compared to mineral soil, and the large proportion (volumetric) of air spaces. Similarly, soil texture can influence the ability of a soil to conduct and store energy. Compiling results from several studies, Jury et al. (1991) reported that wet peat had a thermal conductivity (average) of 0.83 cal cm^{-1} s^{-1} $°C^{-1}$, compared to 4.01 cal cm^{-1} s^{-1} $°C^{-1}$ for wet sand. Since the thermal conductivity of most mineral components of the solid phase is similar (Smith and Byers 1938), differences between mineral soils are primarily due to water content and bulk density.

The hydrologic cycle is a very important driver of soil temperature. When soil is bare, color and moisture content are the main determinants of albedo (Hanks and Ashcroft 1980). Dark soils absorb more energy than lighter ones, and wetting the soil effectively darkens it. As mentioned above, the amount of snow cover exerts a dominant effect on albedo. The albedo of snow- and ice-covered surfaces ranges from 0.45 to 0.90, contributing greatly to the gradient of global albedo (Strahler and Strahler 1983), with energy available to soil decreasing at higher latitudes. However, snow also acts as an effective insulator, retarding heat loss from the soil. At mid-latitudes there is an inverse relationship between depth of snowpack and depth of soil freezing (Bertrand et al. 1994; Stadler et al. 1996). In areas of extremely heavy snowfall, such as in the lake-effect snow-belt region of the Great Lakes, soil freezing can be prevented entirely under very deep snow (Isard and Schaetzl 1995). This has important implications for the hydrology of these systems during spring snowmelt, and can influence the breaking of dormancy. Because the specific heat of water (1.0 cal g^{-1}) is so much greater than that of soil minerals (~0.2 cal g^{-1}), moisture content greatly influences soil thermal capacity and diffusivity, and therefore temperature (Kohnke 1968). Evapotranspiration of water from the soil surface removes latent heat from soil, thereby cooling it. Sumrall et al. (1991) observed maximum dry soil temperatures at 1-cm depth of ~60 °C in burned semi-desert grassland in Arizona, but of only ~42 °C after the summer rainy season had wet the soil. Direct transfer of heat from the atmosphere to soil in rainwater has received some attention from hydrologic modelers (e.g., Bristow et al. 1986), but little from ecologists. Finally, in low-lying systems such as bogs and fens, groundwater can have a dominant influence on soil temperature (Nichols 1998).

Other drivers of soil temperature largely result from land-use practices that influence plant cover, soil moisture status, and cultivation of soil. Maximum daily soil temperature (1-cm depth) in a clear-cut exceeded 50 °C compared to 16 °C in an adjacent mature, western hemlock-fir forest in coastal British Columbia, and had much greater amplitude (Spittlehouse and Stathers 1990). The cumulative warming of soil 19 years after forest removal in western Australia increased soil temperature (relative to intact forest) to many tens of meters in depth, and it was estimated that it would require 208 years to return to an upward heat flux (Taniguchi et al. 1998). Mounding and trenching, forms of site preparation widely practiced in cold, wet soils of northern forests, have been shown to increase soil temperature, significantly increasing tree seedling survival and growth (Weber et al. 1995; Paterson and Mason 1999). Other forms of cultivation, such as blade scarification and plowing, have been shown to increase soil temperature due to the removal of vegetation and improved thermal conductivity from mixing of mineral and organic horizons

(Spittlehouse and Stathers 1990). Use of dark-colored mulches, such as water-permeable plastic, decreases soil albedo and significantly increases soil temperature. As with mounding and trenching, this can be an important way to extend growing seasons at high latitudes. The use of light-colored mulches composed of organic residues has the opposite effect, cooling soil and reducing water loss, which is important in hot climates (Olasantan 1999).

10.2.2 Spatial and Temporal Variation in Soil Temperature

The diurnal and seasonal progression of soil temperature exhibits patterns characteristic of the prevailing climate (Fig. 10.2). During a day in the growing season, soil temperature close to the surface is lowest between the hours of midnight and sunrise. As the solar angle increases, so too does the soil temperature, reaching a peak at midday and declining thereafter. With increasing depth, soil temperature decreases and the amplitude of the daily signal diminishes. Peak temperature is reached progressively later in the day due to a time lag imposed by the thermal conductivity inherent to the soil (Fig. 10.2A). At sufficient depth, diurnal temperature variation is completely dampened, and the temperature of the soil remains near that of the mean annual air temperature (Strahler and Strahler 1983). In winter, the soil temperature profile is inverted, the warmest soil being found at depth. Soil temperature decreases toward the surface, and the diurnal signal has very little amplitude. Over the course of the year, soil temperature exhibits a pattern very similar to that of the diurnal signal. Lowest temperatures are reached in late winter, and rise rapidly during the spring as the cumulative heat sum increases with the changing season (Fig. 10.2B). Soil temperature peaks in late summer and declines for the remainder of the year. As with the diurnal signal, the annual peak temperature is reached progressively later in the year with increasing depth, and the amplitude of the seasonal signal decreases.

The US System of Soil Taxonomy (Soil Survey Staff 1975) recognizes soil temperature classes based on the mean annual temperature at 50-cm depth (Table 10.1). These temperature classes can be thought of as a rough approximation of the latitudinal (or altitudinal) distribution of the different soil thermal regimes. Early investigations of global net primary production (Lieth 1975) found very strong relationships between mean annual air temperature, precipitation, and plant growth. Undoubtedly, some of the ecosystem responses to temperature are responses to soil temperature. Soil temperature has been shown to be a major determinant of the length of the growing season in cold ecosystems (Chapin et al. 1979), and it influences community composition and structure (Van Cleve et al. 1981; Farnsworth et al. 1995; Thompson and Naeem 1996). It is likely that some of the mechanisms by which soil temperature affects these ecosystem properties are exerted indirectly through effects on nutrient availability and uptake by plants.

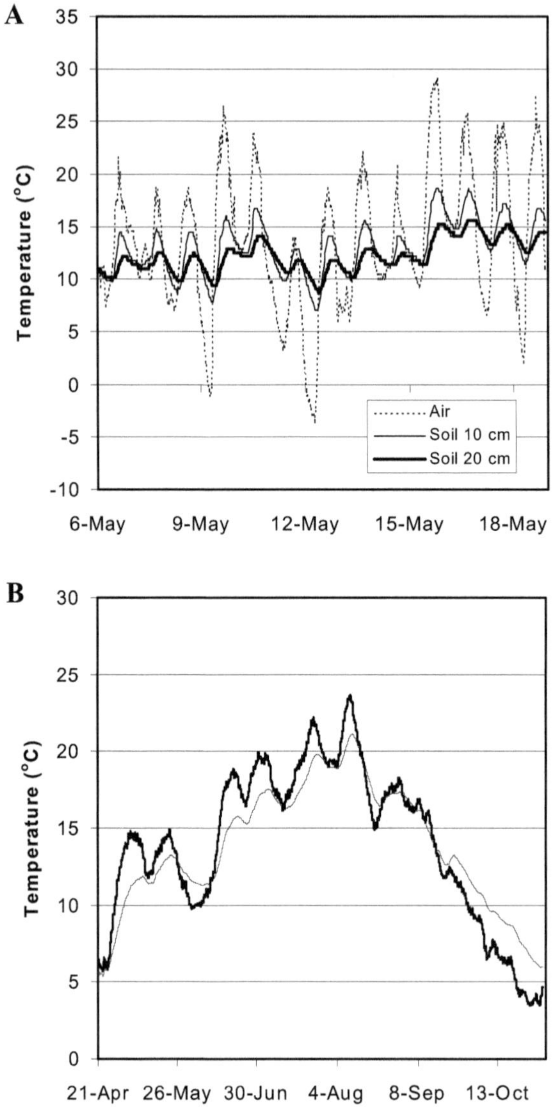

Fig. 10.2A, B. Temperature profile in a young stand of *Populus tremuloides* at the FACTS-II Aspen FACE Project in Rhinelander, Wisconsin, 2001. **A** Several weeks early in the growing season. Note how soil temperature tracks that of the air, with increasing lag time with depth and diminishing amplitude. **B** Ten-day moving average for the entire growing season. *Thick line* Air temperature at 1 m above the ground, *thin line* soil temperature at 20-cm depth. Data were collected using Hobo temperature loggers (Onset Computer Corporation, Pocasset, MA, USA)

We have compiled a list of the predominant biomes of the world, from the poles to the equator, showing peak and mean annual soil temperature, major class of vegetation, and areal extent (Table 10.2). Interestingly, fully 23 % of the land area and 28 % of the biomass occur in the tundra and boreal forest, on soils with mean annual temperatures less than 4 °C (Table 10.2). Although these systems are very cold on average, during the growing season surface soil temperatures can reach 18–20 °C on a regular basis. Temperate forests and steppes comprise approximately 22 % of both the land area and plant biomass, and the soils of these systems exhibit a remarkably wide range of

Table 10.1. Soil thermal regimes recognized by the US System of Soil Taxonomy based on mean annual temperature at 50-cm depth

Soil thermal class	Mean annual temperature (°C)
Frigid	<8
Mesic	8–15
Isomesic[a]	8–15
Thermic	15–22
Hypothermic	>22

[a] The prefix "iso" refers to soils where mean summer and winter temperatures vary by <5 °C

mean annual temperature, from 7.0 to 20.3 °C. Maximum temperatures appear to reach 25–26 °C. While desert ecosystems occur over 16 % of the land area, they only contain 1 % of the plant biomass, an indication of the severe temperature and water stress characteristics of this biome. Mean annual surface soil temperatures are rather moderate, around 16.5 °C, but growing season peak temperatures on bare soil have been reported to be as high as 69.0 °C. However, the presence of a vegetation canopy greatly reduces these extreme temperatures (Table 10.2). Tropical savannas, woodlands, and dry forests occur over 29 % of the land area, and contain approximately 18 % of the plant biomass. Mean annual surface soil temperatures in these systems are around 29.0 °C, and can reach up to 54 °C. Finally, the extremely productive, tropical moist forest covers only 9 % of the land surface, but contains fully 29 % of the total plant biomass. Unless the forest canopy is broken by logging or agriculture, surface soil temperatures experience very little daily or seasonal variation, and remain near the mean annual air temperature of 22–26 °C. Our summary of major world biomes and soil thermal regimes is intended to illustrate the general pattern between plant distributions and soil temperature, and does not capture the obviously much greater variation in soil temperature within each of the biomes. We recommend the measurement and reporting of soil temperature as a regular part of future plant ecological studies. Use of standard depths of measurement, such as 10 and 50 cm, would facilitate comparisons. Systems in particular need of description are tundra and boreal forest, and all tropical ecosystems. Such data will not only allow a better understanding of relationships between primary production and environmental conditions, but will also provide ground-based benchmarks against which to compare satellite and model estimates of changes in soil temperature in response to global warming. General circulation models predict an increase in mean global temperature of 1.4–5.8 °C by the year 2100 as a result of radiative forcing from greenhouse gases (Cubasch et al. 2001). The effects of this on soil thermal regimes is poorly understood, but awareness is growing that responses of soil processes, including nutrient

Table 10.2. Major world biomes with estimates of areal extent, plant biomass, and some peak and mean annual surface soil temperatures. Most values are for depths<15 cm. Annual means have been extracted from original sources by visually estimating plotted values

Biome	Area[a] (10^{12} m^2)	Plant biomass[a] (10^{15} g C)	Peak soil T (°C)	Annual mean soil T (°C)	Latitude
Tundra	11.0	9.0	2–8[b]	≤0	>64°00N[b]
			7–15[c]		64°53′N[c]
Boreal forest	15.0	143.0	8–18[d]		64°48′N[d]
			14[e]		53°42′N[e]
			15[f]	3.9[f]	67°30′N[f]
			15[g]	3.6[g]	52°09′N[g]
			18–20[h]		46°00′N[h]
Temperate forest	9.2	73.3	17[i]	7.0[i]	43°56′N[i]
			16[j]	7.3[j]	43°40′N[j]
			18–20[k]	12.9[k]	35°00′N[k]
			25[l]	20.3[l]	30°39′N[l]
Temperate steppe	15.1	43.8	23[m]	14.4[m]	34°43′S[m]
			21[n]	<5.0[n]	41°25′N[n]
			26[o]	10.5[o]	34°18′N[o]
Desert	18.2	5.9	33[p]	16.3[p]	31°25′N[p]
			41[q]	17.4[q]	
			69.56[r]		33°38′N[r]
Tropical savanna and woodland	7.7	48.8	32.2[s]	28.4[s]	~10°00′N[s]
			~54[t]		13°31′N[t]
Tropical dry forest	24.6	49.7	35[u]	30[u]	8°05′N[u]
Tropical moist forest	10.4	156.0	22–26[v]	22–26[v]	
				26–32[w]	5°15′N[w]

[a] From Schlesinger (1997), after Houghton and Skole (1990)

[b] Mean July soil temperature of the North American High and Low Arctic at 10-cm depth (Bliss 1988)

[c] Late July soil temperature at 10-cm depth for a variety of boreal forest types near Fairbanks, Alaska (147°45′W) (Tryon and Chapin 1983)

[d] July soil temperature at 5-cm depth for open willow, balsam poplar–alder, and mature white spruce systems of the Tanana River floodplain near Fairbanks, Alaska (64°48′N, 147°52′W) (Viereck et al. 1993)

[e] July soil temperature at 10-cm depth in a mature boreal aspen forest in northern Saskatchewan (53°42′N, 106°12′W) as part of the BOREAS project (Russell and Voroney 1998)

[f] Mean annual and late June soil temperature at 5-cm depth in mature Scots pine forest in eastern Finish Lapland (67°30′N, 29°30′E) (Sutinen et al. 1998)
[g] August and mean annual soil temperature at 10-cm depth in a prairie-aspen grove near Saskatoon, Saskatchewan (52°09′N, 106°36′W) (Archibold et al. 1996)
[h] July soil temperature in newly planted red and white pine plantation in the middle of the Great Lakes–St. Lawrence Forest region (46°00′N, 77°33′W) (Weber et al. 1995)
[i] Mean annual and July soil temperature at 8-cm depth in northern hardwood forest at Hubbard Brook Experimental Forest, New Hampshire (43°56′N, 71°45′W) (Federer et al. 1990)
[j] Mean annual and July soil temperature at 15-cm depth in mature sugar maple forests along a latitudinal gradient in Michigan (43°40′N, 86°09′W to 46°52′N, 88°53′W) (Brown et al. 1999)
[k] Mean July soil temperature at 10-cm depth for mixed pine–oak–*Kalmia* forest in southwestern North Carolina (35°00′N, 83°50′W) (Swift et al. 1993)
[l] July soil temperature at 10-cm in mature sweet gum–oak forest on the Neches River floodplain in Tyler County, Texas (30°39′N, 94°50′W) (Londo et al. 1999)
[m] Mean annual and January soil temperature at 40-cm depth in a Mediterranean-climate forest in southwest Australia (34°43′S, 116°04′E) (Stoneman et al. 1995)
[n] Mean annual (estimated from mean annual air temperature of 2.7 °C) and July soil temperature at 10-cm depth in alpine tundra near Saratoga, Wyoming (41°25′N, 107°10′W) (Burke 1989)
[o] Mean annual and July soil temperature at 2-cm depth in semiarid woodland in northern New Mexico (34°18′N, 106°16′W) (Breshears et al. 1998)
[p] Early June and mean annual soil temperature at 1-cm depth in the Chihuahuan Desert in southwestern Texas (31°25′N, 105°40′W) (Scanlon 1994)
[q] "Summer" and mean annual temperature of "wet field seedbeds" of warm-season grasses in southern Arizona (Roundy and Biedenbender 1996)
[r] "Summer" soil temperature at 2-cm depth for bare soil and under a "nurse plant" of *Hilaria rigida*, respectively, at the University of California Desert Research Center at Agave Hill (33°38′N, 116°24′W) (Nobel 1989)
[s] April and mean annual soil temperature at 30-cm depth in the savanna zone of Nigeria (between latitudes 7°30′ and 13°00′N) (Ogigirigi 1977)
[t] Surface temperature of a combination of bare soil and sparse shrub (*Guiera senegalensis*) leaf litter measured with down-facing infrared thermometer in fallow woodland–savannah as part of the HAPEX Sahel project near Niamey, Niger (13°31′N, 2°07′E) (Lhomme et al. 1997)
[u] Dry season and mean annual soil temperature at 15-cm depth at Njala, Sierra Leone (~8°05′N, 12°05′W) (Odell et al. 1974)
[v] Mean surface soil temperature under intact tropical moist forest. This is very close to the mean annual air temperature, with little temporal or spatial variation (Richards 1998)
[w] Diurnal range of soil temperature at 5-cm depth in a grass-dominated clear-cut in an equatorial lowland forest, French Guiana (5°15′N, 52°55′W) (Janssens et al. 1998)

uptake by plants, will be a major driver of future net primary production and climate (Scharpenseel et al. 1990; Zak et al. 1993; Boone et al. 1998; Rosenzweig and Hillel 2000).

10.3 Soil Temperature Effects on Soil Physical and Chemical Properties

Ultimately, the capacity of a soil to supply nutrients for plant growth is a function of the total quantity of nutrients contained in the solid phase (nutrient capital), and a smaller fraction that is in equilibrium between the soil solution and exchange sites of clay particles and organic matter. Soil temperature exerts both long-term and short-term effects on this system. In the long term, soils with a thermic or hyperthermic temperature regime experience accelerated rates of mineral weathering and decomposition, increasing the content of low-activity clays and decreasing organic matter (Sanchez 1976). This has two effects. First, the highly weathered clays (e.g., kaolinite) are composed mainly of iron and aluminum hydroxides, having long since lost most of the elements important to plant nutrition such as Ca and Mg. Therefore, soil nutrient stocks are severely depleted. Second, these clays have low cation exchange capacity, and therefore low capacity to retain nutrients that may be made available from the mineralization of organic matter or atmospheric deposition. In the short term, soil temperature influences plant nutrient uptake through effects on soil water, rates of chemical reactions, and nutrient transport.

10.3.1 Soil Water

Since most chemical reactions and nutrient transport occur in water, how soil water is affected by soil temperature directly impacts nutrient uptake. It has been estimated that only 1 % of the nutrients reaching the surface of plant root systems is due to direct interception, while the remainder is transported to the roots by mass flow (transpiration and hydrodynamic dispersion) and diffusion (Jungk 1996), although interception may be much more important for immobile nutrients such as P (Barber et al. 1989). The most obvious effect of soil temperature on soil water is increased rates and depth of evaporation with increasing soil temperature, especially in situations where the supply of water may be limited. Cortina and Vallejo (1994) reported a decline in litter decomposition due to soil drying associated with higher soil temperature in a clear-felled *Pinus radiata* stand in Mediterranean ecosystems of northeastern Spain. Paláez et al. (1992) reported a very strong inverse relationship between soil temperature and soil water potential (5-cm depth) in *Prosopis* ecosystems in the semiarid region of Argentina. Not only does dry soil prevent mass flow and diffusion of nutrients, but it may also lead to increased mechanical impedance to root growth (Bennie 1991), thereby limiting nutrient interception.

10.3.2 Nutrient Transport

Soil factors that determine nutrient fluxes are treated in Chapter 1 (this Vol.), but here we discuss how soil temperature can modify some of these factors. For example, soil temperature can have a significant effect on the viscosity of water. The viscosity of water is inversely related to its temperature in a nearly linear fashion ($r^2=0.94$) over the range of temperatures experienced in soil in ecosystems around the world (0 to ~70 °C), from 1.15×10^{-3} to 0.41×10^{-3} N s m^{-2} (Brown and LeMay 1981). The increased viscosity at low temperatures is known to decrease rates of water uptake by roots and transport within the plant (Johnson and Thornley 1985; Kramer and Boyer 1995; Wan et al. 2001), and therefore reduces the rate of nutrient transport to the roots in mass flow. Similarly, the transport of nutrient ions from areas of high to low concentration by the process of diffusion is directly influenced by soil temperature. Diffusion of gaseous species in soil atmosphere or solutes in the aqueous phase is usually described by some version of Fick's first law (e.g., Hillel 1998):

$$J_d = -D_s \theta \mathrm{d}c / \mathrm{d}x \tag{1}$$

where J_d is the flux rate of ion d, D_s is the diffusion coefficient of that ion accounting for the tortuosity of the particular soil pore space, θ is the fractional water volume, and dc/dx is the concentration gradient of species d. In most discussions of diffusion, the magnitude of the concentration gradient is emphasized as the main driver of the flux rate. Temperature influences J_d because D_s increases linearly with increasing absolute temperature (Weast 1978). Therefore, for nutrients transported mainly by the process of diffusion (e.g., PO_4^{3-}), we would expect much greater mobility in warm soil.

10.3.3 Chemical Reactions

In general, the rate at which nutrients are taken up by roots is proportional to their concentration in soil solution near the root (Marschner 1995). Therefore, anything that affects nutrient concentrations in soil solution greatly influences nutrient uptake and growth by plants. All chemical reactions that occur in soil, including mineral weathering (Sparks 1995), biologically mediated nitrogen transformations (Paul and Clark (1996), and most reactions involving nutrient ions in soil solution (Sposito 1994), are strongly influenced by temperature. The nonlinear relationship between the rate of chemical reactions and temperature was described in 1888 by Arrhenius (Brown and LeMay 1981):

$$k = A_f e^{-Ea/RT} \tag{2}$$

where k is the reaction rate constant, A_f is the frequency of molecular collisions, e is the base of the natural logarithms, E_a is the energy of activation for the reaction, R is the gas constant, and T is absolute temperature. The linear form of this relationship, $\ln k = \ln A_f - E_a/RT$, has been widely exploited to describe relationships between temperature and various processes that may affect nutrient uptake (Johnson and Thornley 1985), including water uptake (Feng et al. 1990; Kramer and Boyer 1995), root respiration (Gliński and Lipiec 1990), and the oxidation of NH_4^+ and NO_2^- by *Nitrosomonas* and *Nitrobacter*, respectively (Wong-Chong and Loehr 1978).

Thus, the myriad of soil processes, both biotic and abiotic, that control nutrient concentrations in soil solution occur at accelerated rates in warmer soils. We can attribute this to a combination of increased rates of chemical reactions involved in mineral weathering, decomposition of organic matter (see Chap. 2, this Vol.), and exchange reactions between solid and solution phases of the soil. In addition, reduced viscosity of water increases rates of mass flow, and nutrient transport by diffusion probably also increases at higher temperatures. Joslin and Wolfe (1993) reported that warmer soils associated with the sunny side of a clear-felled, high-elevation red spruce stand exhibited higher mean seasonal solution concentrations of NO_3^-, Mg^{2+}, and Al^{3+} compared to cooler soils on the shaded side. Though nutrient availability increased in the short term, the seasonal export (leaching) of NO_3^- and Mg^{2+} from the site was 30–33 % greater due to soil warming, which the authors speculated could *adversely* affect forest nutrition over time. Kelly (1993) reported higher concentrations of total N, NH_4^+-N, and PO_4^{3-}-P in soil solution from the Oa and A horizons of a mixed-age spruce forest as soil temperature was increased from 4 to 24 °C. Concentrations of Ca^{2+}, Mg^{2+}, Na^+, and NO_3^--N *decreased* with increasing soil temperature, however, indicating that it cannot be assumed that all ions respond equally.

10.4 Soil Temperature Effects on Root Biology

Along with chemical and physical effects on soil nutrient concentrations and transport, soil temperature influences nutrient uptake through direct effects on the growth and physiology of plant root systems (Fig. 10.1). The amount and duration of root growth, root morphology, and root spatial distributions can all be affected by soil temperature. In addition, soil temperature can alter specific rates of ion uptake, root respiration, cell membrane permeability, and rates of transport in the xylem.

10.4.1 Root Growth and Morphology

Roots systems, especially in long-lived woody perennials, have functionally distinct fractions with coarse roots being important for anchorage, transport and storage, and small-diameter "fine" roots being most important for uptake of water and nutrients (Eissenstat 1992; Kramer and Boyer 1995; Fitter 1996). There is a large body of evidence demonstrating that, all other factors remaining equal, root growth is enhanced with increasing soil temperature (see reviews by Lyr and Hoffmann 1967; Bowen 1991; Kaspar and Bland 1992; McMichael and Burke 1996, 1998; Pregitzer et al. 2000). Typically, production of root biomass and length commences at some low, limiting temperature, increases with rising temperature to an optimum, and then declines with further increases in temperature (Fig. 10.3). Such temperature response data are common for agronomic plants (e.g., Gliński and Lipiec 1990), and a few forest tree species (Lyr and Hoffmann 1967; Teskey and Hinckley 1981; Kuhns et al. 1985). The range of soil temperatures at which root growth begins, peaks, and then declines varies by species and is related to the environmental conditions under which the plants have evolved (Chapin 1974a, b). Root growth of plants adapted to cold environments usually responds to a lower range of temperatures than those from warmer climates, and vice versa (Gliński and Lipiec 1990). As pointed out by Lyr and Hoffmann (1967), such data generally repre-

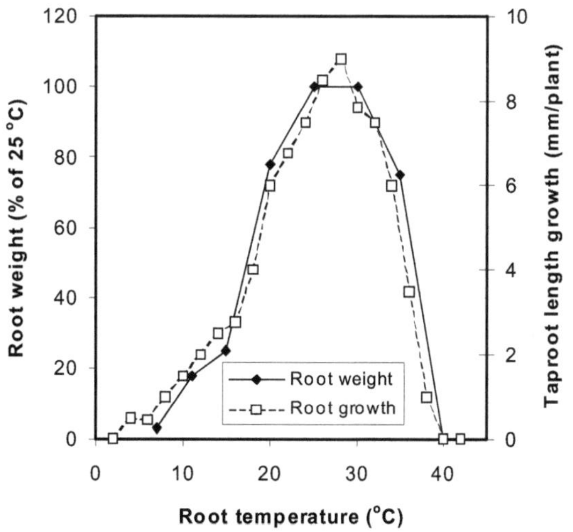

Fig. 10.3. Soil temperature response of maize root dry weight 24 days after germination, and pecan seedling taproot length after 4-day growth at treatment temperatures. (Redrawn from Figs. 3 and 4 in Kaspar and Bland 1992, with kind permission from Lippincott, Williams and Wilkins Publishers)

sent the physiological optimum rather than the ecological optimum, because temperatures employed in controlled-environment experiments are rarely experienced in the soil. The same could be said of "square-wave" temperature treatments, as high temporal variation in soil temperature is a fundamental characteristic of many ecosystems (Pregitzer et al. 2000). Few studies have examined the effects of soil temperature on the allocation of carbon between coarse roots and the very fine feeder roots, which is a particularly important trait affecting nutrient uptake. King et al. (1996) reported that elevated soil (and air) temperature (+5 °C) resulted in small increases in the partitioning of biomass to the smallest-diameter root fraction in *Pinus taeda* and *Pinus ponderosa* seedlings. However, more research is needed before generalizations can be made about soil temperature effects on biomass allocation between root fractions.

One potential mechanism for enhanced root growth in warmer soils is via source–sink relationships between above- and belowground plant parts. It is known that elevated soil temperature increases rates of photosynthesis (Day et al. 1991; Landhäusser et al. 1996; Schwarz et al. 1997). This is due in part to the relief of water stress by greater rates of water uptake and conductance, allowing greater stomatal conductance. Higher rates of photosynthesis increase the availability of fixed carbon, some of which is translocated belowground to sustain new root growth. Another possible mechanism for enhanced root growth with increasing soil temperature is greater production of growth-regulating substances (e.g., ABA, cytokinins, gibberellins) or altered ratios of these substances (Atkin et al. 1973; Bowen 1991, and references therein; Kramer and Boyer 1995). Evidence for this is limited, however, and more work needs to be done to determine the importance of soil temperature control over the production of growth-regulating substances. Finally, rates of enzymatic reactions, cell division and expansion are directly related to temperature (Taiz and Zeiger 1991; Larcher 1995), and so the capacity of plants to construct new root tissue increases as soil temperature rises.

Although the temperature response of the static measure of root standing crop (biomass and length) has been well characterized, much less is known about how soil temperature affects the dynamic processes of fine root production and mortality (Pregitzer et al. 2000). In agronomic plants, uptake of certain ions (e.g., iron, NH_4^+, NO_3^-) appears to occur at higher rates near the apical (youngest) portions of roots, while uptake of K^+ occurs at equal rates along the root (Clarkson 1996; Colmer and Bloom 1998). Such variation in ion uptake with root age may have important implications for plant nutrition if the average longevity of populations of fine roots is altered by environmental conditions. Hendrick and Pregitzer (1993) observed that fine root longevity of sugar maple was shorter at sites with higher mean annual soil temperature. Forbes et al. (1997) found that *Lolium perenne* had decreased root longevity and maintained lower root biomass at elevated soil temperature. Production and mortality of fine roots in *Populus tremuloides* increased with seasonally elevated soil temperature (Fig. 10.4), with the increase in production being proportionately greater, giving rise to an increase in net root production (King et al. 1999). In contrast to the above, Fitter et al. (1998) reported that root production and turnover in grassland communities along an altitudinal gradient responded to solar radiation, but not to soil temperature. In view of the importance of fine root production and turnover to nutrient acquisition and carbon cycling in natural ecosystems, it is imperative that we gain a better understanding of how soil temperature influences this important process.

Soil temperature also influences the vertical distribution of roots in the soil due to direct effects on growth, and indirectly by effects on water and nutrient availability. This is particularly important in extreme environments such as hot deserts in which surface soil conditions may be lethal to roots, or under

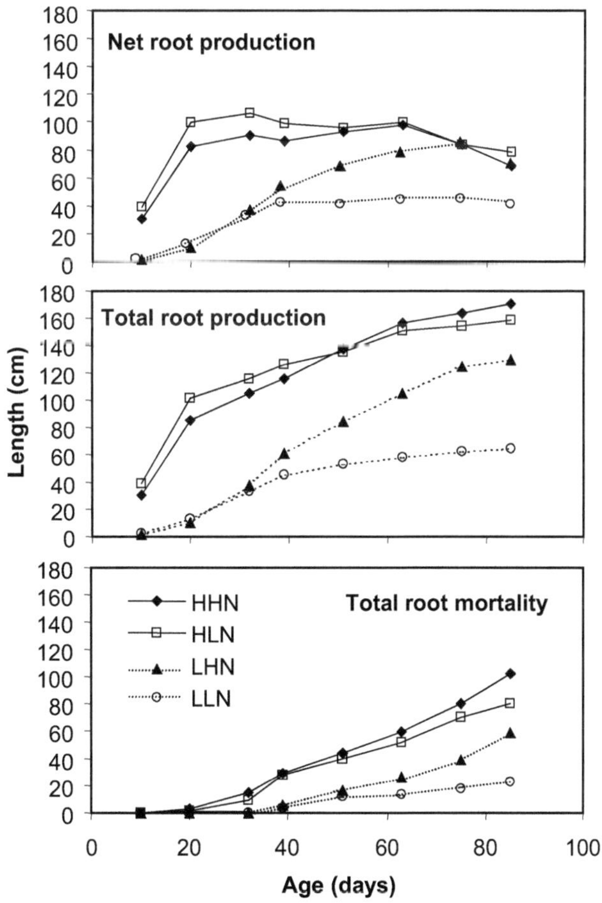

Fig. 10.4. Cumulative root length production and mortality of four clones of aspen grown at two levels of soil N availability and temperature for 98 days at the University of Michigan Biological Station, Pellston, Michigan. *H* and *L* refer to high and low soil temperature, respectively; *HN* and *LN* refer to high and low soil N availability, respectively. (Redrawn from Fig. 4 in King et al. 1999 , with kind permission from Kluwer Academic Publishers)

frigid soil thermal regimes (tundra and high-latitude boreal forest) in which permanently frozen soil limits the depth to which roots can grow. Nobel (1989) reported that temperatures in the top 2 cm of bare soil could exceed 65 °C in the northwestern Sonoran Desert, these temperatures being lethal to the roots of *Agave deserti* and *Ferocactus acanthodes*. However, mean rooting depth of adults was 10 cm, and even less for seedlings of these desert succulents. It was found that shading by *Hilaria rigida* reduced surface soil temperatures by 10 °C, providing safe sites for seedling establishment. Furthermore, in dry years soil water potential *decreased* with depth, indicating that the shallow (but not too shallow) rooting habit is more advantageous in dry years. Nobel and Linton (1997) also reported a mean rooting depth of 10 cm and very few surface roots of *Encelia farinosa* and *Pleuraphis rigidia* in the same desert system. The root response of these and other desert species to soil temperature and water availability controls their spatial distribution and frequency on north- and south-facing slopes. Soils with a frigid thermal regime and permafrost occur under arctic tundra and boreal forest, with surface soils thawing to greater depths with decreasing latitude. Williams and Rastetter (1999) reported that depth of thaw during the growing season in arctic tundra near Prudhoe Bay, Alaska, averaged 0.42 m. Depth of thaw can be as little as 10–20 cm in high-latitude tundra to over 1 m in boreal forest, providing a limited volume of soil for the expansion of root systems. Finally, in mesic systems it is known that roots proliferate in zones of the soil with high resource availability (Pregitzer et al. 1993). Thus, when water is not limiting, roots proliferate near the surface of the soil where higher temperatures do not limit growth and accelerate rates of nutrient mineralization.

At the scale of the individual root, soil temperature can influence morphology in various ways that affect nutrient uptake. A widely reported effect of soil temperature is decreased root diameter at higher temperatures (MacDuff et al. 1986; Rendig and Taylor 1989; Gliński and Lipiec 1990; Bowen 1991), effectively increasing specific root length (SRL). Specific root length is one of most important parameters affecting a plant's capacity to acquire nutrients (Eissenstat 1992; Jungk 1996), and the increase in SRL allows the plant to invest less carbon to meet nutritional demands. Many studies have also reported increased initiation of lateral roots with increasing soil temperature (Miyasaka and Grunes 1990; Sword and Brissette 1993; McMichael and Burke 1996), leading to more highly ramified root systems with a greater abundance of short lateral roots that can exploit a given volume of soil more thoroughly. Decreased root diameter at higher soil temperatures has been attributed to the greater number of small-diameter short roots, rather than to a direct effect on root diameter (Kaspar and Bland 1992), although direct effects on root diameter have indeed been reported (Chapin 1974a; MacDuff et al. 1986). The nutrient content and respiration of tree fine roots increase dramatically below diameters of about 0.5 mm (Pregitzer et al. 1997, 1998). This indicates that the smallest roots (e.g., short lateral roots) are physiologically and functionally distinct from larger roots. Thus, how soil temperature influences the distribution of length among the finest, distal branch orders of roots is crucial to nutrient uptake and nutrient cycling in terrestrial ecosystems, and deserves further study.

Lynch (Chap. 7, this Vol.) has highlighted the role of root architecture in root nutrient acquisition. There is published evidence (albeit scanty) that soil temperature can affect root architecture. For example, it has been reported that the angle at which lateral roots grow from the central axis changes as a function of soil temperature in several agronomic crops, giving rise to root systems of distinct architecture (Rendig and Taylor 1989). Why this occurs, how widespread it is among different taxa (especially forest trees), and the consequences for nutrient uptake are at present poorly understood. There is conflicting information on the effects of soil temperature on the development of root hairs in agronomic plants. Cooper (1973) argued that the number of root hairs generally decreases at higher soil temperature, but several authors have reported increases, decreases, or no effects (MacDuff et al. 1986; Kaspar and Bland 1992). Root hairs provide intimate contact with the soil, reducing resistance to water uptake, and are therefore important to nutrient uptake. Finally, based on very scanty evidence, both Gliński and Lipiec (1990) and Bowen (1991) concluded that the length of "unsuberized" root behind the growing apex decreases at low soil temperature, implying reduced capacity for nutrient uptake. However, the evidence is still too limited to draw firm conclusions on the effects and implications for nutrient uptake of soil temperature on root branching angle, number of root hairs, and degree of suberization, and all should receive further study.

10.4.2 Root Physiology

Changes in soil temperature directly affect root physiology in several ways that, in combination with soil-mediated controls on nutrient availability (diffusion, mass flow, mineralization), determine nutrient uptake. The influx of nutrients into living cells of roots occurs by transport across the plasmalemma through high- and low-affinity transporter proteins that operate at low and high concentrations, respectively, and that are thought to be more or less specific to each ionic species (Marschner 1995; Nissen 1996; BassiriRad 2000). Nutrient uptake rate increases with increasing external concentration until a saturation level is reached, above which uptake is independent of concentration (see Chap. 6, this Vol.). The kinetics of nutrient uptake are suggestive of the operation of enzyme–substrate systems, and for most essential ions can be described reasonably well by the Michaelis-Menten model. Experimental evidence suggests that nutrient uptake capacity is directly related to temperature (Chapin 1974a, b; Gosselin and Trudel 1986; MacDuff et al. 1994; Weih and Karlsson 1999; BassiriRad 2000; Tinker and Nye 2000; Dong et al.

2001). This short-term physiological response has been shown to exhibit Q_{10} values ≥2 between 10 and 30 °C, and lasts for several hours (Deane-Drummond and Glass 1983; Glass 1989). Given the high degree of diurnal variation in soil temperature (Fig. 10.2A), this may confer upon plants the ability to rapidly exploit nutrients made available as soils warm throughout the day. On longer timescales (days), nutrient uptake rates at different temperatures converge, so that uptake appears progressively less sensitive to temperature (Glass and Siddiqu 1985; Toselli et al. 1999), but the reasons for this apparent "acclimation" are poorly understood. In a recent review, BassiriRad (2000) suggests caution when generalizing nutrient uptake responses across a wide range of species and soil temperatures. Earlier work demonstrated that PO_4^{3-} uptake by *Eriophorum vaginatum* increased as soil temperature rose from 5 to 15 °C, but decreased with further increases in temperature (BassiriRad et al. 1996). In addition, other factors such as plant demand for nutrients (growth), root system surface area, and previous N nutrition are known to modify the response of nutrient uptake to changes in soil temperature (MacDuff and Wild 1989; Bowen 1991; BassiriRad et al. 1993; MacDuff et al. 1994).

The mechanisms for increased nutrient uptake with rising soil temperature are not well understood. Root respiration is known to increase with rising soil temperature (Atkin et al. 2000), in part due to higher availability of carbohydrates from enhanced photosynthesis, providing more energy for active transport. Decreased root hydraulic conductance at low root zone temperature was attributed, in part, to decreased capacity to replenish respiratory substrates in *Populus tremuloides* (Wan et al. 2001). However, the correlation between the rise in nutrient uptake and root respiration breaks down at higher temperatures, indicating that other energy-demanding processes are also changing (BassiriRad 2000). Higher rates of root respiration result in higher concentrations of CO_2 in soil solution, of which the dissociation products, H^+ and HCO_3^-, promote ion exchange reactions at the surface of clay and humic particles, freeing nutrient ions for uptake (Larcher 1995). Formation of carbonic acid as a result of increased respiration (auto- and heterotrophic) can decrease rhizosphere and soil pH, which is widely known to affect the availability and uptake of essential ions, especially micronutrients (Tisdale and Nelson 1975; Brady 1990; Marschner 1995). The effect of enhanced root and microbial respiration on soil solution chemistry and plant nutrition is determined to a large extent by the buffering capacity of the soil.

Properties of cell membranes also change with soil temperature, affecting nutrient uptake. There is a decrease in water uptake at low soil temperatures, caused primarily by increased resistance to water movement within the root due to higher viscosity of water and reduced permeability of cell membranes (Johnson and Thornley 1985; Kramer and Boyer 1995; Wan et al. 2001). This decreases mass flow of nutrients to the root surface in soil water. After extended periods at low root zone temperature (3–5 days), the content of unsaturated fatty acids in cell membranes of new roots has been observed to increase, and has been implicated in the acclimation response of nutrient uptake (Markhart et al. 1980; Osmond et al. 1982). Increases in unsaturated fatty acids, decreases in sterols, and altered ratios of linolenic:linoleic acid may be adaptive by offsetting the tendency toward greater membrane viscosity in the cold (White et al. 1990). Further, change in membrane fatty-acid composition at low temperature in roots of a variety of species has been related to abrupt increases in the activation energies (E_a) of membrane-bound ATPases and K^+ influx, suggesting that acclamatory responses occur at the membrane level and are related to lipid composition and transporter activity (Clarkson et al. 1988).

It must noted that root growth and physiology may exhibit distinct, possibly compensatory responses to changes in soil temperature (Fig. 10.1), resulting in variation in nutrient uptake by plant species and for ions of differing mobility. Chapin et al. (1986) found that nutrient uptake [(PO_4^{3-}, NH_4^+, NO_3^-, Rb^+ (analog of K^+), Cl^-] was least sensitive to temperature in slow-growing species such as black spruce (*Picea mariana*), compared to the more rapidly growing poplar (*Populus balsamifera*) and aspen (*Populus tremuloides*). They attributed this in part to the fast-growing species occupying warmer sites, but also more fertile sites where high uptake capacity would confer a selective advantage. It has been suggested that uptake capacity and root growth could act as compensatory responses to changes in soil temperature (Clarkson 1985; Clarkson et al. 1988). Thus, if root growth diminished relative to shoot growth at higher soil temperature, then the specific rate of nutrient uptake would increase, whereas an increase in root:shoot ratio would cause specific uptake rates to decrease. In a 9-week growth chamber experiment with *Pinus sylvestris* seedlings, Domisch et al. (2001) observed increased total biomass as soil temperature increased from 5 to 17 °C, but little change in allocation between roots and shoots. The above considerations underscore the complexity of plant nutrient uptake as influenced by soil temperature, and illustrate the difficulty in making generalizations about the multiple, interacting processes (Fig. 10.1).

10.5 Soil Temperature Effects on Soil Biology

The biological activity of soil organisms affects nutrient uptake by plants through the decomposition of organic matter and mineralization of nutrients, and formation of symbiotic microbial associations. All of these processes are sensitive to soil temperature, and affect uptake largely by influencing the amount of available nutrients in the soil or the extent of the absorbing surface area. A more detailed treatment of this aspect is given in Chapter 2 (this Vol.), but here we provide a brief overview.

10.5.1 Decomposition and Nutrient Mineralization

At the global scale, it has long been recognized that temperature is a major controller of decomposition, resulting in the latitudinal gradient of soil organic matter (Olson 1963). Recent interest in global climate change has stimulated ecosystem-level research on soil warming, usually using buried heating cables, to assess the effects of the predicted warmer climate on soil carbon storage and cycling (Van Cleve et al. 1990; Peterjohn et al. 1994; McHale et al. 1998; Rustad and Fernandez 1998). Universally, these studies have found that moderate warming of soil results in more rapid mass loss of decomposing litter, greater efflux of CO_2 from soil, and greater levels of nutrient availability. For example, after 2 years of decomposition in a northern hardwood forest, mass remaining of American beech leaf litter was approximately 60, 50, and 42% for ambient, ambient+5 °C, and ambient+7.5 °C soil temperature treatments, respectively (McHale et al. 1998). Interestingly, the decomposition of sugar maple litter was unaffected by the soil temperature treatments. Similarly, after 30 months of decay in soil heated approximately 5 °C above ambient in a spruce-fir forest, mass loss of red spruce foliar litter increased 19% but that of red maple was unaffected (Rustad and Fernandez 1998). In a mixed deciduous forest, Peterjohn et al. (1994) observed that heating soil 5 °C doubled daily rates of net N mineralization in both mineral soil and forest floor, which persisted for the length of the growing season. Heating soil 5 °C increased the estimated annual soil C efflux from 712 to 1,250 g m^{-2}, and a strong exponential relationship between soil temperature and CO_2 efflux was observed, in good agreement with the Arrhenius equation. Van Cleve et al. (1990) attributed higher foliar N content and greater rates of photosynthesis (and production) to greater nutrient availability from higher rates of decomposition in soils heated 8–10 °C above ambient in a black spruce forest. Ruark (1993) found that decomposition of lateral and tap roots of loblolly pine seedlings was more rapid at 25 than at 15 °C, suggesting that belowground wood decomposition will also increase in response to soil warming.

The temperature dependence of net N mineralization has been reported for a range of ecosystems, including longleaf pine–wiregrass (Wilson et al. 1999), alpine tundra (Fisk and Schmidt 1995), oak woodland–annual grassland (Leirós et al. 1999), arctic dwarf-shrub heath (Schmidt et al. 1999), and a range of communities along an altitudinal gradient (Powers 1990). It is important to note, however, that other factors such as substrate chemistry, soil moisture and site preparation may have greater impacts than soil temperature on mineralization rates (Paré and Van Cleve 1993; Van Cleve et al. 1993; Schmidt et al. 1999). In addition, rates of microbial immobilization, ammonium oxidation, and denitrification are sensitive to temperature, which may limit the increase in net N mineralization (Stark and Firestone 1996; Pelletier et al. 1999; Schmidt et al. 1999). Currently, it is unknown whether increased C and N mineralization in response to soil warming will sustain long-term increases in productivity, or represents a short-term response of increased metabolism of labile substrates (Bonan and Van Cleve 1991; Peterjohn et al. 1994).

Root-free laboratory incubations of forest soil have been used to characterize the temperature response of microbial mineralization of C and nutrients (Ellert and Bettany 1992; Kirschbaum 1995; MacDonald et al. 1995; Bowden et al. 1998; Ross et al. 1999). These studies indicate that microbial respiration and the mineralization of N and sulfur (S) increase over the range of temperatures most commonly experienced in natural soils, and this response has traditionally been described as exponential (i.e., the Arrhenius equation). However, microbial sensitivity to temperature diminishes at higher temperatures, indicating the exponential model may not always be appropriate. In a survey of the incubation literature, Kirschbaum (1995) determined the Q_{10} for decomposition to be ~8.0 at 0 °C, 4.5 at 10 °C, 2.5 at 20 °C, and ~1.0 at 35 °C. The temperature sensitivity of mineralization of N and S has also been shown to decrease with increasing temperature, and several authors have suggested alternate models (e.g., quadratic) to describe the temperature response of microbial metabolism (Ellert and Bettany 1992; MacDonald et al. 1995). The decrease in Q_{10} at higher temperatures may result from changes in substrate pool size or composition, microbial community composition, or transport processes such as diffusion (Zak et al. 1999). None-the-less, these findings suggest that soil temperature effects on nutrient availability are greatest when soil temperature is low and changes on short timescales (hours to days), such as in temperate systems in spring and fall, or high-latitude soils during the growing season. Increased nutrient availability on short timescales as soils warm may provide the selective pressure for the ability to rapidly increase rates of nutrient uptake.

Altered kinetics of decomposition and nutrient mineralization with changing soil temperature result, in part, from functional shifts in soil microbial communities. Microbial biomass varies seasonally and this is related, at least in part, to seasonal variation in soil temperature (e.g., Fisk and Schmidt 1995). In a literature review, Wardle (1998) concluded that temporal variation in microbial biomass C in forest, grassland, and arable ecosystems was best described by a model (R^2=0.6) incorporating soil pH, soil C, and latitude. High-latitude systems exhibited greater seasonal variation in microbial biomass (greater turnover) due to higher inter-seasonal variations in temperature. Lipson et al. (2000) demonstrated that fall/winter increases in microbial biomass in dry alpine meadows were followed by spring declines related to a sustained increase in soil temperature, and a corresponding decrease in soluble organic C. Microbial biomass was similar in grassland soil incubated for 240 days at 15 and 25 °C, but declined precipitously after 50 days when incubated at 35 °C (Joergensen et al. 1990). Microbial biomass remained low in the highest-temperature treatment, and biomass turnover times were 4, 62, and

139 days for the 35, 25, and 15 °C temperature treatments, respectively. The rapid microbial turnover at high temperature resulted in much higher cumulative CO_2 evolution and N mineralization. These results are consistent with those of Zogg et al. (1997), who also found the lowest microbial biomass and highest CO_2 evolution in soil from a sugar maple forest incubated at 25 °C, compared to incubations at 5 and 15 °C. Using molecular techniques to examine changes in membrane lipid profiles (PLFA and LPS-OHFA), Zogg et al. (1997) concluded that microbial community composition had shifted at higher soil temperature, possibly conferring the ability to metabolize C sources unavailable to the lower-temperature communities. Carreiro and Koske (1992) similarly reported shifts in microfungal community composition in mixed deciduous leaf litter incubated in microcosms at 0, 10, and 20 °C. Litter decay rates increased with increasing soil temperature, and Zygomycete species richness declined while that of Deuteromycetes increased, indicating that microbial community composition was strongly influenced by thermal environment.

10.5.2 Soil Fauna

Soil fauna (micro-, meso-, and macro-) play an important role in maintaining nutrient availability to plants (Paul and Clark 1996). Direct effects include belowground herbivory that can reduce fine root absorbing surface area, or facilitation of root growth through earthworm channels. Indirect effects include improvement of soil physical properties, population-level interactions with microflora (e.g., predation), and the comminution and digestion of litter. Experimental evidence indicates that nutrient mineralization increases significantly in soil systems containing soil fauna, compared to those devoid of fauna (Anderson et al. 1981; Cárcamo et al. 2001). Evidence for direct effects of soil temperature on the soil fauna is, however, limited. Shouten et al. (1998) reported that the abundance of bacterial-feeding nematodes declined in surface litter of a *Pinus sylvestris* forest as soils warmed in spring and summer, and was related to the drying of the litter. Badejo (1990) quantified the seasonal abundance of 14 species of soil mites in secondary forest and a cassava plantation in Nigeria. Most species exhibited unimodal seasonal abundance, peaking at times of warming soil and high soil moisture content (i.e., right after the rainy season), and then declining as soil temperature continued to rise, resulting in the drying of the soil. Direct negative effects of high soil temperature on mite reproduction (reduced sperm vitality and egg laying) were also suggested as possible causes for declining population size. Similar responses to soil temperature and moisture were reported for termites and microarthropod communities in the northern Chihuahuan Desert (Mackay et al. 1986). In contrast, higher soil temperatures associated with areas of deeper snowpack resulted in greater activity of soil microarthropods in tundra and subalpine forest in the Front Range of Colorado (Addington and Seastedt 1999). Abundance of several nematode species has been shown to increase or be unimodal as soil temperature rises (Sohlenius 1968, 1973, 1985; Anderson and Coleman 1982; Bird et al. 1993), suggesting population dynamics of some soil fauna exhibit cardinal temperatures that are probably strongly dependent on soil water content.

10.5.3 Symbiotic Relationships

Of the 320 Tg (10^{12} g) of N input to terrestrial ecosystems annually, fully 175 Tg comes from the biological fixation of dinitrogen (Paul and Clark 1996). Symbiotic relationships between plant hosts and either nitrogen-fixing bacteria (see Chap. 5, this Vol.) or mycorrhizal fungi (see Chap. 9, this Vol.) are known to be sensitive to soil temperature. Indeed, the distribution of plant species in a nitrogen-limited environment may be determined by the temperature relations of the symbiotic N-fixing system (Bowen 1991). Soil temperature optima for intact (e.g., non-excised nodules) symbiotic N fixation in tropical systems appear to be between 25 and 30 °C (Reddell et al. 1985; Daniel 1993), and N fixation may be limited in systems experiencing higher soil temperatures. In a series of flowing solution culture experiments (MacDuff et al. 1989; MacDuff and Dhanoa 1990; Svenning and MacDuff 1996), N fixation in white clover increased with rising temperature from 3 to ~15 °C, but changed little with further increases in solution temperature, up to 25 °C. The increased N fixation at lower temperatures was due to increases in both total nodule mass and specific rates of N fixation (mmol N g^{-1} nodule day^{-1}). Inorganic N nutrition inhibited symbiotic N fixation, and the temperature sensitivity of NO_3^- uptake was much greater than that of N fixation. This caused the proportion of total N nutrition attributable to N fixation to decline from 77 % at 5 °C to 11 % at 25 °C. The finding that symbiotic N fixation in white clover of a Norwegian ecotype had a much lower temperature optimum than that of the tropical *Casuarina* sp. (Reddell et al. 1985) suggests that, as with other aspects of nutrient acquisition, the temperature response is strongly influenced by the habitat of origin.

Growth of mycelial fans of the mycorrhizal fungus *Thelophora terrestris* on roots of Sitka spruce was highly correlated to the seasonal progression of soil temperature, slowing down to a minimum in winter, and reaching a maximum rate of 3 m day^{-1} in mid-summer (July; Coutts and Nicoll 1990). Another species, *Laccaria proxima*, apparently was not perennial and disappeared in winter. Dodd and Jeffries (1989) observed that accumulated thermal time, along with soil moisture, explained much of the variation in vesicular-arbuscular mycorrhizal infection of autumn-sown cereals, and was closely related to overall root growth. Delucia et al. (1997) noted declines in mycorrhizal infection and oxalate levels in soils of ponderosa pine grown in pots at day/night air temperatures of 30/15 °C, compared to higher values recorded at 25/10 °C. Care should be taken when interpreting this result, however, due to the square-wave nature of the temperature treatments that were probably much higher than the average soil temperatures experienced by ponderosa pine at its eastern Sierra Nevada provenance site. Finally, soil temperature effects on spore germination and hyphal growth vary among species, and can thus influence the formation of plant–host relationships, and competitive outcomes between different fungi (Wilson and Tommerup 1992).

10.6 Summary and Areas of Future Research

Soil temperature influences plant nutrient uptake through a multitude of chemical, physical, and biological processes that interact over a wide range of spatial and temporal scales. A fundamental characteristic of most ecosystems is that of constantly changing soil temperature in diurnal and seasonal cycles, which strongly influences the dynamics of nutrient availability and uptake by plants. When water is not limiting, nutrient availability generally increases with rising temperature due to accelerated rates of chemical reactions that liberate nutrients from primary minerals, exchange sites of clay particles, and organic matter. These processes may increase nutrient availability in the short term, but over longer periods of time soil nutrient stocks can be depleted. *A key priority for future research is to determine whether there will be a sustained increase in nutrient availability due to global warming, and how this will affect productivity.*

The structure and function of plant root systems is in part an adaptation to variation in nutrient availability caused by the soil thermal environment. Root biomass and length generally increase with rising temperature, allowing the plant to exploit a greater volume of soil. Root morphology, including specific root length, branching angle, and root hairs, may be directly affected by soil temperature, but experimental evidence is limited. Likewise, the effects of soil temperature on the allocation of C to the root fraction most important for nutrient uptake, and its annual turnover are very poorly understood. *A key priority for future research is to determine the extent to which rising soil temperature will affect the morphology, carbon allocation, and turnover of the ephemeral root fraction.*

In the short term, specific rates of nutrient uptake increase with rising soil temperature, which may be an evolutionary response by plants to acquire nutrients that are suddenly made available. Over a period of days, however, it appears that roots acclimate to changes in soil temperature through changes in membrane fatty-acid composition and possibly transporter activity. *A key area for future research is to characterize the response of specific nutrient uptake to rising soil temperature in a wide range of taxa grown in the field to determine the extent and mechanisms of acclimation.*

Finally, soil temperature has a strong influence on the community composition and activity of the soil fauna and microflora, affecting rates of organic matter decomposition and nutrient mineralization. Abundance and activity of soil fauna generally increase with rising soil temperature, especially in cold soils, but can be limited by drying of the soil at high temperature. *A key area of future research is to determine how rising soil temperature affects community composition and substrate utilization of soil fauna and microflora communities, and the resulting effects on nutrient availability.* Symbiotic nitrogen fixation appears to increase with rising soil temperature, especially at low temperatures, due to increasing nodule mass and activity. However, laboratory studies indicate that rates of inorganic N uptake may increase to a greater extent, dominating total N nutrition. *A key area of future research is to determine to what extent symbiotic N fixation is stimulated by rising soil temperature in the field, for systems in which this is an important part of overall N nutrition.* Growth rates and community composition of mycorrhizae have also been shown to be sensitive to changes in soil temperature, but how these changes affect the acquisition of nutrients and C cycling in forest ecosystems is not well understood. *A key area of future research is to determine the extent to which mycorrhizal nutrient uptake and turnover are influenced by rising soil temperature in natural ecosystems.*

In summary, soil temperature is one of the key factors regulating nutrient availability and uptake in terrestrial ecosystems. Changes in soil temperature are inextricably linked to changes in soil moisture at the ecosystem level. Changes in plant and microbial function will be very tightly coupled to changes in soil temperature and moisture in the future, and these ecological interactions will play critical roles in nutrient availability and net ecosystem productivity.

Acknowledgments. The National Science Foundation, Department of Energy Office of Biological and Environmental Research, and NRI Competitive Grants Program/USDA supported this synthesis. We are grateful for their support.

References

Addington RN, Seastedt (1999) Activity of soil microarthropods beneath snowpack in alpine tundra and subalpine forest. Pedobiologia 43:47–53

Anderson RV, Coleman DC (1982) Nematode temperature responses: a niche dimension in populations of bacterial-feeding nematodes. J Nematol 14:69–76

Anderson RV, Coleman DC, Cole CV (1981) Effects of saprotrophic grazing on net mineralization. In: Clark FE, Rosswall T (eds) Terrestrial nitrogen cycles. Ecol Bull (Stockholm) 33:201–215

Archibold OW, Ripley EA, Bretell DL (1996) Comparison of the microclimate of a small aspen grove and adjacent prairie in Saskatchewan. Am Midl Nat 136:248–261

Atkin RK, Barton GE, Robinson DK (1973) Effect of root-growing temperature on growth substances in xylem exudates of *Zea mays*. J Exp Bot 24:475–487

Atkin OK, Edwards EJ, Loveys BR (2000) Response of root respiration to changes in temperature and its relevance to global warming. New Phytol 147:141–154

Badejo MA (1990) Seasonal abundance of soil mites (*Acarina*) in two contrasting environments. Biotropica 22:382–390

Barber SA, Mackay AD, Kuchenbuch RO, Barraclough PB (1989) Effects of soil temperature and water on maize root growth. Dev Plant Soil Sci 36:231–233

BassiriRad H (2000) Kinetics of nutrient uptake by roots: responses to global change. New Phytol 147:155–169

BassiriRad H, Caldwell MM, Bilbrough C (1993) Effects of soil temperature and nitrogen status on kinetics of $^{15}NO_3^-$ uptake by roots of field-grown *Agropyron desertorum* (Fisch. ex Link) Schult. New Phytol 123:485–489

BassiriRad H, Tissue DT, Reynolds JF, Chapin FS III (1996) Response of *Eriophorum vaginatum* to CO_2 enrichment at different soil temperatures: effects on growth, root respiration and PO_4^{3-} uptake kinetics. New Phytol 133:423–430

Bennie ATP (1991) Growth and mechanical impedance. In: Waisel Y, Eshel A, Kafkafi U (eds) Plant roots: the hidden half. Marcel Dekker, New York, pp 393–414

Bertrand A, Robitaille G, Nadeau P, Boutin R (1994) Effects of soil freezing and drought stress on abscisic acid content of sugar maple sap and leaves. Tree Physiol 14:413–425

Billings WD, Luken OJ, Mortensen DA, Peterson KM (1982) Arctic tundra: a source or sink for atmospheric carbon dioxide in a changing environment? Oecologia 53:7–11

Bird AF, De Ley P, Bird J (1993) Morphology, oviposition and embryogenesis in an Australian population of *Acrobeloides nanus*. J Nematol 25:607–615

Bliss LC (1988) Arctic tundra and polar desert biome. In: Barbour MG, Billings WD (eds) North American terrestrial vegetation. Cambridge University Press, Cambridge, pp 2–32

Bonan GB, Van Cleve K (1991) Soil temperature, nitrogen mineralization, and carbon source-sink relationships in boreal forests. Can J For Res 22:629–639

Boone RD, Nadelhoffer KJ, Canary JD, Kaye JP (1998) Roots exert a strong influence on the temperature sensitivity of soil respiration. Nature 396:570–572

Bowden RD, Newkirk KM, Rullo GM (1998) Carbon dioxide and methane fluxes by a forest soil under laboratory-controlled moisture and temperature conditions. Soil Biol Biochem 30:1591–1597

Bowen GD (1991) Soil temperature, root growth, and plant function. In: Waisel Y, Eshel A, Kafkafi U (eds) Plant roots: the hidden half. Marcel Dekker, New York, pp 309–330

Brady NC (1990) The nature and properties of soils, 10th edn. MacMillan, New York

Breshears DD, Nyhan JW, Heil CE, Wilcox BP (1998) Effects of woody plants on microclimate in a semiarid woodland: Soil temperature and evaporation in canopy and intercanopy patches. Int J Plant Sci 159:101–1017

Bristow KL, Campbell GS, Papendick RI, Elliot LF (1986) Simulation of heat and moisture transfer through a surface residue-soil system. Agric For Meteorol 36:193–214

Brown TL, LeMay HE Jr (1981) Chemistry: the central science. Prentice-Hall, Englewood Cliffs

Brown SE, Pregitzer KS, Reed DD, Burton AJ (1999) Predicting daily mean soil temperature from daily mean air temperature in four northern hardwood forest stands. For Sci 46:297–301

Burke IC (1989) Control of nitrogen mineralization in a sagebrush steppe landscape. Ecology 70:1115–1126

Cárcamo HA, Prescott CE, Chanway CP, Abe TA (2001) Do soil fauna increase rates of litter breakdown and nitrogen release in forests of British Columbia, Canada? Can J For Res 31:1195–1204

Carreiro MM, Koske RE (1992) Effect of temperature on decomposition and development of microfungal communities in leaf litter microcosms. Can J Bot 70:2177–2183

Chapin FS III (1974a) Morphological and physiological mechanisms of temperature compensation in phosphate absorption along a latitudinal gradient. Ecology 55:1180–1198

Chapin FS III (1974b) Phosphate absorption capacity and acclimation potential in plants along a latitudinal gradient. Science 183:521–523

Chapin FS III, Van Cleve K, Chapin MC (1979) Soil temperature and nutrient cycling in the tussock growth form of *Eriophorum vaginatum*. J Ecol 67:169–189

Chapin FS III, Van Cleve K, Tryon PR (1986) Relationship of ion absorption to growth rate in taiga trees. Oecologia 69:238–242

Clarkson DT (1985) Factors affecting mineral nutrient acquisition by plants. Annu Rev Plant Physiol 36:77–115

Clarkson DT (1996) Root structure and sites of ion uptake. In: Waisel Y, Eshel A, Kafkafi U (eds) Plant roots: the hidden half, 2nd edn. Marcel Dekker, New York, pp 483–510

Clarkson DT, Earnshaw MJ, White PJ, Cooper HD (1988) Temperature dependent factors influencing nutrient uptake: an analysis of responses at different levels of organization. In: Long SP, Woodward FI (eds) Plants and temperature. Symp 42, Society for Experimental Biology, Cambridge, pp 281–330

Colmer TD, Bloom AJ (1998) A comparison of NH_4^+ and NO_3^- net fluxes along roots of rice and maize. Plant Cell Environ 21:240–246

Cooper AJ (1973) Root temperature and plant growth – a review. Res Rev no 4, Commonwealth Bureau of Horticulture and Plantation Crops, Commonwealth Agricultural Bureau, Farnham Royal, England

Cortina J, Vallejo VR (1994) Effects of clearfelling on forest floor accumulation and litter decomposition in a *Radiata* pine plantation. For Ecol Manage 70:299–310

Coutts MP, Nicoll BC (1990) Growth and survival of shoots, roots, and mycorrhizal mycelium in clonal Sitka spruce during the first growing season after planting. Can J For Res 20:861–868

Cubasch U, Meehl GA, Boer GJ, Stouffer RJ, Dix M, Noda A, Senior CA, Raper S, Yap KS (2001) Projections of future climate. In: Houghton JT, Ding Y, Griggs DJ, Noguer M, van der Linden PJ, Dai X, Maskell K, Johnson CA (eds) Climate change 2001: the scientific basis. Cambridge University Press, Cambridge, pp 525–582

Daniel JN (1993) Biological fixation and transfer of nitrogen by trees in agroforestry systems. Range Manage Agrofor 14:185–194

Day TA, Heckathorn SA, Delucia EH (1991) Limitations of photosynthesis in *Pinus taeda* L. (loblolly pine) at low soil temperatures. Plant Physiol 96:1246–1254

Dean-Drummond CE, Glass ADM (1983) Compensatory changes in ion fluxes into barely (*Hordeum vulgare* L. cv. Betzes) seedlings in response to differential root/shoot growth temperatures. J Exp Bot 34:1711–1719

Delucia EH, Callaway RM, Thomas EM, Schlesinger WH (1997) Mechanisms of phosphorus acquisition for ponderosa pine seedlings under high CO_2 and temperature. Ann Bot 79:111–120

Dodd JC, Jeffries P (1989) Effect of over-winter environmental conditions on vesicular-arbuscular mycorrhizal infection of autumn-sown cereals. Soil Biol Biochem 21:453–455

Domisch T, Finér L, Lehto T (2001) Effects of soil temperature on biomass and carbohydrate allocation in Scots pine (*Pinus sylvestris*) seedlings at the beginning of the growing season. Tree Physiol 21:465–472

Dong S, Scagel CF, Cheng L, Fuchigami LH, Rygiewicz PT (2001) Soil temperature and plant growth stage influence nitrogen uptake and amino acid concentration of apple during early spring growth. Tree Physiol 21:541–547

Eissenstat DM (1992) Costs and benefits of constructing roots of small diameter. J Plant Nutr 15:763–782

Ellert BH, Bettany JR (1992) Temperature dependence of net nitrogen and sulfur mineralization. Soil Soc Am J 56:1133–1141

Farnsworth EJ, Nunez-Farfan J, Careaga SA, Bazzaz FA (1995) Phenology and growth of three temperate forest life forms in response to artificial soil warming. J Ecol 83:967–977

Federer CA, Flynn LD, Martin CW, Hornbeck JW, Pierce RS (1990) Thirty years of hydrometeorologic data at the Hubbard Brook Experimental Forest, New Hampshire. USDA Forest Service, Radnor, PA, Gen Tech Rep NE-141

Feng Y, Li X, Boersma L (1990) The Arrhenius equation as a model for explaining plant responses to temperature and water stresses. Ann Bot 66:237–244

Fisk MC, Schmidt SK (1995) Nitrogen mineralization and microbial biomass nitrogen dynamics in three alpine tundra communities. Soil Sci Soc Am J 59:1036–1043

Fitter A (1996) Characteristics and functions of root systems. In: Waisel Y, Eshel A, Kafkafi U (eds) Plant roots: the hidden half, 2nd edn. Marcel Dekker, New York, pp 1–20

Fitter AH, Graves JD, Self GK, Brown TK, Bogie DS, Taylor K (1998) Root production, turnover and respiration under two grassland types along an altitudinal gradient: influence of temperature and solar radiation. Oecologia 114:20–30

Forbes PJ, Black KE, Hooker JE (1997) Temperature-induced alteration to root longevity in *Lolium perenne*. Plant Soil 190:87–90

Geiger R (1965) The climate near the ground. Harvard University Press, Cambridge

Glass ADM (1989) Plant nutrition: an introduction to current concepts. Jones and Bartlett, Boston

Glass ADM, Siddiqui MY (1985) Nitrate inhibition of chloride influx in barley: implications for a proposed chloride homeostat. J Exp Bot 36:556–566

Gliński J, Lipiec J (1990) Soil physical conditions and plant roots. CRC Press, Boca Raton

Gosselin A, Trudel MJ (1986) Root-zone temperature effects on pepper. J Am Soc Hortic Sci 111:220–224

Hall FG, Sellers PJ, Williams DI (1996) Initial results from the Boreal Ecosystem-Atmosphere Experiment, BOREAS. Silva Fenn 30:109–212

Hanks RJ, Ashcroft GL (1980) Applied soil physics. Springer, Berlin Heidelberg New York

Hendrick RL, Pregitzer KS (1993) Patterns of fine root mortality in two sugar maple forests. Nature 361:59–61

Hillel D (1998) Environmental soil physics. Academic Press, San Diego

Houghton RA, Skole DL (1990) Carbon. In: Turner BL, Clark WC, Kates RW, Richards JF, Mathews JT, Meyer WB (eds) The Earth as transformed by human action. Cambridge University Press, Cambridge, pp 393–408

Isard SA, Schaetzl RJ (1995) Estimating soil temperatures and frost in the lake effect snowbelt region, Michigan, USA. Cold Reg Sci Tech 23:317–332

Janssens IA, Têtè Barigah S, Ceulemans R (1998) Soil CO_2 efflux rates in different tropical vegetation types in French Guiana. Ann Sci For 55:671–680

Joergensen RG, Brookes PC, Jenkinson DS (1990) Survival of the soil microbial biomass at elevated temperatures. Soil Biol Biochem 22:1129–1136

Johnson IR, Thornley JHM (1985) Temperature dependence of plant and crop processes. Ann Bot 55:1–24

Joslin JD, Wolfe MH (1993) Temperature increase accelerates nitrate release from high-elevation red spruce soils. Can J For Res 23:756–759

Jungk AO (1996) Dynamics of nutrient movement at the soil-root interface. In: Waisel Y, Eshel A, Kafkafi U (eds) Plant roots: the hidden half, 2nd edn. Marcel Dekker, New York, pp 529–556
Jury WA, Gardner WR, Gardner WH (1991) Soil physics, 5th edn. Wiley, New York
Kaspar TC, Bland WL (1992) Soil temperature and root growth. Soil Sci 154:290–299
Kelly JM (1993) Temperature affects solution-phase nutrient concentrations and subsequent calculations of supply parameters. Soil Sci Soc Am J 57:527–531
King JS, Thomas RB, Strain BR (1996) Growth and carbon accumulation in root systems of *Pinus taeda* and *Pinus ponderosa* seedlings as affected by varying CO_2, temperature and nitrogen. Tree Physiol 16:635–642
King JS, Pregitzer KS, Zak DR (1999) Clonal variation in above- and belowground growth responses of *Populus tremuloides* Michaux: influence of soil warming and nutrient availability. Plant Soil 217:119–130
Kirschbaum MU (1995) The temperature dependence of soil organic matter decomposition, and the effect of global warming on soil organic C storage. Soil Biol Biochem 27:753–760
Kohnke H (1968) Soil physics. McGraw-Hill, New York
Kramer PJ, Boyer JS (1995) Water relations of plants and soils. Academic Press, San Diego
Kuhns MR, Garret HE, Teskey RO, Hinckley TM (1985) Root growth of black walnut trees related to soil temperature, soil water potential, and leaf water potential. For Sci 31:617–629
Landhäusser SM, Wein RW, Lange P (1996) Gas exchange and growth of three arctic treeline tree species under different soil temperature and drought preconditioning regimes. Can J Bot 74:686–693
Larcher W (1995) Plant physiological ecology. Springer, Berlin Heidelberg New York
Leirós MC, Trasar-Cepeda C, Seoane S, Gil-Sotres F (1999) Dependence of mineralization of soil organic matter on temperature and moisture. Soil Biol Biochem 31:327–335
Lhomme J-P, Troulfeau D, Monteny B, Chehbouni A, Bauduin S (1997) Sensible heat flux and radiometric surface temperature over sparse Sahelian vegetation II. A model for the kB^{-1} parameter. J Hydrol 188/189:839–854
Lieth H (1975) Modeling the primary productivity of the world. In: Lieth H, Whittaker RH (eds) Primary productivity of the biosphere. Springer, Berlin Heidelberg New York, pp 237–263
Lipson DA, Näsholm T (2001) The unexpected versatility of plants: organic nitrogen use and availability in terrestrial ecosystems. Oecologia 128:305–316
Lipson DA, Schmidt SK, Monson RK (2000) Carbon availability and temperature control the post-snowmelt decline in alpine soil microbial biomass. Soil Biol Biochem 32:441–448
Londo AJ, Messina MG, Schoenholtz SH (1999) Forest harvesting effects on soil temperature, moisture, and respiration in a bottomland hardwood forest. Soil Sci Soc Am J 63:637–644
Lyr H, Hoffmann G (1967) Growth rates and growth periodicity of tree roots. Int Rev For Res 2:181–236
MacDonald NW, Zak DR, Pregitzer KS (1995) Temperature effects on kinetics of microbial respiration and net nitrogen and sulfur mineralization. Soil Soc Am J 59:233–240
MacDuff JH, Dhanoa MS (1990) N_2 fixation and nitrate uptake by white clover swards in response to root temperature in flowing solution culture. Ann Bot 65:325–335
MacDuff JH, Wild A (1989) Interactions between root temperature and nitrogen deficiency influence preferential uptake of NH_4^+ and NO_3^- by oilseed rape. J Exp Bot 40:195–206

MacDuff JH, Wild A, Hopper MJ, Dhanoa MS (1986) Effects of temperature on parameters of root growth relevant to nutrient uptake: measurements on oilseed rape and barley grown in flowing nutrient solution. Plant Soil 94:321–332
MacDuff JH, Gordon AJ, Ryle GJA, Powell CE (1989) White clover N_2 fixation in response to root temperature and nitrate. J Exp Bot 40:517–526
MacDuff JH, Jarvis SC, Cockburn JE (1994) Acclimation of NO_3^- fluxes to low root temperature by *Brassica napus* in relation to NO_3^- supply. J Exp Bot 45:1045–1056
Mackay WP, Silva S, Lightfoot DC, Pagani MI, Whitford WG (1986) Effect of increased soil moisture and reduced soil temperature on a desert soil arthropod community. Am Midl Nat 116:45–56
Markhart AH, Fiscus EL, Naylor AW, Kramer PJ (1980) Low temperature acclimation of root fatty acid composition, leaf water potential, gas exchange and growth of soybean seedlings. Plant Cell Environ 3:435–441
Marschner H (1995) Mineral nutrition of higher plants, 2nd edn. Academic Press, London
McHale PJ, Mitchell MJ, Bowles FP (1998) Soil warming in a northern hardwood forest: trace gas fluxes and leaf litter decomposition. Can J For Res 28:1365–1372
McMichael BL, Burke JJ (1996) Temperature effects on root growth. In: Waisel Y, Eshel A, Kafkafi U (eds) Plant roots: the hidden half, 2nd edn. Marcel Dekker, New York, pp 383–396
McMichael BL, Burke JJ (1998) Soil temperature and root growth. Hortic Sci 33:947–951
Miyasaka SC, Grunes DL (1990) Root temperature and calcium level effects on winter wheat forage. I. Shoot and root growth. Agron J 82:236–242
Näsholm T, Ekblad A, Nordin A, Giesler R, Högberg M, Högberg P (1998) Boreal forest plants take up organic nitrogen. Nature 392:914–916
Nichols DS (1998) Temperature of upland and peatland soils in a north central Minnesota forest. Can J Soil Sci 78:493–509
Nissen P (1996) Uptake mechanisms. In: Waisel Y, Eshel A, Kafkafi U (eds) Plant roots: the hidden half, 2nd edn. Marcel Dekker, New York, pp 511–527
Nobel PS (1989) Temperature, water availability, and nutrient levels at various soil depths-consequences for shallow-rooted desert succulents, including nurse plant effects. Am J Bot 76:1486–1492
Nobel PS, Linton MJ (1997) Frequencies, microclimate and root properties for three codominant perennials in the northwestern Sonoran desert on north- vs. south-facing slopes. Ann Bot 80:731–739
Odell RT, Dijkerman JC, van Vuure W, Melsted SW, Beavers AH, Sutton PM, Kurtz LT, Miedema R (1974) Characteristics, classification, and adaptation of soils in selected areas in Sierra Leone, West Africa. Agricultural Experiment Station, College of Agriculture, University of Illinois at Urbana-Champaign, Bull 748, and Njala University College, University of Sierra Leone, Bull 4
Ogigirigi MA (1977) Climate of the Guinea and Sudan savannas of West Africa. In: Savanna aforestation in Africa. Food and Agriculture Organization of the United Nations, Rome, pp 9–19
Olasantan FO (1999) Effect of time of mulching on soil temperature and moisture regime and emergence, growth and yield of white yam in western Nigeria. Soil Till Res 50:215–221
Olson JS (1963) Energy storage and the balance of producers and decomposers in ecological systems. Ecology 44:322–331
Osmond DL, Wilson RF, Raper CD Jr (1982) Fatty acid composition and nitrate uptake of soybean roots during acclimation to low temperature. Plant Physiol 70:1689–1693
Paláez DV, Bóo RM, Elia OR (1992) Emergence and seedling survival of caldén in the semiarid region of Argentina. J Range Manage 45:564–568

Paré D, Van Cleve K (1993) Soil nutrient availability and relationships with aboveground biomass production on postharvested upland white spruce sites in interior Alaska. Can J For Res 23:1223–1232
Paterson DB, Mason WL (1999) Cultivation of soils for forestry. Forestry Commission Bull 119, Purely Print, Blandford Forum, Dorset
Paul EA, Clark FE (1996) Soil microbiology and biochemistry, 2nd edn. Academic Press, San Diego
Pelletier F, Prévost D, Laliberté, van Bochove E (1999) Seasonal response of denitrifiers to temperature in a Quebec cropped soil. Can J Soil Sci 79:551–556
Peterjohn WT, Melillo JM, Steudler PA, Newkirk KM, Bowles FP, Aber JD (1994) Responses of trace gas fluxes and N availability to experimentally elevated soil temperatures. Ecol Appl 4:617–625
Powers RF (1990) Nitrogen mineralization along an altitudinal gradient: Interactions of soil temperature, moisture, and substrate quality. For Ecol Manage 30:19–29
Pregitzer KS, Hendrick RL, Fogel R (1993) The demography of fine roots in response to patches of water and nitrogen. New Phytol 125:575–580
Pregitzer KS, Kubiske ME, Yu CK, Hendrick RL (1997) Relationships among root branch order, carbon, and nitrogen in four temperate species. Oecologia 111:302–308
Pregitzer KS, Laskowski MJ, Burton AJ, Lessard VC, Zak DR (1998) Variation in sugar maple root respiration with root diameter and soil depth. Tree Physiol 18:665–670
Pregitzer KS, King JS, Burton AJ, Brown SE (2000) Responses of tree fine roots to temperature. New Phytol 147:105–115
Reddell P, Bowen GD, Robson AD (1985) The effects of soil temperature on nodulation and nitrogen fixation in *Casuarina cunninghamiana*. New Phytol 101:441–450
Rendig VV, Taylor HM (1989) Principles of soil-plant interrelationships. McGraw-Hill, New York
Richards PW (1998) The tropical rainforest, 2nd edn. Cambridge University Press, Cambridge
Rosenzweig C, Hillel D (2000) Soils and global climate change: challenges and opportunities. Soil Sci 165:47–56
Ross DJ, Kelliher FM, Tate KR (1999) Microbial processes in relation to carbon, nitrogen and temperature regimes in litter and a sandy mineral soil from a central Siberian *Pinus sylvestris* L. forest. Soil Biol Biochem 31:757–767
Roundy BA, Biedenbender SH (1996) Germination of warm-season grasses under constant and dynamic temperatures. J Range Manage 49:425–431
Ruark GA (1993) Modeling soil temperature effects on *in situ* decomposition rates of fine roots of loblolly pine. For Sci 39:118–129
Russell CA, Voroney RP (1998) Carbon dioxide efflux from the floor of a boreal aspen forest. I. Relationship to environmental variables and estimates of C respired. Can J Soil Sci 78:301–310
Rustad LE, Fernandez IJ (1998) Soil warming: consequences for foliar litter decay in a spruce-fir forest in Maine, USA. Soil Sci Am J 62:1072–1080
Sanchez PA (1976) Properties and management of soils in the tropics. Wiley, New York
Scanlon BR (1994) Water and heat fluxes in desert soils. 1. Field studies. Water Resour Res 30:709–719
Scharpenseel HW, Schomaker M, Ayoub A (eds) (1990) Soils on a warmer earth: Effects of expected climate change on soil processes, with emphasis on the tropics and subtropics. Elsevier, Amsterdam
Schlesinger WH (1997) Biogeochemistry: an analysis of global change, 2nd edn. Academic Press, San Diego
Schmidt IK, Jonasson S, Michelsen A (1999) Mineralization and microbial immobilization of N and P in arctic soils in relation to season, temperature and nutrient amendment. Appl Soil Ecol 11:147–160

Schwarz PA, Fahey TJ, Dawson TE (1997) Seasonal air and soil temperature effects on photosynthesis in red spruce (*Picea rubens*) saplings. Tree Physiol 17:187–194
Shouten AJ, van Esbroek MLP, Alkemade JRM (1998) Dynamics and stratification of functional groups of nematodes in the organic layer of a Scots pine forest in relation to temperature and moisture. Biol Fertil Soils 26:293–304
Smith WO, Byers HG (1938) The thermal conductivity of dry soil of certain of the great soil groups. Soil Sci Soc Am Proc 3:13–19
Sohlenius B (1968) Influence of microorganisms and temperature upon some rhabditid nematodes. Pedobiologia 8:137–145
Sohlenius B (1973) Growth and reproduction of a nematode *Acrobeloides* sp. cultivated on agar. Oikos 24:64–72
Sohlenius B (1985) Influence of climatic conditions on nematode co-existence: a laboratory experiment with coniferous forest soil. Oikos 44:430–438
Soil Survey Staff (1975) Soil taxonomy: a basic system of soil classification for making and interpreting soil surveys. USDA Soil Conservation Service, Washington, DC
Sparks DL (1995) Environmental soil chemistry. Academic Press, San Diego
Spittlehouse DL, Stathers RJ (1990) Seedling microclimate. British Columbia Ministry of Forests, Victoria, Land Management Rep no 65
Sposito G (1994) Chemical equilibria and kinetics in soils. Oxford University Press, New York
Stadler D, Wunderli H, Auckenthaler A, Flühler H (1996) Measurement of frost-induced snowmelt runoff in a forest soil. Hydrol Proc 10:1293–1304
Stark JM, Firestone MK (1996) Kinetic characteristics of ammonium-oxidizer communities in a California oak woodland-annual grassland. Soil Biol Biochem 28:1307–1317
Stoneman GL, Dell B, Turner NC (1995) Growth of *Eucalyptus marginata* (Jarrah) seedlings in Mediterranean-climate forest in south-west Australia in response to overstorey, site and fertiliser application. For Ecol Manage 79:173–184
Strahler AN, Strahler AH (1983) Modern physical geography, 2nd edn. Wiley, New York
Sumrall LB, Roundy BA, Cox JR, Winkel VK (1991) Influence of canopy removal by burning or clipping on emergence of *Eragrostis lehmanniana* seedlings. Int J Wildland Fire 1:35–40
Sutinen M-L, Ritari A, Holappa T, Kujala K (1998) Seasonal changes in soil temperature and in the frost hardiness of Scots pine (*Pinus sylvestris*) roots under subarctic conditions. Can J For Res 28:946–950
Svenning MM, MacDuff JH (1996) Low root temperature retardation of the mineral nitrogen induced decline in N_2 fixation by a northern ecotype of white clover. Ann Bot 77:615–621
Swift LW, Elliot KJ, Ottmar RD, Vihnanek RE (1993) Site preparation burning to improve southern Appalachian pine-hardwood stands: fire characteristics and soil erosion, moisture, and temperature. Can J For Res 23:2242–2254
Sword MA, Brissette JC (1993) Effect of root zone temperature and water availability on shortleaf pine lateral root morphology. In: Brissette JC (ed) Proc 7th Biennial Southern Silvicultural Research Conf, USDA Forest Service, Southern Forest Experiment Station, New Orleans, pp 377–382
Taiz L, Zeiger E (1991) Plant physiology. Benjamin/Cummings, New York
Tajchman SJ, Minton CM (1986) Soil temperature regime in a forested Appalachian watershed. Can J For Res 16:624–629
Taniguchi M, Williamson DR, Peck AJ (1998) Estimations of surface temperature and subsurface heat flux following forest removal in the south-west of Western Australia. Hydrol Proc 12:2205–2216

Teskey RO, Hinckley TM (1981) Influence of temperature and water potential on root growth of white oak. Physiol Plant 52:363–369
Thompson LJ, Naeem S (1996) The effects of soil warming on plant recruitment. Plant Soil 182:339–343
Tinker PB, Nye PH (2000) Solute movement in the rhizosphere. Oxford University Press, New York
Tisdale SL, Nelson WL (1975) Soil fertility and fertilizers, 3rd edn. MacMillan, New York
Toselli M, Flore JA, Marangoni B, Masia A (1999) Effects of root-zone temperature on nitrogen accumulation by non-bearing apple trees. J Hortic Sci Biotech 74:118–124
Tryon PR, Chapin FS III (1983) Temperature control over root growth and root biomass in taiga forest trees. Can J For Res 13:827–833
Van Cleve K, Barney R, Schlentner R (1981) Evidence of temperature control of production and nutrient cycling in two interior Alaska black spruce ecosystems. Can J For Res 11:258–273
Van Cleve K, Oechel WC, Hom JL (1990) Response of black spruce (*Picea mariana*) ecosystems to soil temperature modification in interior Alaska. Can J For Res 20:1530–1535
Van Cleve K, Yarie J, Erickson R, Dyrness CT (1993) Nitrogen mineralization and nitrification in successional ecosystems on the Tanana River floodplain, interior Alaska. Can J For Res 23:970–978
Viereck LA, Van Cleve K, Adams PC, Schlentner RE (1993) Climate of the Tanana river floodplain near Fairbanks, Alaska. Can J For Res 23:899–913
Wan X, Zwiazek JJ, Lieffers VJ, Landhausser M (2001) Hydraulic conductance in aspen (*Populus tremuloides*) seedlings exposed to low root temperatures. Tree Physiol 21:691–696
Wardle DA (1998) Controls of temporal variability of the soil microbial biomass: A global-scale synthesis. Soil Biol Biochem 30:1627–1637
Weast RC (1978) CRC handbook of chemistry and physics, 58th edn. CRC Press, West Palm Beach
Weber MG, McAlpine RS, Wotton BM, Donnelly JG, Hobbs MW (1995) Prescribed burning and disk trenching effects on early plantation performance in eastern Ontario, Canada. For Ecol Manage 78:159–171
Weih M, Karlsson PS (1999) The nitrogen economy of mountain birch seedlings: implications for winter survival. J Ecol 87:211–219
White PJ, Cooke DT, Earnshaw J, Clarkson DT, Burden RS (1990) Does plant growth temperature modulate the membrane composition and ATPase activities of tonoplast and plasma-membrane fractions from rye roots? Phytochemistry 29:3385–3393
Williams M, Rastetter EB (1999) Vegetation characteristics and primary productivity along an arctic transect: implications for scaling-up. J Ecol 87:885–898
Wilson JM, Tommerup IC (1992) Interactions between fungal symbionts: VA mycorrhizae. In: Allen MF (ed) Mycorrhizal functioning: an integrative plant-fungal process. Chapman & Hall, New York, pp 199–248
Wilson CA, Mitchell RJ, Hendricks JJ, Boring LR (1999) Patterns and controls of ecosystem function in longleaf pine–wiregrass savannas. II. Nitrogen dynamics. Can J For Res 29:752–760
Wong-Chong GM, Loehr RC (1978) Kinetics of microbial nitrite nitrogen oxidation. Water Res 12:605–609
Woodward A (1998) Relationships among environmental variables and distribution of tree species at high elevation in the Olympic mountains. Northwest Sci 72:10–22

Zak DR, Pregitzer KS, Curtis PS, Teeri JA, Fogel R, Randlett DL (1993) Elevated atmospheric CO_2 and feedback between carbon and nitrogen cycles. Plant Soil 151:105–117

Zak DR, Holmes WE, MacDonald NW, Pregitzer KS (1999) Soil temperature, matric potential, and the kinetics of microbial respiration and nitrogen mineralization. Soil Sci Am J 63:575–584

Zogg GP, Zak DR, Ringelberg DB, MacDonald NW, Pregitzer KS, White DC (1997) Compositional and functional shifts in microbial communities due to soil warming. Soil Sci Soc Am J 61:475–481

11 Nutrient Acquisition of Terrestrial Plants in a Changing Climate

D.G. BIELENBERG and H. BASSIRIRAD

11.1 Introduction

The previous chapters highlighted the many ways by which nutrient acquisition can regulate plant response to the environment. Today, ecosystems and ecological processes are impacted by a suite of anthropogenic agents that will define plant ecology in ways that were not imagined even as little as five decades ago. Changes in land use and anthropogenic production of environmental pollutants have all but devastated many native systems of the world and will remain a growing danger to the health of the remaining native systems. Ecologists, particularly those working with plants and ecosystems, bear the intellectual task of reliably predicting the potential consequences of global change. Such knowledge will be critical in management and policy measures designed to halt further deterioration of native systems. In this chapter, we argue that reliable predictions of the fate of terrestrial plant communities, and thereby the policy and management choices of the future, will require a clear understanding of plant nutrient uptake in a changing climate.

By the end of this century, most of the globe will be experiencing atmospheric CO_2 concentrations twice as high as those encountered during the mid-20th century, while the global temperature is likely to rise by as much as 5 °C during the same period. In addition to their direct effects, the higher concentration of CO_2 and other greenhouse gasses will impact terrestrial ecosystems via dramatic changes in climate pattern, including the occurrence of extreme drought and flooding. Major biomes of the earth will also experience above-average input of atmospheric N, ozone (stratospheric ozone will decrease and tropospheric levels increase), and UV-B radiation. Either individually or collectively, alterations in these environmental factors will change plant nutrient demand. Unless these demands are met via adjustments in nutrient availability and/or uptake, the productivity and composition of plant communities may change irreversibly.

Ecological Studies, Vol. 181
H. BassiriRad (Ed.) Nutrient Acquisition by Plants
An Ecological Perspective

Obviously, we will not be able to cover all the components of global change in this chapter. Chapter 10 (this Vol.) has covered soil temperature as a component of global change. Here, we will focus on the impacts of three major components of global change, and examine the state of the knowledge on how these factors affect plant nutrient uptake. The components of global change focused on here are elevated tropospheric ozone, increased incident UV-B radiation, and elevated atmospheric CO_2 concentration.

11.2 Elevated Tropospheric Ozone or Elevated UV-B Radiation

11.2.1 Extent of the Problem and Relevance to Nutrient Uptake

At first consideration, elevated levels of tropospheric ozone (O_3) and elevated levels of UV-B solar radiation would appear to have little similarity. Nonetheless, these topics will be considered together because of the shared themes that emerge from the literature: both have direct effects upon aboveground plant organs, both have the potential to generate oxidative damage, and both have the potential to alter biomass accumulation and partitioning. Fundamentally, however, little is known about the effect of either of these stresses on nutrient acquisition from the soil environment.

Tropospheric O_3 is the major phytotoxic constituent of photochemical smog, and was first identified as damaging to plants in the 1950s (Richards et al. 1958). Large areas of Asia, Africa, Australia, and the Americas have high potential for the production of elevated tropospheric O_3, based on predicted energy use patterns and global solar radiation patterns (Kickert and Krupa 1990). Ozone reduces photosynthesis, growth, and yield, increases respiration, causes accelerated foliar senescence, reduces root:shoot ratios, and can increase plant susceptibility to pathogens (Runeckles and Krupa 1994).

Elevated UV-B, resulting from stratospheric O_3 depletion, has the potential to affect a wide range of ecosystems and plant types. In some sensitive species, laboratory experiments have demonstrated that UV-B can reduce photosynthesis, leaf area, water use efficiency, growth, and yield as well as altering many aspects of plant morphology (Runeckles and Krupa 1994). However, these impacts of UV-B exposure are minimal in many plant species, and by and large the evidence available to date predicts no impact for most species at UV-B levels that reflect realistic environmental dosages (Searles et al. 2001).

11.2.2 Direct Effects on Nutrient Uptake

11.2.2.1 Changes in Root-to-Shoot Ratios, Morphology and Architecture

A shift in biomass allocation to the shoots (a decrease in root-to-shoot ratios, RSR) has been observed in a wide range of plant species in response to O_3 exposure (Cooley and Manning 1987). The shift in partitioning to the shoot is generally ascribed to increased use of carbohydrates for repair of O_3-induced damage in the leaves, and O_3-induced disruption of phloem loading (Cooley and Manning 1987). However, when assessing the effects of environmental treatments upon RSR shifts, it is important to evaluate the impact of changes in plant size as a causal factor in altered RSR (Ledig et al. 1970; Coleman et al. 1994). Coleman and McConnaughay (1995) emphasized the need for caution with their reexamination of a study of biomass partitioning involving the air pollutant SO_2. When they accounted for treatment-induced changes in plant size, they eliminated differences in RSR that previously appeared to support a conclusion of altered biomass partitioning. Similarly, when Bielenberg et al. (2002) investigated the effects of O_3 on biomass partitioning in hybrid poplar, shifts in RSR occurred as a result of treatment-induced changes in plant size, and not from specific alterations in biomass partitioning.

The root:shoot ratio can also be affected by UV-B exposure (Sullivan 1997), although the results from different experiments have been variable. Increased, decreased and unchanged RSR have all been reported in response to UV-B exposure (Sullivan 1997; Weih et al. 1998). Alterations in RSR induced by UV-B can be subtle and sometimes transient. UV-B increased RSR in the first year of a 2-year study by Sullivan et al. (1994). No differences in RSR were detected in the second year of the study (Sullivan et al. 1994). Ziska et al. (1993) reported a strong reduction of RSR in cassava by UV-B, without changes in whole-plant biomass. Reduced biomass allocation to roots may decrease nutrient uptake, as well as reducing the yield of economically important plant parts (Ziska et al. 1993). It remains to be determined whether most UV-B-induced changes in RSR result from altered ontogeny or from altered partitioning priorities.

Root system architecture is also affected by exposure to O_3, although specific changes in architecture appear variable and highly species-specific. In an O_3-sensitive loblolly pine selection, O_3 was shown to decrease both total root surface area and root surface area per unit root mass (Qiu et al. 1992). Reduced root surface area may limit nutrient uptake. However, Taylor and Davies (1990) noted an increase in specific root length (length per unit mass) in response to O_3 in beech seedlings. Temple et al. (1993) reported an increase in fine root mass, and a decrease in coarse root mass in a field study with ponderosa pine. Nutrient acquisition is facilitated by increased specific root length and increased fine root numbers. Finally, no impact of

O_3 on the proportion of coarse and fine root biomass, or the specific root length of the coarse and fine fractions was reported in birch (Maurer and Matyssek 1997).

11.2.2.2 Changes in Root Respiration and Carbohydrate Supply to the Roots

Root respiration is generally reduced by O_3 stress. Theoretically, such reduction can limit the energy supply required for active nutrient uptake. In some cases, decreases in total root system respiration are attributable to decreases in total root mass (Coleman et al. 1996), while in other cases a reduction in respiration on a specific root mass basis has been suggested (Edwards 1991). A large fraction of total root respiration is related to nutrient uptake (Lambers 1987), but nutrient uptake was not measured in either of these studies.

A number of studies using isotopic tracers have shown that proportionately less photosynthate is translocated from leaves to roots in O_3-stressed plants (Nouchi et al. 1995; Pausch et al. 1996a; Samuelson and Kelly 1996). The decrease in translocation of C to the roots is viewed as the cause of decreased root mass (Cooley and Manning 1987). By decreasing both root growth and root respiration, it is likely that O_3 also limits nutrient acquisition from the soil environment.

11.2.2.3 Changes in Root Uptake Kinetics

The preceding sections indicate O_3 has the potential to alter nutrient uptake by reducing carbohydrate availability, decreasing root biomass, and altering root architecture. Direct investigations of the effect of O_3 on nutrient uptake are few, and indicate species-specific differences in response. Nouchi et al. (1991) reported a specific reduction in ammonium uptake in roots excised from O_3-treated rice plants. Pausch et al. (1996b) showed a shift in N acquisition away from N-fixing nodules to N acquisition from the root zone in O_3-exposed soybean, and ascribed this reduction to reduced photosynthate translocation to roots from the shoots (Pausch et al. 1996a). However, Bielenberg et al. (2002) showed no O_3 impact on labeled N uptake by hybrid poplars on a whole-plant basis at two levels of N availability. Similarly, in one of the few studies that included other mineral nutrients besides N, chronic O_3 exposure did not affect tissue concentrations or shoot uptake of a number of these (N, P, K, S, Ca, Mg, Fe, Mn, and Zn) in spring wheat (Fangmeier et al. 1997). However, a later study in potato found that tuber concentrations of N and Mg significantly increased in response to O_3 exposure, but this effect may be attributable to reduced biomass, rather than increased nutrient uptake (Fangmeier et al. 2002).

Few studies to date have addressed the effect of UV-B exposure on root system uptake of nutrients from the soil environment (Baker and Allen 1994; Sullivan 1997; Caldwell et al. 1998; Zepp et al. 1998). A small number of studies have investigated the effect of UV-B on whole-plant accumulation of nutrients. These studies have used nutrient content (mass nutrient per plant or plant part; Murali and Teramura 1985; Yue et al. 1998) or concentration (mass nutrient per dry weight of plant or plant part; Elawad et al. 1985; Larsson et al. 1998; Moorthy and Kathiresan 1998; de la Rosa et al. 2001; Shukula and Kakkar 2002) in the standing biomass (shoot, leaves, or whole plant) as an indication of whole-plant uptake. Nutrients increase in concentration (Larsson et al. 1998), or remain unchanged (de la Rosa et al. 2001) in plant tissues in response to UV-B exposure. In the studies where nutrient content was measured along with concentration (Murali and Teramura 1985; Yue et al. 1998), plant content was usually unaffected or very slightly reduced in response to UV-B exposure. Therefore, UV-B appears to have very little or no affect on nutrient accumulation on a whole-plant basis. Plant nutrient concentration changes appear to result from UV-B-induced biomass reductions, rather than changes in the physiology of nutrient acquisition.

11.2.2.4 Changes in Foliar Uptake/Leaching

Solutes, particularly cations and small-molecular weight molecules, can be taken up by the leaf through the waxy cuticle that forms the outer boundary of the leaf surface. Atmospheric deposition of nitrogen and sulfur onto leaf surfaces can result in significant foliar uptake of these nutrients (Lambers et al. 1998). The ease with which uptake occurs is inversely related to cuticle thickness. UV-B has been shown to increase the thickness of the cuticle layer surrounding the leaf (Grammatikopoulos et al. 1998), potentially limiting the uptake of nutrients such as N and S deposited from the atmosphere.

Similarly, UV-B-induced thickening of leaf cuticles may reduce leaching of foliar nutrients. Foliar leaching can result in considerable loss of nutrients, particularly Ca and Mg, in moist climates (Marschner 1995). Foliar leaching may negatively impact the plant by removing essential nutrients, but also may increase the soil availability of nutrients for new growth that are inefficiently remobilized from older leaves (Marschner 1995). In areas of acidic deposition, however, foliar leaching and pH-induced soil leaching may result in significant losses of cations from the plant–soil system (Edwards et al. 1995).

11.2.3 Indirect Effects on Nutrient Uptake

Chronic O_3 stress has been demonstrated to cause accelerated senescence of leaves in many species of plants (Reich and Lassoie 1985; Sanders et al. 1992;

Wiltshire et al. 1993; Sicher and Bunce 1998; Pell et al. 1999). UV-B has also been shown to trigger early senescence of cotyledons and leaves (Ambler et al. 1975; Ernst et al. 1997). Additionally, UV-B impaired remobilization of Zn (Ambler et al. 1975), and Fe, Mg, and Mn (Ernst et al. 1997) from cotyledons of seedlings. Early leaf loss will alter the schedule of litter-fall availability to the decomposer community, and may change temporal patterns of soil nutrient cycling and availability.

In addition to altering the timing of leaf fall, O_3 affects the tissue quality and decomposition of abscised foliage (Kickert and Krupa 1990). In a study of the effect of O_3 on leaf loss and decomposition in poplar, Jones et al. (1994) demonstrated that leaf litter resulting from O_3-induced accelerated senescence contained greater amounts of N and bound phenolic compounds. Contrary to the expectation that increased N would speed decomposition, this change in litter quality resulted in slowed decomposition, presumably as a result of phenolic-N complexation. Additionally, naturally senescing leaves from the same O_3-exposed plants had reduced N content, compared to leaves abscised from non-O_3-exposed plants (Jones et al. 1994).

UV-B exposure frequently induces the synthesis of UV-absorbing phenylpropanoid compounds (Zepp et al. 1998). Grammatikopoulos et al. (1998) reported increased phenolics and tannins, and decreased N content in leaves of *Laurus nobilis* and *Ceratonia siliqua*. In contrast, no effects of UV-B exposure upon leaf litter quality (nutrient or phenolic contents) were found in a study of *Quercus robur* (Newsham et al. 2001). These compounds can affect nutrient cycling by retarding the rate of decomposition and/or affecting the availability of other nutrients in leaf litter (Zepp et al. 1998). A slowed rate of decomposition could result in lower turnover of organic matter and mineral nutrients in the soil, potentially reducing productivity as more and more nutrients are sequestered. However, no direct effects of UV-B exposure on litter decomposition were found in a dune grassland system (Verhoef et al. 2000).

Far more attention has been paid to the effects of O_3 than of UV-B on root characteristics and function. Much of the O_3 literature is contradictory and indicates that, depending upon species, there may be a wide range of responses to elevated tropospheric O_3. Additional studies involving a wider range of functional types are needed to predict possible community and ecosystem alterations resulting from O_3-impaired nutrient acquisition. The little UV-B literature that exists indicates that only limited, if any, effects on direct nutrient uptake can be expected. Both stresses have potential for indirect effects upon ecosystem nutrient cycling in the long term. More studies are needed to assess the potential for species- and community-level feedbacks of elevated O_3 or UV-B on nutrient cycling that result from alterations in phenology, litter quality, or decomposition processes.

11.3 Elevated Atmospheric CO2 Concentration

11.3.1 Extent of the Problem and Relevance to Nutrient Uptake

The advent of the industrial revolution, increased combustion of fossil fuel, and deforestation have substantially accelerated the rate of CO_2 production on earth, and thereby its accumulation in the atmosphere. Data from ice-core gasses indicate that between 1750 and the pre-industrial revolution, the atmospheric CO_2 concentration remained constant at roughly 280 ppm (Houghton et al. 2001). Since then, however, the average CO_2 concentration has risen to 370 ppm. More importantly, direct atmospheric CO_2 measurements from Mauna Loa, Hawaii, indicate that more than 60 % of this dramatic rise in CO_2 has occurred in the last 40 years (Keeling and Whorf 2002). At this rate, it is expected that by the end of this century, the CO_2 concentration of the atmosphere will be double the current levels.

One of the most common plant responses to increased CO_2 concentration is the reduction in foliage N and many other nutrients (BassiriRad et al. 2001). This is perhaps why most researchers in this field suggest that while short-term exposure to CO_2 enrichment stimulates growth and photosynthetic C fixation, in most managed and natural ecosystems long-term growth and photosynthetic responses may be confined by the limited availability of mineral nutrients, particularly nitrogen (N) and phosphorus (P; Kramer 1981; Eamus and Jarvis 1989; Bazzaz 1990; Conroy 1992; Sinclair 1992; McKee and Woodward 1994; Wolfenden and Diggle 1995; Lloyd and Farquhar 1996). Therefore, factors that may affect the availability and uptake of nutrients are critical in determining plant and ecosystem responses to high CO_2. Here, we distinguish between responses in plant ability to take up nutrients (direct effects), as opposed to ecosystem availability of nutrients (indirect effects).

11.3.2 Direct Effects of CO2 on Plant Nutrient Uptake

11.3.2.1 Changes in Root-to-Shoot Ratios

Most of the early work in this area was based on the assumption that plant responses to high CO_2 would be driven by the functional balanced model. Such theoretical underpinning predicts that at high availability of C, biomass allocation to the roots should increase. Over the past three decades, considerable debate has emerged as to whether high CO_2 stimulates root biomass allocation (Stulen and den Hertog 1993). Although earlier literature indicated that elevated CO_2 almost universally increased root-to-shoot ratios (Eamus and Jarvis 1989; Bazzaz 1990; Poorter 1993), more recent literature reveals a less pronounced effect (Norby 1994; Rogers et al. 1994). Ceulemans and Mousseau

(1994) reviewed the literature between 1989 and 1993 for woody plants, and confirmed the earlier conclusions that more biomass is allocated to roots in response to high CO_2, but such a shift in biomass was less likely when supplemental N was added to the experimental plants. In contrast, in a survey of 224 observations of woody plants grown at high CO_2, Wullschleger et al. (1995) found a significant increase in root-to-shoot ratio in only 6% of the cases. Norby (1994) also concluded that elevated CO_2 does not lead to a significant shift in biomass allocation, regardless of soil N availability. The conclusion that elevated CO_2 will not alter biomass allocation to roots is further confirmed by results of a meta-analysis by Curtis and Wang (1998), and the earlier review by McGuire et al. (1995). More recently, Poorter and Nagel (2000) reviewed 170 observations from 80 publications, and concluded that there would be no significant changes in response to a doubling of CO_2 concentration.

The survey by Poorter and Nagel (2000) is particularly relevant to our discussion here, because they highlighted the importance of other environmental factors, e.g., light, water and nutrients, in regulating plant biomass allocation responses to CO_2. Therefore, caution must be exercised when the results of experiments with dissimilar protocols are compared. When comparable experiments are considered, it is often observed that changes in root-to-shoot ratio in response to CO_2 enrichment depend on soil nutrient availability, i.e., increased root-to-shoot ratio is often associated with nutrient limitation (McDonald et al. 1991; Ericsson et al. 1992) – a response that could be suppressed when nutrient limitation is avoided (Bazzaz 1990; Pettersson and McDonald 1992; Ceulemans and Mousseau 1994). In their survey of more than 40 tree species, McGuire et al. (1995) reported that as N availability increased, the CO_2-induced increase in biomass allocation to roots was substantially decreased, and overall the effect was not statistically significant. However, they attributed the lack of a statistical difference in root-to-shoot ratio at high vs. low N availability to an inadequate sample size.

In addressing the effects of elevated CO_2 on root growth, particularly as it relates to plant nutrient uptake, the relative distribution of biomass between fine and coarse roots may be particularly important. For example, a mere increase in root-to-shoot ratio may not be highly informative in terms of plant nutrient acquisition, although it is an important parameter in terms of plant C allocation, particularly if biomass is allocated into taproots or other highly suberized components of the root system not involved in nutrient uptake. Even when it is used to evaluate C partitioning in response to CO_2, root-to-shoot ratio may be a poor indicator of root carbon allocation since it does not take into account biomass lost by fine root turnover (Pregitzer et al. 1995) or root exudation. Therefore, a more relevant indicator of root characteristics that could potentially enhance plant nutrient capture is the proportion of biomass allocated to fine roots (BassiriRad et al. 1996a, 1997a, b). The fine root ratio, defined as fine roots relative to total plant biomass, is sensitive to nutrient availability (Körner and Renhardt 1987), but depending upon the species, it may or may not be sensitive to CO_2 levels (BassiriRad et al. 1997b).

Fine root biomass allocation responses to high CO_2 provide a robust estimate of plant nutrient acquisition only if it is complimented with information about root turnover. In many deciduous forest species, fine root turnover can account for as much as 80 % of annual NPP (Ceulemans and Mousseau 1994). Pregitzer et al. (1995) showed that growth in high CO_2 increased root turnover in *Populus tremuloides*. Elevated CO_2 has also been shown to increase root turnover in a number of grassland species (Fitter et al. 1996, 1997; Hungate et al. 1997). Matamala and Schlesinger (2000) distinguished between live fine root (LFR) and dead fine root (DFR) in a loblolly pine stand under ambient and elevated CO_2, and found that after one season of fumigation LFR increased by 86 % in response to CO_2 enrichment, but DFR was relatively unaffected by the CO_2 treatment. In the same system, Pritchard et al. (2001) used mini-rhizotron observations and reported only modest changes in root turnover, but found a significant CO_2 by depth interaction on root mortality. It is therefore necessary that biomass allocation studies consider spatial and temporal dynamics of root growth responses to elevated CO_2.

11.3.2.2 Changes in Root Morphology and Architecture

Alterations in root morphology (e.g., root length and radius) and architecture (e.g., branching pattern) are also effective mechanisms that influence plant nutrient acquisition (Caldwell 1987; Fitter and Hay 1987). Finer roots can confer greater nutrient uptake per unit root mass. Quantitative assessments of such root characteristics, particularly in natural ecosystems, are difficult, which is perhaps why very little information is available as to how these parameters change in response to high CO_2. In a growth chamber study, Rogers et al. (1992) demonstrated that elevated CO_2 doubled root length and increased root diameter by 27 %, but had no effect on the number of first-order laterals. Ferris and Taylor (1995) found that elevated CO_2 had contrasting effects on root morphological and architectural characteristics among four native chalk grassland species. After 100 days of treatment, the root-to-shoot ratio was unchanged but root length was significantly higher in three out of four species grown at high vs. ambient CO_2. In contrast, specific root length (m/g) significantly increased in response to CO_2 enrichment in only one of the four species. Berntson and Woodward (1992) examined the root branching pattern of *Senecio vulgaris*, a common fast-growing annual in Britain, and found that elevated CO_2 resulted in longer roots and increased root branching. However, they found that root density (root length per volume of soil) was unaffected by growth at high CO_2. More recently, Berntson and Bazzaz (1997) examined a number of architectural parameters in roots of yellow birch, and found no elevated CO_2 effects on specific root length and specific root number.

It is also clear that mycorrhizal associations will be an important component of the root system responses to CO_2 that may regulate nutrient acquisition capacity of the whole plant. We have not discussed this topic here, but the reader is encouraged to consult the following works for a more comprehensive treatment of that topic: O'Neill (1994), Fitter et al. (2000), and BassiriRad et al. (2001). The growth and morphological characteristics discussed here are all important factors in plant nutrient acquisition, and many of them appear to respond to CO_2 enrichment. However, studies of root morphology and architecture have seldom, if ever, shown a direct link between these structural characteristics and nutrient uptake responses to high CO_2. Future studies in this field must make such linkages in order to establish a clear cause-and-effect relationship between form and function.

11.3.2.3 Changes in Root Uptake Kinetics

There are at least three theoretical reasons why kinetics of root nutrient uptake should respond to changes in atmospheric CO_2 concentration. First, given the energy requirement for the active transport of nutrients, one would expect that higher availability of carbohydrates under CO_2 would result in up-regulation of root nutrient transporters. Second, elevated CO_2, at least in the short term, has been shown to accelerate growth. Higher growth rate should then increase plant nutrient demand and uptake capacity (Lee 1982; Clarkson 1985). Third, models of whole-plant carbon nutrient balance (Bloom et al. 1985; Johnson 1985; Robinson 1986; Luo et al. 1994; Gutschick and Kay 1995) predict that resources of abundant availability – C – must be allocated to increase the uptake of the most limiting resource – nutrients. A large number of CO_2 studies have demonstrated both a short-term stimulation of growth and an increased supply of root respiratory substrates (Cruz et al. 1993; Tschplinski et al. 1993; Norby et al. 1995). Yet to date, the data on root uptake kinetics show no consistent pattern of uptake response to high CO_2.

The earliest direct measurements of root uptake responses to CO_2 enrichment were reported by Newbery et al. (1995) who examined the rates of N, P and K uptake in *Agrostis capallaris*, at three different external concentrations, and found no significant CO_2 effect. Using field-grown loblolly pine saplings that had been exposed to 18 months of CO_2 treatment, BassiriRad et al. (1996b) found no significant changes in V_{max} of NH_4^+ uptake, but elevated CO_2 significantly enhanced V_{max} for NO_3^- uptake. Jackson and Reynolds (1996) also found no effect of elevated CO_2 on NH_4^+ uptake rates of six grassland annuals, but NO_3^- uptake rate was inhibited in response to CO_2 enrichment. Other studies (see BassiriRad 2000; BassiriRad et al. 2001) also produced equally inconsistent patterns. Even in cases where similar species are examined, the effect of CO_2 on the kinetics of NH_4^+ varies from one experiment to the other (BassiriRad et al. 1996a, b).

The observed variability in root nutrient uptake responses to high CO_2 is consistent with that observed in other plant characteristics such as root growth (Tingey et al. 2000), shoot growth (Norby et al. 1999) as well as water use efficiency, photosynthetic rate, tissue N concentration, and nonstructural carbohydrates (Luo et al. 1999). Until we find a unified mechanism to explain such variability, our predictive capacity for plant and ecosystem responses to elevated CO_2 will remain limited. Some of the variations in response to high CO_2 may simply result from differences in experimental protocols. However, Zerihun and BassiriRad (2001) examined NH_4^+ and NO_3^- uptake kinetic responses of six broad-leaf and conifer tree species grown under identical conditions, and found that root NH_4^+ uptake kinetic responses to high CO_2 are clearly species-dependent.

Interspecific variation in root physiological characteristics may explain why some species do not exhibit a commonly observed decline in tissue nutrient concentration at high CO_2. In a study of three desert species, BassiriRad et al. (1997b) showed that elevated CO_2 decreased foliar N and P concentrations in *Larrea tridentata* and *Prosopis glandulosa*, but not in *Bouteloua eriopoda* that was the only species that had up-regulated its root uptake rate for both N and P. Species-specific responses of N uptake kinetics can also be one of the potential mechanisms by which elevated CO_2 may affect competitive balance among co-occurring species, thereby affecting ecosystem productivity and composition (Berntson et al. 1998).

It is important to note that on their own, changes in root uptake kinetics will not be a meaningful index of plant nutrient acquisition responses to elevated CO_2. Often, in cases where kinetics of uptake are severely down-regulated, the overall plant growth response to CO_2 is still positive. This is partly because root uptake kinetic response is only one potential mechanism that can lead to changes in nutrient uptake. Compensatory adjustments in other root characteristics, such as morphology and life span, along with increased nutrient use efficiency, can prevent the need for a positive adjustment in uptake kinetics. Jackson and Reynolds (1996) examined root physiological uptake capacity for N, and root biomass responses of a mixed community of annual grasses and forbs. They reported that N uptake responses to elevated CO_2 were more closely correlated with root physiological capacity under low N availability, but correlated more strongly with root biomass under relatively high N availability. Zerihun et al. (2000) reported that the relative contribution of root physiological uptake capacity in determining the overall plant responses to CO_2 enrichment must be evaluated in conjunction with other root mechanisms that can also influence nutrient acquisition. Therefore, future studies addressing the relative contribution of active ion uptake to plant nutrient status and growth responses to high CO_2 will require experimental designs that monitor changes in these mechanisms simultaneously (Berntson and Bazzaz 1996, 1997; Berntson et al. 1998).

11.3.3 Indirect Effects of CO2 on Nutrient Availability at the Root Surface

Indirectly, elevated CO_2 can alter nutrient availability at the root surface via a number of mechanisms. Increased C supply to the soil can enhance microbial activities (Dhillion et al. 1996), accompanied by higher mineralization and plant uptake (Zak et al. 1993; Rice et al. 1994), although this effect is not universally observed (Diaz et al. 1993). A higher microbial activity was also reported for a grassland system by Hungate et al. (1997), but they argued that this enhancement was caused by reduced transpiration and improved soil moisture, rather than improved C supply to the soil. Reduced transpiration response to high CO_2 has also been invoked as a possible indirect mechanism that suppressed mass flow of K^+ to the root surface in wheat, but this effect was offset by a higher diffusion rate (Van Vuuren et al. 1997). Finally, availability of nutrients such as N in many ecosystems could also increase due to chronic deposition of atmospheric N (Ollinger et al. 1993; Lovett 1994; Galloway et al. 1995; Vitousek et al. 1997; Norby 1998). Whether ecosystem availability of nutrients in response to rising CO_2 concentration will keep pace with increased plant nutrient demand is not the focus of this chapter. However, such information is important for models designed to reliably predict CO_2 responses in natural ecosystems.

11.4 Summary

There are many facets of global climate change that could synergistically affect native vegetation. Whenever possible, experimental approaches must consider the consequences for plants under realistic, multiple stress conditions. Models designed to predict plant and ecosystem responses to global change should integrate more thoroughly the role of plant nutrient acquisition, and how it may respond to a changing climate. Our current mechanistic models are largely based on plant gas-exchange parameters. We can significantly improve the robustness of these models by incorporating parameters of nutrient uptake, e.g., root system characteristics. We also highlight the need to examine the root system controllers of nutrient uptake collectively, rather than individually. It is not reasonable to assume that the knowledge of changes in just one parameter, e.g., root-to-shoot ratio, can elucidate the mechanism and/or the extent of the effects of global change on plant nutrient uptake, much the same way that we do not expect that the knowledge of changes in leaf area alone will be sufficient to reliably predict canopy gas-exchange in response to climate change. Within this context, we recommend that future studies should pay more attention to the collective effects of root morphology, mycorrhizal association (Chap. 9, this Vol.), kinetics (Chap. 6, this Vol.) and architecture (Chap. 7, this Vol.), as well as life span (Chap. 8, this Vol.) in determining whole-plant nutrient responses to global change.

Acknowledgments. The corresponding author, H. BassiriRad, would like to acknowledge support from the National Science Foundation for this project.

References

Ambler JE, Krizek DT, Semeniuk P (1975) Influence of UV-B radiation on early seedling growth and translocation of ^{65}Zn from cotyledons in cotton. Physiol Plant 34:177–181

Baker JT, Allen LH (1994) Assessment of the impact of rising carbon-dioxide and other potential climate changes on vegetation. Environ Pollut 83:223–235

BassiriRad H (2000) Kinetics of nutrient uptake by roots: responses to global change. New Phytol 147:150–169

BassiriRad H, Griffin KL, Strain BR, Reynolds JF (1996a) Effects of CO_2 enrichment on root $^{15}NH_4$ uptake kinetics and growth in seedlings of loblolly and ponderosa pine. Tree Physiol 16:957–962

BassiriRad H, Thomas RB, Reynolds JF, Strain BR (1996b) Differential responses of root uptake kinetics of and to enriched atmospheric CO_2 in field grown loblolly pine. Plant Cell Environ 19:367–371

BassiriRad H, Griffin KL, Reynolds JF, Strain BR (1997a) Changes in root and absorption rates of loblolly and ponderosa pine in response to CO_2 enrichment. Plant Soil 190:1–9

BassiriRad H, Reynolds JF, Virginia RA, Brunelle MH (1997b) Growth and root and $PO_4^{\ 3}$ uptake capacity of three desert species in response to atmospheric CO_2 enrichment. Aust J Plant Physiol 24:353–358

BassiriRad H, Gutschick V, Lussenhop J (2001) Root system adjustments: regulation of plant nutrient uptake and its control of growth responses to elevated CO_2. Oecologia 126:305–320

Bazzaz FA (1990) The response to natural ecosystems to the rising global CO_2 levels. Annu Rev Ecol Syst 21:167–196

Berntson GM, Bazzaz FA (1996) Belowground positive and negative feedbacks on CO_2 growth enhancement. Plant Soil 187:119–131

Berntson GM, Bazzaz FA (1997) Nitrogen cycling in microcosms of yellow birch exposed to elevated CO_2: simultaneous positive and negative below-ground feedbacks. Global Change Biol 3:247–258

Berntson GM, Woodward FI (1992) The root system architecture and development of *Senicio vulgaris* in elevated carbon dioxide and drought. Funct Ecol 6:324–333

Berntson GM, Rajakaruna N, Bazzaz FA (1998) Species-level and community-level growth and nitrogen uptake in elevated CO_2 atmospheres in an experimental annual community. Global Change Biol 4:101–120

Bielenberg DG, Lynch JP, Pell EJ (2002) Nitrogen dynamics during O_3-induced accelerated senescence and related compensation in hybrid poplar. Plant Cell Environ 25(4):501–512

Bloom AJ, Chapin FS, Mooney HA (1985) Resource limitation in plants – an economic analogy. Annu Rev Ecol Syst 16:363–392

Caldwell MM (1987) Plant architecture and resource competition. In: Schulze E-D, Zwolfer H (eds) Ecological Studies, vol 61. Springer, Berlin, Heidelberg New York, pp 164–179

Caldwell MM, Bjorn LO, Bornman JF, Flint SD, Kulandaivelu G, Teramura AH, Tevini M (1998) Effects of increased solar ultraviolet radiation on terrestrial ecosystems. J Photochem Photobiol B Biol 46:40–52

Ceulemans R, Mousseau M (1994) Effects of elevated atmospheric CO_2 on woody plants. New Phytol 127:425–446

Clarkson DT (1985) Factors affecting mineral nutrient acquisition by plants. Annu Rev Plant Physiol 36:77–115

Coleman JS, McConnaughay KDM (1995) A non-functional interpretation of a classical optimal-partitioning example. Funct Ecol 9:951–954

Coleman JS, McConnaughay KDM, Ackerly DD (1994) Interpreting phenotypic variation in plants. Trends Ecol Evol 9:187–191

Coleman MD, Dickson RE, Isebrands JG, Karnosky DF (1996) Root growth and physiology of potted and field-grown trembling aspen exposed to trophospheric ozone. Tree Physiol 16:145–152

Conroy JP (1992) Influence of elevated atmospheric CO_2 concentrations on plant nutrition. Aust J Bot 40:445–456

Cooley DR, Manning WJ (1987) The impact of ozone on assimilate partitioning in plants: a review. Environ Pollut 47:95–113

Cruz C, Lips SH, Martins-Loução MA (1993) The effect of nitrogen source on photo-synthesis of carob at high CO_2 concentrations. Physiol Plant 89:552–556

Curtis PS, Wang X (1998) A meta-analysis of elevated CO_2 effects on woody plant mass, form and physiology. Oecologia 113:299–313

de la Rosa TM, Julkunen-Tiitto R, Lehto T, Aphalo PJ (2001) Secondary metabolites and nutrient concentrations in silver birch seedlings under five levels of daily UV-B exposure and two relative nutrient addition rates. New Phytol 150:121–131

Dhillion SS, Roy J, Abrams M (1996) Assessing the impact of elevated CO_2 on soil microbial activity in a Mediterranean model ecosystem. Plant Soil 187:333–342

Diaz S, Grime JP, Harris J, McPherson E (1993) Evidence of a feedback mechanism limiting plant response to elevated carbon dioxide. Nature 364:616–617

Eamus D, Jarvis PG (1989) The direct effects of increase in the global atmospheric CO_2 concentration on natural and commercial temperate trees and forests. Adv Ecol Res 19:5l–55

Edwards NT (1991) Root and soil respiration responses to ozone in *Pinus taeda* L. seedlings. New Phytol 118:315–321

Edwards GS, Sherman RE, Kelly JM (1995) Red spruce and loblolly-pine nutritional responses to acidic precipitation and ozone. Environ Pollut 89:9–15

Elawad SH, Allen LH, Gascho GJ (1985) Influence of UV-B radiation and soluble silicates on the growth and nutrient concentration of sugarcane. Soil Crop Sci Soc Fla Proc 44:134–141

Ericsson T, Rytter L, Linder S (1992) Nutritional dynamics and requirements of short rotation forests. In: Mitchel CP, Ford-Robertson JB, Hinckley T, Sennerby-Forsse L (eds) Ecophysiology of short rotation forest crops. Elsevier, London, pp 35–65

Ernst WHO, van de Staaij JWM, Nelissen HJM (1997) Reaction of savanna plants from Botswana on UV-B radiation. Plant Ecol 128:162–170

Fangmeier A, Gruters U, Hogy P, Vermehren B, Jager HJ (1997) Effects of elevated CO_2, nitrogen supply and tropospheric ozone on spring wheat. II. Nutrients (N, P, K, S, Ca, Mg, Fe, Mn, Zn). Environ Pollut 96:43–59

Fangmeier A, De Temmerman L, Black C, Persson K, Vorne V (2002) Effects of elevated CO_2 and/or ozone on nutrient concentrations and nutrient uptake of potatoes. Eur J Agron 17(4):353–368
Ferris R, Taylor G (1995) Contrasting effects of elevated CO_2 on the root and shoot growth of four native herbs commonly found in chalk grassland. New Phytol 125:855–866
Fitter AH, Hay RKM (1987) Environmental physiology of plants, 2nd edn. Academic Press, New York, pp 99–106
Fitter AH, Self GK, Wolfenden J, Vuuren MMI van, Brown TK, Williamson L, Graves JD, Robinson D (1996) Root production and mortality under elevated atmospheric carbon dioxide. Plant Soil 187:299–306
Fitter AH, Graves JD, Wolfenden J, Self GK, Brown TK, Bogie D, Mansfield TA (1997) Root production and turnover and carbon budgets of two contrasting grasslands under ambient and elevated atmospheric carbon dioxide concentrations. New Phytol 137:247–256
Fitter AH, Heinemeyer A, Staddon PL (2000) Mycorrhizal fungi have a potential role in soil carbon storage under elevated CO_2 and nitrogen deposition. New Phytol 147:189–200
Galloway JN, Schlesinger WH, Levy H II, Michaels A, Schnoor JL (1995) Nitrogen fixation: anthropogenic enhancement-environmental response. Global Biogeochem Cycles 9:235–252
Grammatikopoulos G, Kyparissis A, Drilias P, Petropoulou Y, Manetas Y (1998) Effects of UV-B radiation on cuticle thickness and nutritional value of leaves in two Mediterranean evergreen sclerophylls. J Plant Physiol 153:506–512
Gutschick VP, Kay LE (1995) Nutrient-limited growth rates: quantitative benefits of stress responses and some aspects of regulation. J Exp Bot 46:995–1009
Houghton JT, Ding Y, Griggs DJ, Noguer M, van der Linden PJ, Xiaosu D (eds) (2001) Climate change 2001. The scientific basis. Cambridge University Press, Cambridge
Hungate BA, Holland EA, Jackson RB, Chapin FS III, Mooney HA, Field CB (1997) The fate of carbon in grasslands under carbon dioxide enrichment. Nature 388:576–579
Jackson RB, Reynolds HL (1996) Nitrate and ammonium uptake for single-and mixed-species communities grown at elevated CO_2. Oecologia 105:74–80
Johnson IR (1985) A model of the partitioning of growth between the shoots and the roots of vegetative plants. Ann Bot 55:421–431
Jones CG, Coleman JS, Findlay S (1994) Effects of ozone on interactions between plants, consumers and decomposers. In: Alscher RG, Wellburn AR (eds) Plant responses to the gaseous environment: molecular, metabolic and physiological aspects. Chapman & Hall, New York, pp 339–363
Keeling CD, Whorf TP (2002) Atmospheric CO_2 records from sites in the SIO air sampling network. In: Trends: a compendium of data on global change. Carbon Dioxide Information Analysis Center, Oak Ridge National Laboratory, US Department of Energy, Oak Ridge, TN (http://cdiac.ornl.gov/trends/CO_2/sio-mlo.htm)
Kickert RN, Krupa SV (1990) Forest responses to tropospheric ozone and global climate change – an analysis. Environ Pollut 68:29–65
Körner CH, Renhardt U (1987) Dry matter partitioning and root length/leaf area ratios in herbaceous perennial plants with diverse altitudinal distribution. Oecologia 74:411–418
Kramer PJ (1981) Carbon dioxide concentration, photosynthesis, and dry matter production. BioScience 31:29–33
Lambers H (1987) Growth, respiration, exudation and symbiotic association: the fate of carbon translocated to roots. In: Gregory PJ, Lake JV, Rose DA (eds) Root development and function. Cambridge University Press, London, pp 125–143

Lambers H, Pons TL, Chapin FS (1998) Plant physiological ecology. Springer, Berlin Heidelberg New York

Larsson EH, Bornman JF, Asp H (1998) Influence of UV-B radiation and Cd^{2+} on chlorophyll fluorescence, growth and nutrient content in *Brassica napus*. J Exp Bot 49:1031–1039

Ledig FT, Bormann FH, Wenger KF (1970) The distribution of dry matter growth between shoot and roots in loblolly pine. Bot Gaz 131:349–359

Lee RB (1982) Selectivity and kinetics of ion uptake by barley plants following nutrient deficiency. Ann Bot 50:429–449

Llyod J, Farquhar GD (1996) The CO_2 dependence of photosynthesis, plant growth responses to elevated atmospheric CO_2 concentrations and their interaction with soil nutrient status. I. General principles and forest ecosystems. Funct Ecol 10:4–32

Lovett GM (1994) Atmospheric deposition of nutrients and pollutants in North America: an ecological perspective. Ecol Appl 4:629–650

Luo Y, Field CB, Mooney HA (1994) Predicting responses of photosynthesis and root fraction to elevated carbon dioxide. Interactions among carbon, nitrogen and growth. Plant Cell Environ 17:1195–1204

Luo Y, Reynolds J, Wang Y, Wolfe D (1999) A search for predictive understanding of plant responses to elevated [CO_2]. Global Change Biol 5:143–156

Marschner H (1995) Mineral nutrition of higher plants. Academic Press, San Diego

Matamala R, Schlesinger W (2000) Effects of elevated atmospheric CO_2 on fine root production and activity in a temperate forest ecosystem. Global Change Biol 6:967–980

Maurer S, Matyssek R (1997) Nutrition and the ozone sensitivity of birch (*Betula pendula*) 2. Carbon balance, water-use efficiency and nutritional statue of the whole plant. Trees Struct Funct 12:11–20

McDonald AJS, Ericsson T, Ingestad T (1991) Growth and nutrition of tree seedlings. In: Raghavendra AS (ed) Physiology of trees. Wiley, New York, pp 199–220

McGuire AD, Melillo JM, Joyce LA (1995) The role of nitrogen in the response of forest net primary production to elevated atmospheric carbon dioxide. Annu Rev Ecol Syst 26:473–503

McKee IF, Woodward FI (1994) CO_2 enrichment responses of wheat: interactions with temperature, nitrate and phosphate. New Phytol 127:447–453

Moorthy P, Kathiresan K (1998) Effects of UV-B irradiance on biomass and uptake of nutrients in mangrove seedlings of *Rhizophora apiculata* (Rhizophorales: Rhizophoraceae). Indian J Mar Sci 27:239–242

Murali NS, Teramura AH (1985) Effects of ultraviolet-B irradiance on soybean. 7. Biomass and concentration and uptake of nutrients at varying P-supply. J Plant Nutr 8:177–192

Newbery RM, Wolfenden J, Mansfield TA, Harrison AF (1995) Nitrogen, phosphorus and potassium uptake and demand in *Agrostis capillaris*: the influence of elevated CO_2 and nutrient supply. New Phytol 130:565–574

Newsham KK, Splatt P, Coward PA, Greenslade PD, McLeod AR, Anderson JM (2001) Negligible influence of elevated UV-B radiation on leaf litter quality of *Quercus robur.* Soil Biol Biochem 33:659–665

Norby RJ (1994) Issues and perspectives for investigating root responses to elevated atmospheric carbon dioxide. Plant Soil:165:9–20

Norby RJ (1998) Nitrogen deposition: a component of global change analysis. New Phytol 139:189–200

Norby RJ, Wullschleger SD, Gunderson CA, Johnson DW, Ceulemans R (1999) Tree responses to rising CO_2 in field experiments: implications for the future forests. Plant Cell Environ 22:683–714

Nouchi I, Ito O, Harazono Y, Kobayashi K (1991) Effects of chronic ozone exposure on growth, root respiration and nutrient uptake of rice plants. Environ Pollut 74:149–164

Nouchi I, Ito O, Harazono Y, Kouchi H (1995) Acceleration of ^{13}C-labelled photosynthate partitioning from leaves to panicles in rice plants exposed to chronic ozone at the reproductive stage. Environ Pollut 88:253–260

Ollinger SV, Aber JD, Lovett GM, Millham SE, Lathrop RG, Ellis JM (1993) A spatial model of atmospheric deposition for the northeastern United States. Ecol Appl 3:459–472

O'Neill EG (1994) Responses of soil biota to elevated atmospheric carbon dioxide. Plant Soil 165:55–65

Pausch RC, Mulchi CL, Lee EH, Forseth IN, Slaughter LH (1996a) Use of ^{13}C and ^{15}N isotopes to investigate O_3 effects on C and N metabolism in soybeans. I. C fixation and translocation. Agric Ecosyst Environ 59:69–80

Pausch RC, Mulchi CL, Lee EH, Meisinger JJ (1996b) Use of ^{13}C and ^{15}N isotopes to investigate O_3 effects on C and N metabolism in soybeans. II. Nitrogen uptake, fixation, and partitioning. Agric Ecosyst Environ 60:61–69

Pell EJ, Sinn JP, Brendley BW, Samuelson L, Vinten-Johansen C, Tien M, Skillman J (1999) Differential response of four tree species to ozone-induced acceleration of foliar senescence. Plant Cell Environ 22:779–790

Pettersson R, McDonald AIS (1992) Effects of elevated carbon dioxide concentration on photosynthesis and growth of small birch plants (*Betula pendula* Roth.) at optimal nutrition. Plant Cell Environ 15:911–919

Poorter H (1993) Interspecific variation in the growth response of plants to an elevated ambient CO_2 concentration. Vegetatio 104/105:77–97

Poorter H, Nagel O (2000) The role of biomass allocation in growth response of plants to different levels of light, CO_2, nutrients and water: a quantitative review. Aust J Plant Physiol 27:595–607

Pregitzer KS, Zak DR, Curtis PS, Kubiske ME, Teeri JA, Vogel CS (1995) Atmospheric CO_2, soil nitrogen and turnover of fine roots. New Phytol 129:579–585

Pritchard S, Rogers H, Davis M, Van Santen E, Prior S, Schlesinger W (2001) The influence of elevated atmospheric CO_2 on fine root dynamics in an intact temperate forest. Global Change Biol 7:829–837

Qiu Z, Chappelka AH, Somers GL, Lockaby BG, Meldahl RS (1992) Effects of ozone and simulated acidic precipitation on above- and below-ground growth of loblolly pine (*Pinus taeda*). Can J For Res 22:582–587

Reich PB, Lassoie JP (1985) Influence of low concentrations of ozone on growth, biomass partitioning and leaf senescence in young hybrid poplar plants. Environ Pollut 39:39–51

Rice CW, Garcia FO, Hampton CO, Owensby CE (1994) Soil microbial response in tallgrass prairie to elevated CO_2. Plant Soil 165:67–74

Richards BL, Middleton JT, Hewitt WB (1958) Air pollution with relation to agronomic crops. V. Oxidant stipple of grape. Agron J 50:559–561

Robinson D (1986) Compensatory changes in the partitioning of dry matter In relation to nitrogen uptake and optimal variations in growth. Ann Bot 58:841–848

Rogers HH, Peterson CM, McCrimmon JN, Cure JD (1992) Response of plant roots to elevated atmospheric carbon dioxide. Plant Cell Environ 15:749–752

Rogers HH, Runion GB, Krupa SV (1994) Plant responses to atmospheric CO_2 enrichment with emphasis on roots and the rhizosphere. Environ Pollut 83:155–189

Runeckles VC, Krupa SV (1994) The impact of UV-B radiation and ozone on terrestrial vegetation. Environ Pollut 83:191–213

Samuelson LJ, Kelly JM (1996) Carbon partitioning and allocation in northern red oak seedlings and mature trees in response to ozone. Tree Physiol 16:853–858

Sanders GE, Colls JJ, Clark AG (1992) Physiological changes in *Phaseolus vulgaris* in response to long-term ozone exposure. Ann Bot 69:123–133

Searles PS, Flint SD, Caldwell MM (2001) A meta analysis of plant field studies simulating stratospheric ozone depletion Oecologia 127:1–10

Shukula UC, Kakkar P (2002) Effect of dual stress of ultraviolet-B radiation and cadmium on nutrient uptake of wheat seedlings. Commun Soil Sci Plant Anal 33(11/12):1737–1749

Sicher RC, Bunce JA (1998) Evidence that premature senescence affects photosynthetic decline of wheat flag leaves during growth in elevated carbon dioxide. Int J Plant Sci 159:798–804

Sinclair TR (1992) Mineral nutrition and plant growth response to climate change. J Exp Bot 253:1141–1146

Stulen I, den Hertog J (1993) Root growth and functioning under atmospheric CO_2 enrichment. Vegetatio 104/105:99–115

Sullivan JH (1997) Effects of increasing UV-B radiation and atmospheric CO_2 on photosynthesis and growth: implications for terrestrial ecosystems. Plant Ecol 128:194–206

Sullivan JH, Teramura AH, Dillenburg LR (1994) Growth and photosynthetic responses of field-grown sweetgum (*Liquidambar Styraciflua* Hamamelidaceae) seedlings to UV-B radiation. Am J Bot 81:826–832

Taylor G, Davies WJ (1990) Root growth of *Fagus sylvatica*: impact of air quality and drought at a site in southern Britain. New Phytol 116:457–464

Temple PJ, Riechers GH, Miller PR, Lennox RW (1993) Growth responses of ponderosa pine to long-term exposure to ozone, wet and dry acidic deposition, and drought. Can J For Res 23:59–66

Tingy DT, Phillips DL, Johnson MG (2000) Elevated CO_2 and conifer roots: effects on growth, life span and turnover. New Phytol 147:87–104

Tschaplinski TJ, Norby RJ, Wullschleger SD (1993) Responses of loblolly pine seedlings to elevated CO_2 and fluctuating water supply. Tree Physiol 13:283–296

Van Vuuren MMI, Robinson D, Fitter AH, Chasalow SD, Williamson L, Raven JA (1997) Effects of elevated carbon dioxide and soil water availability on root biomass, root length and N, P and K uptake in wheat. New Phytol 135:455–465

Verhoef HA, Verspagen JMH, Zoomer HR (2000) Direct and indirect effects of ultraviolet-B radiation on soil biota, decomposition and nutrient fluxes in dune grassland soil systems. Biol Fertil Soils 31:366–371

Vitousek PM, Aber J, Howarth RW, Likens GE, Matson PA, Schindler DW, Schlesinger WH, Tilman GD (1997) Human alteration of the global nitorgen cycle: sources and consequences. Ecol Appl 7:737–750

Weih M, Johanson U, Gwynn-Jones D (1998) Growth and nitrogen utilization in seedlings of mountain birch (*Betula pubescens* ssp. *tortuosa*) as affected by ultraviolet radiation (UV-A and UV-B) under laboratory and outdoor conditions. Trees Struct Funct 12:201–207

Wiltshire JJJ, Wright CJ, Unsworth MH, Craigon J (1993) The effects of ozone episodes on autumn leaf fall in apple. New Phytol 124:433–437

Wolfenden J, Diggle PJ (1995) Canopy gas exchange and growth of upland pasture swards in elevated CO_2. New Phytol 130:369–380

Wullschleger SD, Post WM, King AW (1995) On the potential for a CO_2 fertilization effect in forests: estimates of the biotic growth factor based on 58 controlled-exposure studies. In: Woodwell GM, Mackenzie FT (eds) Biotic feedbacks in the global climatic system: will warming feed the warming? Oxford University Press, New York, pp 85–107

Yue M, Li Y, Wang XL (1998) Effects of enhanced ultraviolet-B radiation on plant nutrients and decomposition of spring wheat under field conditions. Environ Exp Bot 40:187–196

Zak DR, Pregitzer KS, Curtis PS, Teeri JA, Fogel R, Randlett RL (1993) Elevated atmospheric CO_2 and feedback between carbon and nitrogen cycles. Plant Soil 51:105–117

Zepp RG, Callaghan TV, Erickson DJ (1998) Effects of enhanced solar ultraviolet radiation on biogeochemical cycles. J Photochem Photobiol B Biol 46:69–82

Zerihun A, BassiriRad H (2001) Interspecific variation in nitrogen uptake kinetic responses of temperate forest species to elevated CO_2: potential causes and consequences. Global Change Biol 7:211–222

Zerihun A, Gutschick VP, BassiriRad H (2000) Relative roles of photosynthetic N-use efficiency and N uptake in determining plant growth response to elevated CO_2: evaluation using the functional balance model approach. Ann Bot 86:723–730

Ziska LH, Teramura AH, Sullivan JH, McCoy A (1993) Influence of ultraviolet-B (UV-B) radiation on photosynthetic and growth-characteristics in field-grown cassava (*Manihot-Esculentum* Crantz). Plant Cell Environ 16:73–79

12 From Molecular Biology to Biogeochemistry: Toward an Integrated View of Plant Nutrient Uptake

H. BassiriRad

12.1 Introduction

In most ecosystems of the world, one or more of the essential minerals can be expected to limit plant growth and development. As a consequent, natural selection may have favored a number of plant traits that lend themselves to efficient uptake and assimilation of limited nutrients. Yet, a comprehensive and integrative treatment of these traits, how they respond to the environment, and whether there are tradeoffs among these traits have not been recently addressed.

Today, more than ever before, technical and theoretical developments have made it possible to reliably quantify plant traits that are critical to nutrient acquisition under various realistic scenarios. The older models of plant nutrient uptake can be markedly improved by these new findings. For example, we now know that in most terrestrial ecosystems, the availability of nutrients is highly heterogeneous in time and space. Moreover, plants may be able to meet a considerable portion of their nutrient requirement from organic as opposed to inorganic sources. There is a timely need to synthesize and incorporate these concepts into new models of nutrient uptake. Therefore, in this volume, we focused on (1) recent findings in root traits that control nutrient uptake, (2) integration of these traits into models of plant nutrient acquisition, and (3) a road map for future studies. We use Fig. 12.1 as a guideline to illustrate the factors that received wider attention in this volume as important controllers of nutrient uptake. The diagram encapsulates the content of most of the chapters, with the exception of two chapters that dealt with consequences of global change on nutrient acquisition (see below).

Ecological Studies, Vol. 181
H. BassiriRad (Ed.) Nutrient Acquisition by Plants
An Ecological Perspective

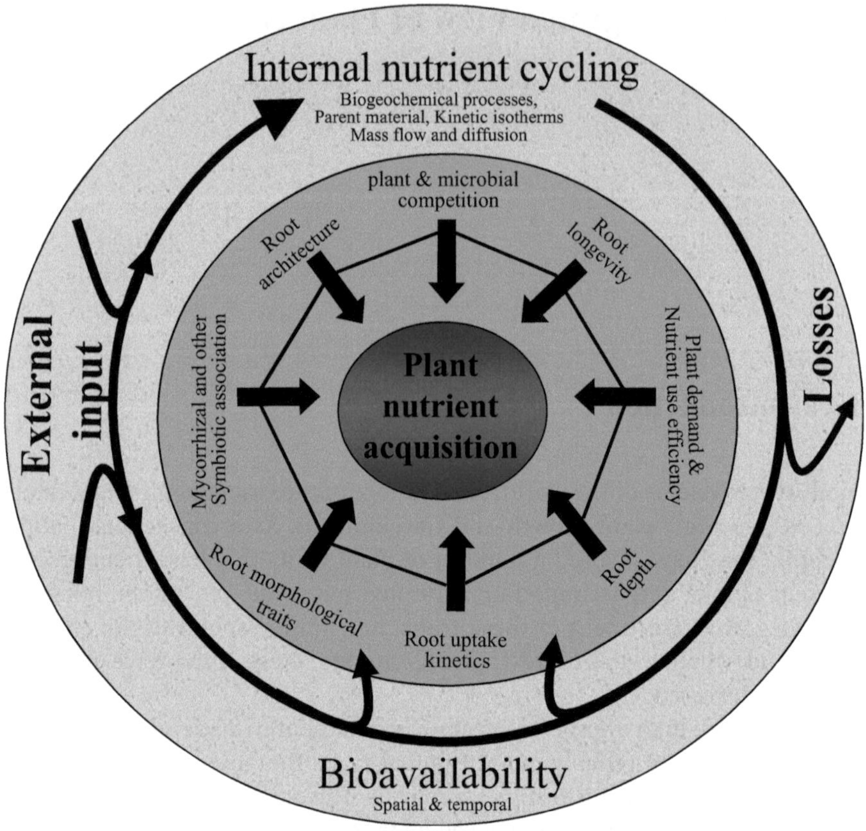

Fig. 12.1. A conceptual model depicting factors that control nutrient acquisition by plants. The innermost shell highlights rhizospheric traits including root morphology, architecture, kinetics, life span, and mycorrhizal colonization that received the most coverage in this volume

12.2 Nutrient Bioavailability

Understanding plant nutrient acquisition requires robust knowledge of processes that control bioavailability of nutrients, decomposition, mineralization, and soil physiochemical characteristics (Fig. 12.1). In Chapter 1, Comerford highlighted that sorption and desorption isotherms of nutrients from parent material are major factors in determining the bioavailability of many macro- and micronutrients. More importantly, Chapter 1 also emphasized the critical role of soil water content, hydraulic properties, mass flow and diffusion in transporting nutrients from the site of production to the site of consumption by plant roots. In this chapter, a brief treatment of root exudates as a mechanism that can increase nutrient availability was also given. Despite the rather important role that they play in influencing labile pools of nutrients, soil physiochemical characteristics are still poorly incorporated into studies designed to understand nutrient dynamics in native systems. Integration of these principles will be indispensable for a better understanding of nutrient dynamics and plant competition in native habitats.

In contrast to soil physiochemical characteristics, internal cycling of nutrients from organic matter has received considerably more attention in ecological studies. Internal cycling of nutrients is the focus of Chapter 2 by Prescott. Admittedly, processes of nutrient release from organic matter have a high leverage on the availability of minerals such as N, P and S, with N dominating most of the studies in this field. While temperature, precipitation, and aspects of tissue quality such as lignin content and C:N ratio are the strong predictors of decomposition and mineralization, there is emerging evidence (Silver and Miya 2001) that less commonly determined traits such as root Ca^{2+} concentration may be the strongest predictor of decomposition.

An important characteristic of soil bioavailability is that nutrient concentration in the bulk soil solution is heterogeneous in time and space, which was the main focus of Chapter 3 by Robinson. During the last two decades, the growing recognition that soil patches and pulses of nutrients are extremely common in natural habitats has been matched by an equal recognition that plant roots exhibit a wide array of plasticity in response to these patches and pulses. Collectively, these observations have provided a fresh perspective in explaining plant–plant competition for nutrients that would have otherwise been inconsistent with the view of homogenous soil nutrient availability. A considerable amount of work has yet to be accomplished in developing a cohesive theory of nutrient acquisition that would routinely take into account uptake by a non-uniform root system.

The outer shell in Fig. 12.1 also indicates that nutrient bioavailability within a system can be influenced by input from external sources. External input is best exemplified in systems where chronic atmospheric deposition in the form of acid rain and N deposition can significantly alter S and N availability to plants. There is no specific chapter devoted to this topic here, but chronic N and S deposition is the focus of a growing body of ecological literature. Nutrients from external sources can be either directly taken up by plants or they may indirectly affect the rate of internal cycling of these ions. For example, in the case of chronic nitrate deposition, an important indirect effect is excess cation loss from the system (Aber et al. 1989, 1998; Shultze 1989). In this volume, free-living and symbiotic N_2 fixation are assumed to be an external input for N and were discussed in Chapter 5. There, Sprent makes a compelling case that the amount of N_2 fixed may be small, but it is enough to enhance broader ecosystem outcomes such as biodiversity. However, in some habitats, introduction of exotic legumes is a major cause of decreased diversity.

12.3 Plant Nutrient Demand and Use Efficiency

One of the major challenges in modeling plant nutrient acquisition is how to incorporate important drivers of uptake, e.g., "nutrient demand" and "nutrient use efficiency". In Chapter 4, Gutschick and Pushnik present a road map showing how a functional definition can be developed for "demand" as well as "use efficiency". According to the most commonly used definition, nutrient use efficiency is simply an inverse of concentration. This definition of efficiency is a good first step to develop a broader categorization of plant types, but lacks the necessary rigor to predict and/or explain the vast differences in nutrient acquisition capacities among species with the same nutrient concentration. Gutschick and Pushnik suggest a functionally based definition for N use efficiency based on photosynthetic utility of N. They also address the need for a functionally meaningful definition of "demand". Too often, plant growth rate is assumed to be a surrogate for nutrient demand. The fact that so many plants of native and cultivated habitats exhibit luxurious consumption of nutrients, when supply exceeds the growth demand, argues strongly against the usefulness of a term based on growth rate alone.

12.4 Root System Traits That Regulate Nutrient Acquisition

The challenges associated with a mechanistically oriented definition of "demand" is also raised in Chapter 6. There, Glass argues that, at least in the case of N, P and S, specific internal pools of these minerals (including organic by-products of the assimilatory pathways) are important feedback regulators of uptake. Therefore, root or vascular tissue concentrations of intermediary metabolites may provide a quantitative measure of the demand. For example, there is growing recognition that root glutamine concentration, and the rate of glutamine transport to the root via phloem are strong regulators of NH_4^+ and NO_3^- uptake. In this chapter, Glass also focuses on the kinetics of root nutrient acquisition, and the role of membrane transporters in regulating ion uptake. This chapter highlights the fact that transporters of many ions, specifically NO_3^-, NH_4^+, PO_4^{3-} and K^+, have been characterized, and the genes encoding for many of those transporters have been isolated. Therefore, future studies designed to better understand the role of kinetics in the ecology of nutrient acquisition can benefit greatly from molecular approaches, including the use of mutants.

As is shown in Fig. 12.1, one of the important root traits that can influence nutrient acquisition is root system architecture, which is the topic of Chapter 7 by Lynch. Despite the widespread recognition of its importance, root architecture remains a poorly examined aspect of ion uptake. Theoretically, root architectural/topological differences can give rise to differences in nutrient acquisition by influencing the vascular network that carries nutrients from the root to the shoot. Lynch argues, however, that such interpretation of the possible role of root architecture is premature, and requires further testing. Lynch also highlights that the value of an architectural trait for ion uptake depends on its scale. Borrowing from the works of Fitter and coworkers, this chapter discusses architectures that are theoretically more efficient in the delivery of nutrients, based on their diffusive properties in the soil.

In Chapter 8, Eissenstat and Volder cover the many facets by which root life span influences plant nutrient acquisition. In many ways, root life span is, to nutrient acquisition, what leaf life span is to C acquisition. However, compared to leaf life span, which is commonly incorporated into models of C uptake, root life span has received less attention in models of nutrient acquisition. In general, data are limited on root longevity of different species and in different environments, partly because of shortcomings in methodology. From the limited studies that are available, however, the authors highlight the wide range of root life spans that can occur. For example, in fruit trees a sizeable fraction of the fine roots lives only a few weeks while in other species, like sugar maple, the fine roots may live about a year or more. The authors further argue that species with long-lived absorptive roots likely exhibit different strategies for nutrient acquisition than those with short-lived roots. The authors use a cost-benefit analysis to quantitatively assess changes in nutrient acquisition efficiency as root life span changes. In examining the potential effect of aging over the life of a single root, the authors of this chapter provide evidence that root absorption capacity may generally decline as the root ages. Finally, Eissenstat and Volder provide an excellent overview of biotic (e.g., pathogens) and abiotic (e.g., temperature, soil fertility, and CO_2) factors that could influence root life span.

Eissenstat and Volder also highlight the role of one particular biotic factor that could significantly affect root longevity, namely, mycorrhizal colonization. Mycorrhizal control of plant nutrient acquisition is covered in more detail by Finlay in Chapter 9. Here, Finlay covers issues ranging from factors that control root colonization, and implications for nutrient uptake of individual plants, to the effects of the mycorrhizal symbiosis on plant communities. An extremely valuable contribution of this chapter is in providing a comprehensive treatment of mycorrhizal facilitation of uptake of organic sources of nutrients that would otherwise be unavailable to plants. Finlay argues that the general theories of nutrient acquisition by plants must better incorporate the role of mycorrhizae and the uptake of organic sources of nutrients, a point of view that is rapidly gaining acceptance (Lipson and Näshlom 2001; Schimel and Bennet 2004). In this chapter, we also learn that mycorrhizal fungi can influence nutrient availability to the plant either by their direct impact on the weathering of the parent material, or by interaction with other soil organisms involved in nutrient cycling.

12.5 Global Change and Plant Nutrient Acquisition

Root symbiotic association is also prominently covered in Chapter 10 where Pregitzer and King review the effects of soil temperature on various processes in the rhizosphere. Here, the authors argue that an increase in soil temperature is likely to accompany the projected rise in the temperature of ambient air and, therefore, deserves special attention. Pregitzer and King present three areas in which soil temperature can influence nutrient uptake by plants. First, they argue that changes in soil temperature can influence nutrient availability by affecting the biogeochemical cycling of nutrients. Second, alteration in soil thermal regime could result in changes in nutrient uptake by modifications in root growth, longevity, architecture, and morphology. Third, soil temperature significantly influences symbiotic N_2 fixation, the colonization of roots by mycorrhizal fungi, and activities of other soil biota that have an interactive relationship with the rhizospheric nutrient status.

The penultimate chapter of this volume considers plant nutrient acquisition responses to other components of global change. Many ecological theories are based on our understanding of organisms under average environmental conditions for the past decades and centuries. In Chapter 11, Bielenberg and BassiriRad argue, however, that global conditions are changing so rapidly that our ability to predict the future of terrestrial ecosystems may require new and more realistic models. More specifically, they consider the effects of increased O_3, UV-B and CO_2 on nutrient uptake by plants. The ability of terrestrial plants to respond to increased levels of these factors may largely depend on the extent to which nutrient uptake is altered. The authors find that the available literature shows a highly inconsistent pattern of nutrient uptake responses to these factors among species. Zerihun and BassiriRad (2001) propose that some of the inconsistencies of N uptake responses to elevated CO_2 concentrations may be attributed to the fact that results from different experimental protocols were compared. Therefore, they designed an experiment in which seedlings of six tree species were grown for 9 months under identical conditions, except for CO_2 levels. They report that elevated CO_2 resulted in species-specific responses in kinetics of NH_4^+ uptake (Fig. 12.2). More specifically, they found that elevated CO_2 elicited an up-regulation of the kinetics in three species, down-regulation in one species, and no changes in the kinetics of the remaining two species.

Much remains to be done to make sure that the results from these limited, but disparate studies are representative of a larger number of species and other ecosystems. If these species-dependent responses in nutrient acquisition prove to be a general outcome of global change, then such information will be necessary to predict possible consequences for important ecosystem properties such as species diversity.

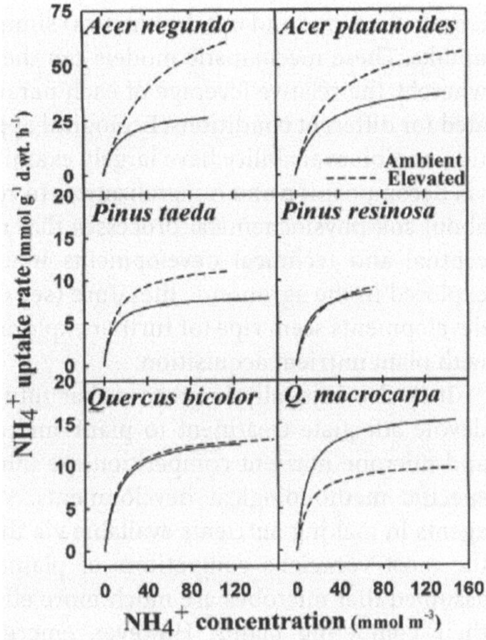

Fig. 12.2. Kinetics of NH_4^+ uptake responses of seedlings of six tree species in response to elevated CO_2. Data from Zerihun and BassiriRad (2001)

12.6 Broader Issues and Directions for Future Research

This volume focused largely on root system characteristics that govern nutrient uptake by plants. We highlighted a number of root traits that are critical in nutrient acquisition, but these remain highly understudied. For example, kinetics of root ion uptake, fine root turnover, longevity, and architecture show an incredible level of plasticity in response to the environment, yet their contribution to the overall nutrient budget of plants are qualitative at best. In contrast, analogous characteristics of the shoot system have been more successfully integrated into quantitative models of carbon uptake. We also need mechanistic approaches that integrate these root parameters collectively, rather than individually. Models that simply use one trait, e.g., using root:shoot ratio as a predictive parameter of nutrient uptake or competition, are grossly inadequate. The information about root morphology, kinetics, longevity, architecture and mycorrhizal association must be collectively incorporated into studies of nutrient uptake if we expect to develop mechanistic understanding of nutrient uptake.

Integration of the root and soil system characteristics into models of nutrient uptake may be aided by using modeling approaches described by Tinker and Nye (2000), or by Barber and coworkers (Barber 1995) who incorporated a number of soil (e.g., mass flow and diffusion) and root parameters (e.g., root length, diameter, and uptake kinetics) simultaneously into models of nutrient uptake. These mechanistic models can then be used for sensitivity analyses, whereby the relative leverage of each parameter can be quantitatively evaluated for different conditions. Ecological approaches to plant nutrient acquisition and bioavailability have largely examined biological release of nutrients via decomposition and mineralization. In contrast, considerably less is known about soil physiochemical processes that regulate nutrient availability. Conceptual and technical developments in this area have been much better explored in the agronomic literature (see Chap. 1, this Vol.). Therefore, these developments seem ripe for further exploration in ecological research dealing with plant nutrient acquisition.

It is also critically important that future studies of nutrient acquisition devote adequate treatment to plant–microbe interactions. Studies of plant and microbe nutrient competition are inherently difficult, and may require specific methodological developments. While microbes are the primary agents in making nutrients available via the mineralization process, they are the most voracious competitors of plants for nutrients. It is traditionally assumed that microbes are much more efficient in absorbing nutrients than their competing plants. However, emerging evidence indicates that this assumption may not be appropriate in many circumstances (Kay and Hart 1997; Korsaeth et al. 2001). Well-designed studies to elucidate the role of microbial competition are likely to contribute significantly to models of plant nutrient acquisition.

It cannot be overstated that the current literature in the ecology of nutrient acquisition is dominated by studies of N and, to a lesser extent, P. Understandably, this "nitrocentric" approach reflects the important and frequent limitation of plant processes by this element. We believe that in many circumstances, however, nutrients other than N and P may regulate plant responses to the environment. In this regard, micronutrients may deserve a greater attention in ecological research. By definition, micronutrients are required in very small quantities, but their importance to plant biology is disproportionately large. For example, in a recent study, Hungate et al. (2004) have shown that biological N_2 fixation of a leguminous vine grown in a scrub–oak ecosystem increased in response to elevated CO_2 during the first year of treatment, but the effect declined and then disappeared in the second and third years. Between the fifth and seventh years, elevated CO_2 significantly decreased N_2 fixation. Hungate et al. (2004) concluded that the decline reflected the onset of soil Mo deficiency.

Perhaps the most fitting way to end this volume is to propose a greater effort to link studies of plant nutrient acquisition to broader issues of plant ecology. Too often, we encounter studies that show enormous adjustments in root traits when nutrient availability changes, e.g., changes in specific root length, uptake kinetics, architecture root turnover, and even the extent of mycorrhizal colonization. While these studies are extremely valuable, they do not clearly establish ecological consequences to individual plants, or to the whole community. The logical next step in such cases is to establish the extent to which these traits contribute to broader ecological issues such as whole-plant nutrient content, vigor, fitness, competitiveness, and even evolution of plant communities.

Acknowledgments. I thank John Lussenhop, Vince Gutschick and Jacqui Charette for their comments and suggestions on an earlier version of this manuscript. I would also wish to acknowledge the National Science Foundation for supporting my research in this area.

References

Aber JD, Nadelhoffer KJ, Steudler P, Melillo JM (1989) Nitrogen saturation in northern forest ecosystems. Bioscience 39(6):378–386

Aber JD, McDowell W, Nadelhoffer K, Magill A, Bernston G, Kamakea M, McNulty S, Currie W, Rustad L, Fernandez I (1998) Nitrogen saturation in temperature forest ecosystems. Bioscience 48:921–934

Barber SA (1995) Soil nutrient bioavailability: a mechanistic approach, 2nd edn. Wiley, New York

Hungate BA, Stiling PD, Dijkstra P, Johnson DW, Ketterer ME, Hymus GJ, Hinkle CR, Drake BG (2004) CO_2 elicits long-term decline in nitrogen fixation. Nature 304:129–129

Kaye JP, Hart SC (1997). Competition for nitrogen between plants and soil microorganisms. Trends Ecol Evol 12:139–143

Korsaeth A, Molstad L, Bakken LR (2001) Modeling the competition for nitrogen between plants and microflora as a function of soil heterogeneity. Soil Biol Biochem 33:215–226

Lipson D, Näshlom T (2001) The unexpected versatility of plants: organic nitrogen use and availability in terrestrial ecosystems. Oecologia 128:305–316

Schimel JP, Bennet J (2004) Nitrogen mineralization: challenges of a changing paradigm. Ecology 85(3):591–602

Schulze ED (1989) Air pollution and forest decline in a spruce (*Picea abies*) forest. Science 224:776–783

Silver WL, Miya RK (2001) Global patterns in root decomposition: comparison of climate and litter quality effects. Oecologia 129:407–419

Tinker PB, Nye PH (2000) Solute movement in the rhizosphere. Oxford University Press, Oxford

Zerihun A, BassiriRad H (2001) Interspecies variation in nitrogen uptake kinetic responses of temperate forest species to elevated CO_2: potential causes and consequences. Global Change Biol 7:211–222

Subject Index

Subject Index

MIX
Papier aus verantwortungsvollen Quellen
Paper from responsible sources
FSC® C105338

FSC
www.fsc.org

If you have any concerns about our products,
you can contact us on
ProductSafety@springernature.com

In case Publisher is established outside the EU,
the EU authorized representative is:
Springer Nature Customer Service Center GmbH
Europaplatz 3, 69115 Heidelberg, Germany

Printed by Libri Plureos GmbH
in Hamburg, Germany